Kazhdan-Lusztig Theory and Related Topics

Recent Titles in This Series

(*Continued in the back of this publication*)

CONTEMPORARY MATHEMATICS

139

Kazhdan-Lusztig Theory and Related Topics

Proceedings of an AMS Special Session
held May 19–20, 1989
at the University of Chicago,
Lake Shore Campus, Chicago, Illinois

Vinay Deodhar

American Mathematical Society
Providence, Rhode Island

The AMS Special Session on Kazhdan-Lusztig Theory and Related Topics was held at the 849th Meeting of the American Mathematical Society on the Lake Shore Campus of Loyola University of Chicago, on May 19–20, 1989.

1991 *Mathematics Subject Classification*. Primary 22E47, 20H15, 20G05, 17B35, 14M15; Secondary 55N35, 51M35, 17B20, 14M20, 14F45, 14E15.

Library of Congress Cataloging-in-Publication Data

Kazhdan-Lusztig theory and related topics: proceedings of an AMS special session held May 19–20, 1989 at the University of Chicago, Lake Shore Campus, Chicago, Illinois/Vinay Deodhar.

p. cm.—(Contemporary mathematics; v. 139)

ISBN 0-8218-5150-0

1. Lie groups—Congresses. 2. Representations of groups—Congresses. 3. Verma modules—Congresses. I. Deodhar, Vinay, 1948– . II. Series: Contemporary mathematics (American Mathematical Society); v. 139.

QA387.K.38 1992 — 92-27738

512′.55—dc20 — CIP

Printed in the United States of America.

The paper used in this book is acid-free and falls within the guidelines established to ensure permanence and durability. ♾

This publication was printed directly from author-prepared copy. Portions of this volume were prepared by the authors using AMS-TEX, the American Mathematical Society's TEX macro system.

10 9 8 7 6 5 4 3 2 1 97 96 95 94 93 92

Contents

Preface

In the years 1979–80, Kazhdan and Lusztig published two papers which dealt with interesting problems from different topics related to Lie Theory such as structure of Verma modules and its relation with the combinatorics of Bruhat ordering in Weyl groups, geometry of Schubert varieties in generalized flag manifolds, and primitive ideals in enveloping algebras. As far as these individual topics are concerned, some noteworthy results had been obtained by mathematicians prior to the work of Kazhdan and Lusztig. However, the link between these seemingly diverse topics was made clear in the papers of Kazhdan–Lusztig. In fact, these papers seem to have started a new topic in Lie Theory and related areas, a topic which can be called Kazhdan–Lusztig Theory. Since then a number of mathematicians have worked on problems related to this circle of ideas and several interesting results have been obtained. The aim of this special volume is to put together articles which deal with new results on different aspects of the Kazhdan–Lusztig Theory and related topics.

This volume has its origin in the special session on this topic at the A.M.S. meeting in Chicago in May 1989. Several mathematicians could not attend this meeting due to conflicts with conferences held elsewhere in the U.S.A. and Europe. Considering the enthusiasm with which this session was received by the audience, it seemed natural to put together a volume based on the proceedings of the special session. In addition to the speakers, several mathematicians were invited to contribute articles. The response was very encouraging. It has been a very rewarding experience for me to edit this volume.

My sincere thanks are due to the authors for their contributions to the volume and their patience with me through several schedule changes, to the referees for their time and careful evaluation of the papers, and to the American Mathematical Society for publishing this volume.

Vinay Deodhar
Bloomington
May 15, 1992.

Contemporary Mathematics
Volume **139**, 1992

A Counterexample to the Gabber-Joseph Conjecture

BRIAN D. BOE

ABSTRACT. A counterexample to the Gabber-Joseph Conjecture on extensions between Verma modules is provided.

Let $\mathfrak{g}$ be a complex simple Lie algebra with Weyl group W, Borel subalgebra $\mathfrak{b}$ and nilradical $\mathfrak{n}$. Fix a regular integral weight λ such that $-\lambda$ is dominant (with respect to $\mathfrak{b}$). For $w \in W$, denote by $M(w)$ the Verma module with highest weight $w\lambda - \rho$ (ρ as usual is the half-sum of positive roots for $\mathfrak{b}$), and by $L(w)$ its unique simple quotient. A family of polynomials $R_{y,w}(q)$, indexed by a pair of elements $y, w \in W$, was introduced by Kazhdan and Lusztig in [**KL**]. They are defined recursively in terms of the Bruhat order on W as follows.

$$R_{e,e} = 1; \qquad R_{y,e} = 0 \quad \text{if} \quad y \neq e \tag{1.1}$$

For $w > e$, choose a simple reflection s such that $sw < w$.

$$R_{y,w} = R_{sy,sw} \quad \text{if} \quad sy < y \tag{1.2}$$

$$R_{y,w} = (q-1)R_{sy,w} + qR_{sy,sw} \quad \text{if} \quad sy > y \tag{1.3}$$

A second family of polynomials $R'_{y,w}(q)$ was defined by Gabber and Joseph in [**GJ**]. These are the "generating functions" for the dimensions of Ext groups between Verma modules:

$$R'_{y,w}(q) = \sum_{i=0}^{\infty} (-1)^{\ell(w)-\ell(y)-i} q^i \dim \operatorname{Ext}^i\big(M(y), M(w)\big). \tag{2}$$

Gabber and Joseph proved that the polynomials $R'_{y,w}$ satisfy the recursion relations (1.1) and (1.2), and gave a necessary and sufficient condition for them to also satisfy (1.3). Although not actually stated in [**GJ**], the following became known as the Gabber-Joseph Conjecture.

1991 *Mathematics Subject Classification*. Primary 17B10; Secondary 22E47.
Partially supported by a University of Georgia Faculty Research Grant.
This paper is in final form and no version of it will be submitted for publication elsewhere.

GABBER-JOSEPH CONJECTURE. *$R_{y,w} = R'_{y,w}$ for all $y, w \in W$.*

The main interest of the Kazhdan-Lusztig theory is in $\mathfrak{n}$-cohomology of the irreducibles $L(w)$; equivalently, $\operatorname{Ext}^*(M(y), L(w))$. The Kazhdan-Lusztig polynomials, $P_{y,w}(q)$, are defined recursively in terms of the R-polynomials via

$$q^{\ell(w)}\, P_{y,w}(q^{-1}) = q^{\ell(y)} \sum_{y \leq z \leq w} R_{y,z}(q)\, P_{z,w}(q). \tag{3}$$

The Kazhdan-Lusztig Conjecture (which is a theorem due to Beilinson-Bernstein and Brylinski-Kashiwara in this context) was shown by D. Vogan to be equivalent to the statement that

$$P_{y,w}(q) = \sum_{i \geq 0} q^i \dim \operatorname{Ext}^{\ell(w)-\ell(y)-2i}(M(y), L(w)).$$

Part of the appeal of the Gabber-Joseph Conjecture was that it would provide a pleasing "symmetry" between the R-polynomials and the P-polynomials – the former would give extensions between pairs of Verma modules while the latter compute extensions between Verma modules and irreducibles. Moreover, (3) would become a simple recursion formula for the latter Ext groups in terms of the former. The fact that the Gabber-Joseph Conjecture is false leaves open the very intriguing question, what **is** the relation between these two types of Ext groups?

For our counterexample, note that a trivial consequence of the conjecture is that successive coefficients of the R polynomials should have opposite signs (since from (2) this is certainly true for the R''s). We simply produce a polynomial $R_{y,w}$ with two successive non-zero coefficients having the same sign.

The "smallest" such example occurs in the Weyl group of type B_4. (Our criteria for "smallness" are first, the rank of $\mathfrak{g}$, and second, $\ell(w) - \ell(y)$.) Label the Dynkin diagram as usual:

and consider the pair

$$w = s_2 s_3 s_4 s_2 s_3 s_4 s_2 s_1 s_2 s_3\ , \qquad y = s_4.$$

Then applying the recursion relations (1.1) – (1.3), one calculates

$$R_{y,w}(q) = -1 + 4q - 6q^2 + 3q^3 + 2q^4 - 2q^5 - 3q^6 + 6q^7 - 4q^8 + q^9.$$

Examples of the same sort (but with larger length difference between y and w) also occur in types D_4 and A_5.

It is perhaps curious that this phenomenon of non-alternating signs does not arise in smaller rank: other non-trivial phenomena certainly do. For example in B_3, there are 4 distinct non-zero Kazhdan-Lusztig polynomials (including 1), and there is a Verma module in which the same irreducible composition factor occurs 3 times. In A_4, there are 5 Kazhdan-Lusztig polynomials, and a Verma module composition factor multiplicity of 4. On the other hand, B_4 has 41 distinct non-zero P polynomials, and a Verma multiplicity of 14, so the complexity has increased dramatically. Evidently the Weyl group structure of B_3 and A_4 is just not rich enough to produce an R polynomial having non-alternating coefficients.

The counterexample was originally discovered with the help of a computer program, which first calculates the Weyl group and its "weak Bruhat order" (the order obtained by using simple reflections only), and then implements the recursion relations (1.1)–(1.3) to compute the R polynomials. It has subsequently been checked by hand by both the author and the referee.

The author would like to thank David Collingwood for suggesting this method of approach to the problem, and Ron Irving for a helpful communication on the status of the conjecture.

References

[GJ] O. Gabber and A Joseph, *Towards the Kazhdan-Lusztig Conjecture*, Ann. Scient. Ec. Norm. Sup. **14** (1981), 261–302.

[KL] D. Kazhdan and G. Lusztig, *Representations of Coxeter Groups and Hecke Algebras*, Inv. Math. **53** (1979), 165–184.

Department of Mathematics, University of Georgia, Athens, GA 30602

E-mail address: brian@joe.math.uga.edu

Contemporary Mathematics
Volume **139**, 1992

EQUIVARIANT INTERSECTION COHOMOLOGY

JEAN-LUC BRYLINSKI

Introduction. There has recently been much progress in the equivariant cohomology of symplectic manifolds, for hamiltonian group actions [**1**], [**5**], [**21**], [**27**]. In particular, Kirwan and Ginsburg have proved very precise fixed point theorems for hamiltonian torus actions. On the other hand, very little is known about torus actions on singular algebraic varieties (see however [**6**], [**12**]). As Goresky and MacPherson have shown, the natural cohomological theory for singular Morse theory is the intersection cohomology (with middle perversity) [**24**]. It seems therefore natural that an equivariant intersection cohomology theory should be introduced to study actions of compact groups on singular varieties. This is done here, for arbitrary perversities. The idea is very simple : let the compact group G act on the pseudo-manifold X; we have a simplicial pseudo-manifold $X \overset{G}{\times} EG$, with smooth face maps $X \times G^n \to X \times G^{n-1}$. The intersection complexes $\underline{IC}^*$ of [**23**] on the $X \times G^n$ organize into a simplicial complex of sheaves on $X \overset{G}{\times} EG$. We define the equivariant intersection cohomology groups

1991 Mathematics Subject Classification. Primary 55N25; Secondary 55N33
Partially supported by a NSF grant.
This paper is in final form and no version of it will be published elsewhere.

$IH^*_G(X)$ to be the hypercohomology groups of this simplicial complex of sheaves. They are the abutment of a spectral sequence with E_2-term $H^*(BG) \otimes IH^*(X)$.

Much of the recent progress in equivariant cohomology is due to the explicit localization theory of Berline and Vergne [**4**], [**5**], which uses an explicit description of $H^*_T(X)$ as the cohomology of the "Cartan complex" $[\Omega^*_X]^T \otimes S(t^*)$ (for X a manifold, T a torus action differentiably on X). We give a similar description of $IH^*_T(X)$, using a complex of differential forms on the smooth part of X, satisfying perversity conditions near the various strata. We borrow this complex of differential forms from unpublished work of Goresky-MacPherson.

A well-known and striking feature of intersection cohomology is Poincaré duality. Equivariant cohomology of smooth compact manifolds with action of a torus T also satisifes Poincaré duality, if one takes as base ring the cohomology ring $H^*(BT)$. This observation, which appears new, is best seen on the Cartan complex mentioned above. In the present paper, we prove Poincaré duality for equivariant intersection cohomology $IH^*_T(X)$, using $H^*(BT) = S(t^*)$ as base ring. This is done by establishing a sheaf-theoretic duality (in the sense of Verdier).

The verification of the duality is local near an orbit, and splits into a duality on the orbit and a duality on a transversal slice, each with its own base ring (see §3.2 for details).

We prove a location theorem, à la Segal-Quillen, for $IH^*_T(X)$. We deduce from it a fixed point theorem for actions of tori on projective singular varieties, using the hard Lefschetz theorem for intersection cohomology, due to Gabber [**18**]. This fixed point theory gives an inequality of sums of intersection cohomology Betti numbers; to get an equality, we have to assume purity for the intersection complex restricted to the fixed point set. The example of the projective cone over a smooth projective variety, with the usual S^1-action, shows that this fixed point theorem is very closely related to the hard Lefschetz theorem for that smooth variety. This might suggest that fixed point theorems in the singular case lie deeper than in the smooth case.

We have not discussed here a model for equivariant middle intersection cohomology of complex analytic spaces, based on the de

Rham complex of a holonomic $\mathcal{D}$-module with regular singularities. We hope to treat that in the future. Also, it would be easy to give a definition, à la Cartan, of equivariant L^2-cohomology. This might be very useful for explicit localization theorems, à la Berline-Vergne.

Equivariant intersection cohomology has been defined independently by Roy Joshua. He works in a slightly different context (algebraic groups acting on algebraic varieties, in arbitrary characteristics), but his construction is essentially the same as ours. His article [**25**] and the present article have (at least) the following overlap:

- construction of a spectral sequence for equivariant intersection cohomology, and a degeneracy theorem for it
- localization theorem
- application to Schubert varieties.

It is a pleasure for us to thank him for his correspondence. In addition, I would like to thank Mark Goresky and Robert MacPherson for enlightening discussions, especially concerning fixed point theorems, and for letting me use some unpublished work of theirs, mentioned earlier. The results of this paper were exposed at the Singularity Seminar of Ecole Polytechnique, in March 1986. I thank the audience of that seminar (in particular Jean-Pierre Brasselet and Claude Sabbah) for their kind interest. I am also grateful to William Browder and Victor Ginsburg for interesting comments on this work.

In addition, I am very grateful to I.H.E.S. for its hospitality during the academic year 1985-86.

§1. D.G.M. Complexes.

§1.1. We consider an n-dimensional topological pseudo-manifold X, equipped with a stratification

$$X = X_n \supset X_{n-2} \supset \cdots \supset X_0 \supset X_{-1} = \phi$$

which satisfies the assumptions of [**22**] (in particular, the stratum $S_i = X_i - X_{i-1}$ is a manifold of dimension i, or is empty). We

---

NOTE: This article, which was first circulated as an IHES preprint in 1986, was revised in 1989.

give ourselves a perversity function $c \mapsto \overline{p}(c)$ as in [**22**], [**23**]. So $\overline{p}$ is a function from $\{2, 3, \cdots\}$ to $\mathcal{N}$ such that both $\overline{p}(c)$ and $\overline{q}(c) =: c - 2 - \overline{p}(c)$ are weakly increasing functions of c. We give ourselves a commutative field k of characteristic 0. The "D.G.M. complex" (an "intersection complex") for the perversity $\overline{p}$ is an object $P^{\bullet}$ of the derived category of bounded complexes of sheaves of k-vector spaces on X. It is defined, using the stratification of X, as follows. For each k, we consider the open set $U_k = X - X_{n-k}$, and we let $j_k : U_k \hookrightarrow U_{k+1}$ be the inclusion (this is denoted i_k in [**2**], §3.1). Notice $U_1 = U_2$. The Deligne construction gives an inductive definition of $P_k^{\bullet} \in D^b(U_k)$. Start from $P_2^{\bullet} = k_{U_2}$ on $U_2 = X - X_2$. Then $P_{k+1}^{\bullet} = \tau_{\leqslant p(k)} \mathbb{R} j_{k,*} P_k^{\bullet}$ for $k \geqslant 1$. For our purposes, we need actual complexes of sheaves (not merely objects of derived categories). We therefore give an inductive construction of a complex of sheaves $P_k^{\bullet}$ on U_k. First introduce a notation for Godement's canonical flasque resolutions [**21**]: if $K^{\bullet}$ is a complex of sheaves on some space, $\underline{Go}(K^{\bullet})$ will be the simple complex of sheaves obtained from the double complex of sheaves, where we take the Godement resolution of each sheaf K^i. We then start with $P_2^{\bullet} = \underline{Go}(k_{U_2})$. Then we set

$$P_{k+1}^{\bullet} = \tau_{\leqslant p(k)} j_{k,*} \underline{Go}(P_k^{\bullet}) \text{ for } k \geqslant 1.$$

Note that our normalization of the D.G.M. complexes differs from [**23**] by a shift of n. Note that each $P_k^{\bullet}$ is zero in degrees $> \overline{p}(k-1)$. We call this $P^{\bullet}$ the canonical flasque D.G.M. complex. In practice, of course, one works with other models of D.G.M. complexes, and one wishes to compare them to $P^{\bullet}$ in as precise a manner as possible. For this, we need

Definition 1.1.1. *A generalized quasi-isomorphism $A^{\bullet} \xrightarrow{\Phi} B^{\bullet}$ (where $A^{\bullet}$, $B^{\bullet}$ are complexes of sheaves) is a diagram*

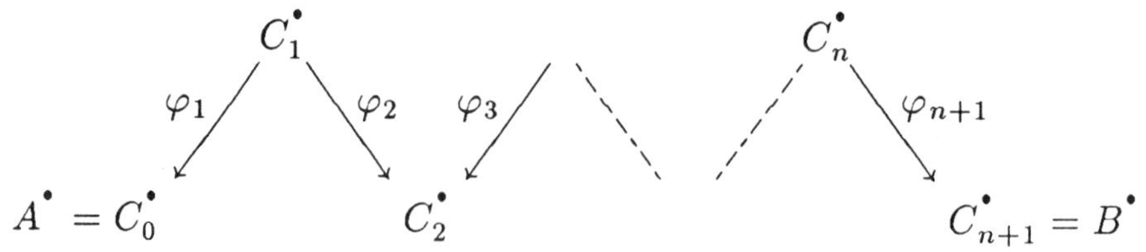

where each φ_j is a quasi-isomorphism of complexes of sheaves.

If $E^{\bullet}$ is another D.G.M. complex with each E^i <u>soft</u>, we will produce a generalized quasi-isomorphism $E^{\bullet} \to P^{\bullet}$. Again, this is done on U_k by induction on k. We let $E_k^{\bullet} = E^{\bullet}_{/U_k}$; the inclusion $\tau_{\leqslant p(k-1)}\, E_k^{\bullet} \hookrightarrow E_k^{\bullet}$ is a quasi-isomorphism. If we find a generalized quasi-isomorphism $E_{k+1}^{\bullet} \to \tau_{\leqslant p(k)} j_{k,*} E_k^{\bullet}$ then we may extend the quasi-isomorphism $\varphi_k : E_k^{\bullet} \to P_k^{\bullet}$ to U_{k+1} as follows. Compose the following generalized quasi-isomorphisms:

(a) $E_{k+1} \to \tau_{\leqslant p(k)} j_{k,*} E_k^{\bullet}$
(b) $\tau_{\leqslant p(k)} j_{k,*} E_k^{\bullet} \to \tau_{\leqslant p(k)} j_{k,*} P_k^{\bullet}$ *deduced from* φ_k
(c) $\tau_{\leqslant p(k)} j_{k,*} P_k^{\bullet} \to \tau_{\leqslant p(k)} j_{k,*} \underline{Go}(P_k^{\bullet}) = P_{k+1}^{\bullet}$ *deduced from the map* $P_k^{\bullet} \to \underline{Go}(P_k^{\bullet})$.

Of course, to start with, on U_2, since $E_2^{\bullet}$ is concentrated in degrees $\leqslant 0$ and $\underline{H}^{\circ}(E_2^{\bullet}) = k_{U_2}$, we have a natural morphism of complexes $\varphi_2 : E_2^{\bullet} \to k_{U_2} \to \underline{Go}(k_{U_2}) = P_2^{\bullet}$

So we are reduced to the following

Construction 1.1.2. *Let X be a topological space, $U \hookrightarrow X$ an open set, $E^{\bullet}$ a complex of <u>soft</u> sheaves on X, which is such that, for some integer ℓ, the map $E^{\bullet} \to j_* j^* E^{\bullet}$ factors (in the derived category) through a quasi-isomorphism $E^{\bullet} \to \tau_{\leqslant \ell} \mathbb{R} j_* j^* E^{\bullet}$. Then we construct a generalized quasi-isomorphism of complexes of sheaves $E^{\bullet} \to \tau_{\leqslant \ell} j_* j^* E^{\bullet}$ (note that $j_* j^* E^{\bullet}$ is a realization of $Rj_* j^* E^{\bullet}$, since $E^{\bullet}$ is a complex of soft sheaves).*

The generalized quasi-isomorphism is given by the diagram:

$$\begin{array}{ccc} & \tau_{\leq l} E^{\bullet} & \\ \swarrow & & \searrow \\ E^{\bullet} & & \tau_{\leq l} j_* j^* E^{\bullet} \end{array}$$

This construction would apply in particular to the variant of the complex of sheaves $IC_{\overline{p}}^{\bullet}$ of [**23, §2.1**], where X is a real-analytic space, the X_i are real-analytic subspaces, and one takes $IC_{\overline{p}}^{-k}$ to consist of <u>sub-analytic</u> k-chains (rather than P.L. chains) which satisfy the dimensionality conditions of loc. cit.

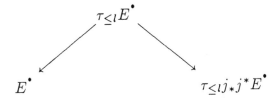

However, we are interested only in D.G.M. complexes $K^\bullet_X$ which have the following property:

(1.1.3) If $f : Y \to X$ is a <u>smooth</u> map of topological pseudo-manifolds, and if Y is endowed with the stratification $Y_j = f^{-1}(X_{j+\dim(X)-\dim(Y)})$, there is a natural inverse image map: $f^{-1}(K^\bullet_X) \to K^\bullet_Y$, which is a quasi-isomorphism.

Certainly the canonical flasque complex $P^\bullet$ has this property. That is not the case for $IC^\bullet_{\overline{p}}$; however, for $\overline{q}$ the dual perversity, consider the following realization $A^\bullet_{\overline{p}}$ of the Grothendieck-Verdier dual complex of $IC^\bullet_{\overline{q}}$:

$$A^\bullet_{\overline{p}}(U) = K\text{-dual of } \Gamma_c(U, IC^\bullet_{\overline{q}}).$$

This definition is licit because the sheaves $IC^i_{\overline{q}}$ are soft, cf. [**9, V.7.6**], and k is a field.

For U an open set of X, there is a natural map

$$\Gamma_c(f^{-1}(U), \underline{IC^\bullet_{\overline{q},Y}}) \longrightarrow \Gamma_c(U, \underline{IC^\bullet_{\overline{q},X}})$$

(push-forward of finite sub-analytic cycles), hence a map $A^\bullet_{\overline{p}}(U) \to A^\bullet_{\overline{p}}(f^{-1}(U))$ as required. In this discussion, it was assumed that $f : Y \to X$ was a real-analytic map.

Definition 1.1.4. : *$A^\bullet_{\overline{p}}[-n]$ is called the <u>cohomological D.G.M. complex</u> (for the perversity $\overline{p}$).*

For this $A^\bullet_{\overline{p}}$, as well as for the complex to be considered in the next section the following general fact will be easily deduced from the details of Construction 1.1.2.

Principle 1.1.5. *For f as in (1.1.3) and $K^\bullet$ satisfying (1.1.3) the generalized quasi-isomorphisms $\Phi : K^\bullet \to P^\bullet$ fit into a commutative diagram*

$$\begin{array}{ccc} f^{-1}(K^\bullet_X) & \xrightarrow{f^{-1}(\Phi_X)} & f^{-1}(P^\bullet_X) \\ \downarrow & & \downarrow \\ K^\bullet_Y & \xrightarrow{\Phi_Y} & P^\bullet_Y \end{array}$$

Finally, note that our construction 1.1.2 provides a generalized quasi-
isomorphism between the canonical flasque D.G.M. complexes relative to different topological stratifications. This is enough to ensure independence of the stratification for all our constructions in §2.

§1.2 A D-G-M-complex of differential forms.

The complex of sheaves we will describe is an unpublished invention of Goresky and MacPherson, which they kindly allowed us to use. All results in this section are due to them.

We need a C^∞-structure on the pseudo-manifold X. We will need Thom-Mather data for the stratification $X = X_n \supset X_{n-2} \supset \cdots \supset X_0 \supset X_{-1} = \phi$ (cf. [**28,** §**8**]). First we assume that all strata S_i are C^∞-manifolds. So for each i, we need an open neighborhood T_i of X_i in X and a continuous retraction π_i of T_i onto X_i; we also need a continuous function $\rho_i : T_i \to (0,\infty)$ such that $X_i = \{x \in T_i : \rho_i(x) = 0\}$. The (T_i, ρ_i) should satisfy the axioms (A8) and (A9) of loc. cit. Particularly noteworthy is the condition that for any i, j the equality $\pi_i \circ \pi_j = \pi_i$ should hold on $T_i \cap T_j$ (we warn the reader that our notations are slightly different from those of Mather). We will assume that for any $i, j, (\pi_i)_{/T_i \cap X_j}$ is a C^∞-map.

These data enter in the definition of the Goresky-MacPherson complex $\Omega^\bullet_{\overline{p}}$ of differential forms (on U_2), which we will give shortly. We first need a filtration of the de Rham complex $\Omega^\bullet_M$ on a C^∞-manifold M, which is obtained from a smooth C^∞ fibration $M \xrightarrow{\pi} B$, where B is another C^∞-manifold.

Definition 1.2.1. *For $k \geqslant 0$, $F_k\Omega^\bullet_M$ is the sub-complex of $\Omega^\bullet_M$ consisting of differential forms ω which satisfy (P_k) if $\xi_1, \cdots, \xi_{k+1}$ are vector fields on M, which are tangent to the fibers of π, then $i(\xi_1) \circ \cdots \circ i(\xi_{k+1})(\omega) = 0$, and such that $d\omega$ satisfies (P_k).*

Assume π has connected fibers. Then $F_\circ\Omega^\bullet_M$, which is the subcomplex of $\Omega^\bullet_M$ consisting of basic differential forms (with respect to π), identifies with $\pi^{-1}(\Omega^\bullet_B)$. The filtration F_k was introduced by Cartan [**13**], who proved:

Proposition 1.2.2. *The filtration of $H^\bullet(M, \Omega^\bullet_M)$ by the images of*

$H^{\bullet}(M, F_k\Omega^{\bullet}_M)$ is the one obtained from the Leray spectral sequence for π, which filters $H^{\bullet}(M, \Omega^{\bullet}_M) = \mathbb{H}^*(B, \mathbb{R}\pi_*\mathbb{C}_M)$ by the images of $\mathbb{H}^*(B, \tau_{\leqslant k}\mathbb{R}\pi_*\mathbb{C}_M)$. In particular, if B is contractible, we have:

$$H^i(M, F_k\Omega^{\bullet}_M) = \begin{cases} H^i(M)\,, & \text{if } i \leqslant k \\ 0\,, & \text{if } i > k \end{cases}.$$

We notice the following elementary properties of this filtration

(1.2.3) *If $\alpha \in F_k\Omega^{\bullet}_M, \beta \in F_\ell\Omega^{\bullet}_M$ then $\alpha \wedge \beta \in F_{k+\ell}\Omega^{\bullet}_M$*

(1.2.4) *Let ξ be a vector field on M, η a vector field on B, such that $\pi_*(\xi) = \eta$ everywhere. Then the interior product $i(\xi)$ maps each $F_k\Omega^{\bullet}_M$ to itself.*

We return now to our pseudo-manifold X, equipped with Thom-Mather data. Let $j = j_2 : U_2 = X - X_{n-2} \hookrightarrow X$.

Definition 1.2.5. *$\Omega^{\bullet}_{\overline{p}}$ is the sub-complex of sheaves of $j_*\Omega^{\bullet}_U$ consisting of differential forms ω so that, for every point of X_{n-c}, the restriction of ω to a neighborhood V of that point, contained in T_{n-c}, belongs to $\Omega^{\bullet}_{\overline{p}(c)}(V \cap U_2)$, with respect to the projection $V \cap U_2 \to X_{n-c}$ induced by π_{n-c}.*

In other words, near X_{n-c}, ω should satisfy $i(\xi_1)\cdots i(\xi_{\overline{p}(c)+1})\omega = 0$, if the ξ_i are vector fields on the smooth open set U_2 which are tangent to the fibers of π_{n-c}.

Proposition 1.2.6. *$\Omega^{\bullet}_{\overline{p}}$ is a D.G.M. complex with respect to the perversity $\overline{p}$.*

Proof. We verify that $\Omega^{\bullet}_{\overline{p}}[n]$ satisfies the axioms [**AX1**] in [**23**, **§3.3**]. Properties (a) and (b) are obvious. The vanishing condition (c) for the stalk cohomology is easy: for $x \in X_{n-k}$, the cohomology of the stalk complex $(\Omega^{\bullet}_{\overline{p}})_x$ is equal to the cohomology of $\Omega^{\bullet}_{\overline{p}}(S)$, for S a contractible normal slice. Now an element ω of $\Omega^k_{\overline{p}}(S)$ satisfies $i(\xi_1)\circ\cdots\circ i(\xi_{\overline{p}(k)+1})\omega = 0$, for any vector fields ξ_i on $U_2 \cap S$. Hence if $l > \overline{p}(k), \omega$ must be zero.

The same reasoning shows that the complex $\Omega^{\bullet}_{\overline{p}}(S)$ is exactly equal to the truncated complex $\tau_{\leq \overline{p}(k)}\Omega^{\bullet}_{\overline{p}}(S - \{x\})$, hence the "attaching map" is an isomorphism in all degrees $\leq \overline{p}(k)$, and condition (d) in [**AX1**] is also verified.

As in §1.1.3, the complex of sheaves $\Omega^\bullet_{\bar{p}}$ is functorial with respect to a smooth map $f : X \to Y$, proved that Y is endowed with the inverse image of the stratification of X, and that all strata of Y are C^∞-manifolds, and the restriction of f to each stratum is smooth. We also have to assume that the retractions π_i for Y are compatible with those for X.

§2. Definition of equivariant intersection cohomology.

§2.1. We assume that a compact topological group G acts continuously on a topological pseudo-manifold X, in such a way that there exists a stratification of X, as in §1.1, which is G-invariant. We first recall the construction of the simplicial space $X \overset{G}{\times} EG$ (cf. [**15**]). First we have the simplicial manifold EG : for each $n \geqslant 0$, we have manifolds $(EG)_n = G^{n+1}$ and we have face maps $d_i : G^{n+1} \to G^n$ such that $d_i(g_0, g_1, \cdots, g_n) = (g_0, g_1, \cdots, \widehat{g}_i, \cdots, g_n)$ (for $0 \leqslant i \leqslant n$). We do not give here the degeneracy maps, since we will not need them. So $X \times EG$ will be a simplicial pseudo-manifold, with $(X \times EG)_n = X \times G^{n+1}$ and face maps $d_i : X \times G^{n+1} \to X \times G^n$ given by $d_i(x; g_0, \cdots, g_n) = (x; g_0, \cdots, \widehat{g}_i, \cdots, g_n)$. Now G acts on the left on $X \times EG$; the action of G on $X \times G^{n+1}$ is the diagonal action: $\gamma.(x; g_0, \cdots, g_n) = (\gamma.x; \gamma g_0, \cdots, \gamma g_n)$. Then $X \overset{G}{\times} EG$ is the quotient of $X \times EG$ by this action. It is a simplicial pseudo-manifold, with $(X \overset{G}{\times} EG)_n = (X \times G^{n+1})/G$. As usual, we have homeomorphisms $X \times G^n \xrightarrow{h_n} (X \times G^{n+1})/G$ such that $h_n(x; g_1, \cdots, g_n) =$ class of $(x; 1, g_1, g_1g_2, \cdots, g_1g_2 \cdots g_n)$. So we may view $X \overset{G}{\times} EG$ as the simplicial space such that $(X \overset{G}{\times} EG)_n = X \times G^n$, with face maps given by:

$$d_0(x; g_1, \cdots, g_n) = (g_1^{-1}x; g_2, \cdots, g_n) \tag{2.1.1}$$

$$d_i(x; g_1, \cdots, g_n) = (x; g_1, \cdots, g_ig_{i+1}, \cdots, g_n) \quad \text{for } 1 \leqslant i \leqslant n-1 \tag{2.1.2}$$

$$d_n(x; g_1, \cdots, g_n) = (x; g_1, \cdots, g_{n-1}). \tag{2.1.3}$$

Now our stratification of X induces a stratification of each $X \times G^n$ by the $X_i \times G^n$, so that each $X_i \times G^n$ is a stratified pseudo-manifold. Since our stratification is G-invariant, all face maps d_i :

$X \times G^{n+1} \to X \times G^n$ are smooth maps, and the inverse image by d_i of the stratification of $X \times G^n$ is the stratification of $X \times G^{n+1}$.

Now take any of the D.G.M. complexes, for a perversity $\overline{p}$, discussed in §1. On each $X \times G^n$, we then get a complex of soft sheaves $K_n^\bullet$ on $X \times G^n$, and, as discussed in 1.1.3, we have for $i \in \{0, \cdots, n\}$ a morphism $d_i^* K_{n-1}^\bullet \to K_n^\bullet$ of complexes of sheaves on $X \times G^n$. These morphisms behave nicely under composition (to see this, one uses the well-known category whose objects $[n]$ are indexed by integers $n \geqslant 0$, and such that the morphisms from $[m]$ to $[n]$ are the injections of $\{0, \cdots, m\}$ into $\{0, \cdots, n\}$, which are increasing). We denote by d_i^* the corresponding map $\Gamma(X \times G^{n-1}, K_{n-1}^\bullet) \to \Gamma(X \times G^n, K_n^\bullet)$. We then have a double complex $L^{i,j} = \Gamma(X \times G^i, K_i^j)$, with the differential $L^{i,j} \xrightarrow{d} L^{i,j+1}$ induced by that of the complex $K_i^\bullet$, and the differential $L^{i,j} \xrightarrow{\delta} L^{i+1,j}$ given by $\delta = d_0^* - d_1^* + d_2^* \cdots$. Let $L^\bullet$ be the associated simple complex.

Definition 2.1.3. *The equivariant intersection cohomology groups* $IH^m_{\overline{p},G}(X) = IH^m_G(X)$ *are the cohomology groups of* $L^\bullet$.

To justify this definition, it is necessary to prove independence of the DGM complex $K^\bullet$. This follows from Principle 1.1.5, applied to the face map d_i.

Let $\overline{X} = X|G$ be the quotient space. Then one may localize the previous construction to produce a complex of sheaves on $\overline{X}$, whose global sections on an open set $\overline{U}$ are equal to the simple complex associated to $\Gamma(U \times G^\bullet, K^\bullet)$, where $U = \pi^{-1}(\overline{U})$ (for $\pi : X \to \overline{X}$ the canonical projection). We will denote by $K_G^\bullet$ this complex of sheaves on $\overline{X}$, which depends on the choice of the D.G.M. complex $K^\bullet$; however, $K_G^\bullet$ is well-defined up to quasi-isomoprhism. $K_G^\bullet$ may be described as the direct image of the simplicial complex of sheaves $(K_n^\bullet)$ on $X \overset{G}{\times} EG$, under the augmentation map $X \overset{G}{\times} EG \to \overline{X}$ (an augmentation map ϵ of a simplicial space (Z_n) to a space S is just a map $\epsilon : Z_o \to S$ such that the two composite maps $\epsilon \circ d_o$ and $\epsilon \circ d_1$, from Z_1 to S, coincide; see for instance **[15]**). Note that for the 0-perversity, $IH^m_{\overline{o},G}(X)$ is just the usual equivariant cohomology $H^m_G(X)$.

2.1.4. Let us examine $IH^*_G(X)$ in special cases.

If G acts trivially on X, $IH^*_G(X) = IH^*(X) \otimes H^*(BG)$. More precisely, the complex of sheaves $K^\bullet_G$ on $\overline{X} = X$ is quasi-isomorphic to a direct sum $\bigoplus_i H^i(BG)[-i]$ of shifted constant sheaves.

Now let G be a Lie group (compact, as always).

If G acts freely on X, it is easily seen that $\overline{X}$ is a pseudo-manifold, with topological stratification given by the X_i/G. Then $IH^m_G(X) = IH^{m+d}(\overline{X})$, where $d = \dim(G)$.

If G is a compact abelian Lie group, and if there exists a closed sub-group H of G such that $G_x = H$ for all $x \in X$, then again $\overline{X}$ is a pseudo-manifold, and we have:

$$\begin{aligned} IH^m_G(X) &= IH^m_H(X) \underset{H^*(BH)}{\otimes} H^*(BG) \\ &= IH^{m+e}(X) \underset{H^*(BH)}{\otimes} H^*(BG) \\ &= IH^{m+e}(X) \underset{k}{\otimes} H^*(B(G/H)) \end{aligned}$$

(for $e = \dim H$).

§**2.2.** It is shown in [**23, §5.2**] that if $\overline{\ell}$, $\overline{m}, \overline{p}$ are perversities, with $\overline{\ell}+\overline{m} \leqslant \overline{p}$, there is a product morphism $\underline{IC}^\bullet_\ell \otimes IC^\bullet_m \to \underline{IC}^\bullet_{\overline{p}}$. In loc. cit., this product is constructed on derived category objects; however, using the canonical flasque complexes of § 1.1, one obtains an actual morphism of complexes $P^\bullet_{\overline{\ell}} \otimes P^\bullet_{\overline{m}} \xrightarrow{U} P^\bullet_{\overline{p}}$; the point is that for complexes of sheaves $K^\bullet$, $L^\bullet$, there is a natural morphism of complexes $\underline{Go}(K^\bullet) \otimes \underline{Go}(L^\bullet) \to \underline{Go}(K^\bullet \otimes L^\bullet)$ we recall that $\underline{Go}$ means "Godement flasque resolution", cf. §1.1). Then for each n, one gets a product morphism of complexes of sheaves on $X \times G^n$; this is compatible with face maps. Now a variant of the shuffle-product, with degeneracies replaced by face maps, allows us to define the product of $\alpha \in \Gamma(X \times G^a, P^\bullet_{\overline{\ell}})$ and $\beta \in \Gamma(X \times G^b, P^\bullet_{\overline{m}})$ as the following element of $\Gamma(X \times G^{a+b}, P^\bullet_{\overline{p}})$:

$$\sum_{(\mu,\nu)} \epsilon(\mu,\nu)(d^*_{\nu_b} \cdots d^*_{\nu_1}\alpha) \cup (d^*_{\mu_a} \cdots d^*_{\mu_1}\beta) \tag{2.2.1}$$

where (μ, ν) runs over all (a, b)-shuffles of $\{1, \dots, a+b\}$, i.e. permutations of $\{1, \dots, a+b\}$, say $(\mu_1, \dots, \mu_a, \nu_1, \dots, \nu_b)$, such that $\mu_1 < \cdots < \mu_a$ and $\nu_1 < \cdots < \nu_b$, and where $\epsilon(\mu, \nu)$ is the sign of the permutation. This product is commutative and associative. It induces a morphism of complexes of sheaves on $\overline{X}$

$$(P^{\bullet}_{\overline{\ell}})_G \otimes (P^{\bullet}_{\overline{m}})_G \to (P^{\bullet}_{\overline{p}})_G. \tag{2.2.2}$$

This induces intersection pairings.

(2.2.3) In particular, for any perversity $\overline{\ell}$, $IH^*_{\overline{\ell},G}(X)$ is a graded module over the ordinary equivariant cohomology ring $H^*_G(X)$.

Another model of D.G.M. complexes for which products are defined is the complex $\Omega^{\bullet}_{\overline{\ell}}$ of §1.2 (defined when X has a certain C^{∞}-structure, which is preserved by the action of G; in particular, we consider here only compact Lie groups G, and the action of G should be C^{∞}, and should preserve Thom-Mather data). Indeed, it follows easily from (1.2.3) that the cup-product on $j_*\Omega^{\bullet}_{U_2}$ induces a product of complexes

$$\Omega^{\bullet}_{\overline{\ell}} \otimes \Omega^{\bullet}_{\overline{m}} \to \Omega^{\bullet}_{\overline{p}} \text{ when } \overline{\ell} + \overline{m} \leqslant \overline{p}.$$

§2.3 We fix a perversity $\overline{m}$, and we analyse the spectral sequence, for
$IH^*_{\overline{m},G}(X)$, for the fibration of simplicial spaces:

$$\begin{array}{ccc} X & \longrightarrow & X \overset{G}{\times} EG \\ & & \Big\downarrow p \\ & & BG \end{array}$$

(recall $BG = \text{pt} \overset{G}{\times} EG$, so that $(BG)_n = G^n$). As in §2.1, we have a simplicial complex of sheaves $(K^{\bullet}_n)$ on $X \overset{G}{\times} EG$, each $K^{\bullet}_n$ being a complex of soft sheaves on $X \times G^n$. The direct image $p_*(K^{\bullet}_n)$ is a simplicial complex of sheaves on BG, made of $p_{n,*}(K^{\bullet}_n)$ on G^n, where $p_n : X \times G^n \to G^n$ is the projection. The cohomology sheaves of each complex $p_{n,*}(K^{\bullet}_n)$ are locally constant sheaves on G^n, of fiber $IH^i_{\overline{m}}(X)$. The Leray spectral sequence is obtained from the filtration of $p_*(K^\bullet_n)$ by the sub-simplicial complex of sheaves which is $\tau_{\leqslant \ell} p_{n,*}(K^\bullet_n)$ on G^n. Hence this spectral sequence may be written $E_2^{p,q} = H^p(BG, IH^q_{\overline{m}}(X)) \Rightarrow IH^{p+q}_{\overline{m},G}(X)$. If G is a compact Lie group, the above simplicial local systems on BG are necessarily constant, so the spectral sequence may be written $E_2^{p,q} = H^p(BG) \underset{k}{\otimes} IH^q_{\overline{m}}(X) \Rightarrow IH^{p+q}_{\overline{m},G}(X)$. For the 0-perversity, this is the Leray spectral sequence considered by Borel in [**8, Chapter IV**]. This spectral sequence has a algebra structure. It follows from §2.2 that the spectral sequence relative to any perversity $\overline{m}$ has a module structure over the latter; on the E_2-terms, this module structure is deduced, by extension of scalars from k to $H^*(BG)$, from the $H^*(X)$-module structure on $IH^*_{\overline{m}}(X)$.

§**2.4** Let X be a topological pseudo-manifold on which the compact group G acts continuously, preserving a stratification, as in §2.1.

We assume that the stratification satisfies the following condition, which we call "G-local triviality". Let x be any point of X, with stabilizer G_x. Let S_i be the stratum containing x. There should exist a G-invariant neighborhood N of x, a G_x-invariant slice Z to the orbit $G.x$ inside S_i, and a compact Hausdorff space L with a continuous action of G_x, together with a G_x-invariant topological stratification

$$L = L_{n-i-1} \supset \cdots \supset L_1 \supset L_0 \supset L_{-1} = \emptyset,$$

and a G-equivariant homeomorphism $\Phi : G \times^{G_x} (S_i \times cone(L)) \to N$, which takes each $G \times^{G_x} (S_i \times cone(L_j))$ homeomorphically to $N \cap X_{i+j-1}$.

Notice this is an equivariant version of the local triviality condition in [**23, §1.1**]. If X is real-analytic and G acts by real-analytic automorphisms, such a stratification which is "G-locally trivial" exists.

Lemma 2.4.1. *If X admits a G-locally trivial stratification, then it admits a G-locally trivial stratification $X = Z_n \supset Z_{n-1} \supset \ldots Z_0 \supset Z_{-1} - \emptyset$ such that, for any x in X, there exists a neigh-*

borhood W of x in its stratum with the property that for y in W, the stabilizer subgroup G_y is conjugate to G_x.

Proof. Start from a G-invariant stratification

$$X = X_n \supset X_{n-1} \cdots \supset X_1 \supset X_0 \supset X_{-1} = \emptyset,$$

which is G-locally trivial. By induction on the integer k, $0 \leq k \leq n$, we construct a G-invariant subspace Z_{n-k} of X in such a way that:

(a)

$$X = Z_n \supset Z_{n-1} \supset \dots Z_{n-k} \supset X_{n-k-1} \cdots \supset X_0 \supset X - 1 = \emptyset$$

is a G-locally trivial stratification;

(b) $Z_{n-i} - Z_{n-i-1}$ is a topological manifold of dimension $n-i$, for $0 \leq i \leq k-1$;

(c) for $0 \leq i \leq k-1$, and for any x in $Z_{n-i} - Z_{n-i-1}$, there exists a neighborhood W of X inside $Z_{n-i} - Z_{n-i-1}$ such that, for $y \in W$, G_y is conjugate to G_x.

Note that we do not assert that $Z_{n-k} - X_{n-k-1}$ is a manifold.

If Z_{n-i} has been constructed for $i \leq k$, in such a way that (a), (b) and (c) are satisfied for integers $\leq k$, we construct $Z_{n-k-1} = X_{n-k-1} \cup Y_1 \cup Y_2$, where Y_1 is the "singular set" of Z_{n-k} (i.e. the set of points near which Z_{n-k} is not a manifold of dimension $n-k$), and Y_2 is the subset of Z_{n-k} consisting of points near which the conjugacy type of the stabilizer subgroup is not locally constant.

Clearly Z_{n-k-1} is G-invariant, (b) and (c) are satisfied, for $0 \leq i \leq k$. To verify (a) for the new stratification

$$X = Z_n \supset Z_{n-1} \supset \dots Z_{n-k} \supset Z_{n-k-1} \supset X_{n-k-2} \cdots \supset X_0 \supset X_1 = \emptyset$$

one simply observes that the local G-homeomorphism Φ for the previous stratification preserves the subspace Z_{n-k-1}, in the sense that Φ maps $G \times^{G_x} (S_i \times cone(A))$ homeomorphically to $N \cap Z_{n-k-1}$, for a suitable G_x-subspace A of L. But this is clear, if one defines $A := L_{n-k-1-i} \cup B_1 \cup B_2$, where B_1 is the singular set of $Z_{n-k} \cup L$, and B_2 the set of x in $Z_{n-k} \cup L$ near which the conjugacy type of the stabilizer subgroup (inside G_x) is not locally constant.

For $k = n$, the stratification obtained has all the required properties.

The advantage of such a stratification is the following. Denote by $\overline{Z}_k = \pi(Z_k)$ the induced stratification of $\overline{X}$.

Proposition 2.4.2. *For $K^\bullet$ a D-G-M complex, the cohomology sheaves of the complex $K^\bullet_G$ are locally constant on each stratum $\overline{Z}_k - \overline{Z}_{k-1}$.*

Proof. We consider the Leray spectral sequence for the augmentation $\epsilon : X \overset{G}{\times} EG \to \overline{X}$, applied to the simplicial complex of sheaves $(K^\bullet_n)$. This gives a spectral sequence, in the category of sheaves of k-vector spaces over $\overline{X}$:

$$E_2^{p,q} = E_2^{p,q} = \mathcal{K}^p \otimes R^q \epsilon_*(k_{X \overset{G}{\times} EG}) \Rightarrow \underline{H}^{p+q}(K^\bullet_G)$$

where $\mathcal{K}^p$ is the constructible sheaf on $\overline{X}$ with stalk at $\overline{x} = \pi(x)$ equal to $H^p(K^\bullet_x)^{G_x}$ (see [**2**] for a discussion of descent to $\overline{X}$ of G-equivariant sheaves on X). Now each $\mathcal{K}^p$ is locally constant on each stratum (because the topology is locally constant, and G_x too). As for $R^q\epsilon_*(kX \overset{G}{\times} EG)$, its restriction to a stratum, associated to the subgroup H (up to conjugacy) is locally constant, of fiber equal to $H^q(BH)$ (if G is abelian, it is even constant, as a quotient of the constant sheaf of fiber $H^q(BG)$). Since the spectral sequence converges, the cohomology sheaves of $K^\bullet_G$ are also locally constant on the stratum.

(2.4.3) Let us note that in the above proof, we have found a spectral sequence $E_2^{p,q} = [\underline{H}^p(K^\bullet_x)]^{G_x} \otimes H^q(BG_x) \Rightarrow \underline{H}^{p+1}(K^\bullet_G)_x$ for $x \in X$, $\overline{x} = \pi(x) \in \overline{X}$. In §3.1, we will give an interpretation of the differential d_2 of this spectral sequence, when G is a compact torus acting differentiably on a C^∞ pseudo-manifold X.

§3. A perverse version of the Cartan equivariant complex.

§3.1. For a torus T acting differentiably on a C^∞-manifold M, Henri Cartan [**13**, §**6**] introduces a complex of differential forms $\Omega^*(M)^T \otimes S(t^*)$, where $t = \mathrm{Lie}(T)$, and an element of T^* is of degree 2 (see also [**13**]).

We consider here a differentiable action of T on a pseudo-manifold X, and we consider the complex $[\Omega^*_{\overline{p}}]^T \otimes S(t^*)$ of sheaves on $\overline{X}$, where $\Omega^*_{\overline{p}}$ is the complex of differential forms studied in §1.2, and we consider the sub-complex of T-invariant forms $[\Omega^*_{\overline{p}}]^T$ to be a complex of sheaves on $\overline{X}$. The differential D of the sheaf of algebras $[\Omega^*_{\overline{p}}]^T \otimes S(t^*)$ is given by the following rules, where (ξ_k) denotes a basis of t, (u_k) the dual basis.

$$D(\omega \otimes 1) = (d\omega) \otimes 1 + \sum_k i(\xi_k)\omega \otimes u_k \, . \tag{3.1.1}$$

$$D(1 \otimes P) = 0 \text{ for } P \in S(t^*) \, . \tag{3.1.2}$$

Note that $i(\xi_k)$ operates on $\Omega^*_{\overline{p}}$ by (1.2.4), hence on $[\Omega^*_{\overline{p}}]^T$. We will now construct an explicit quasi-isomorphism $[\Omega^*_{\overline{p}}]^T \otimes S(t^*) \to (\Omega^*_{\overline{p}})_T$. First we embed $[\Omega^*_{\overline{p}}]^T$ into $\pi_*\Omega^*_{\overline{p},X}$, which sits inside $(\Omega^*_{\overline{p}})_T$ (not as a sub-complex, though); next we send any $u \in t^*$ to the pull-back to $X \times T$ of the differential form du on T; since the shuffle-product in §2.2 defines a (super) commutative algebra structure on the de Rham complex of the simplicial manifold BT, we then know how to map $S(t^*)$ to that algebra; note that de Rham algebra acts on the simplicial algebra $\Omega^*_{\overline{p}}(X \times T^n)$, so we have a map $[\Omega^*_{\overline{p}}]^T \otimes S(t^*) \to (\Omega^*_{\overline{p}})_T$ which is a morphism of sheaves of algebras on $\overline{X}$. It remains to check the compatibility with the differentials. Recall that $\xi \in t$ acts on functions on X by $(\xi.f)(p) = \frac{d}{dt}f(\exp(-t\xi)p)_{|t=0}$. Now take a T-invariant differential form $\omega \in [\Omega^*_{\overline{p}}]^T$ which maps to itself, viewed as a section of $\Omega^*_{\overline{p}}$ over X. The differential of this element is the sum of $d\omega \in \Omega^*_{\overline{p},X}$ and of $d_0^*\omega - d_1^*\omega \in \Omega^*_{\overline{p},X\times T}$. Now, since ω is T-invariant, we have $d_0^*\omega = \omega \otimes 1 + \sum_k i(\xi_k)\omega \otimes u_k$; also $d_1^*\omega = \omega \otimes 1$, therefore $d_0^* - d_1^*\omega = \sum_k i(\xi_k)\omega \otimes u_k$ which fits with (3.1.1). Also $du \in \Omega^1(T)$ is closed and $d_0^*(du) - d_1^*(du) + d_2^*(du) = (-du \otimes 1 + 1 \otimes du) - 1 \otimes du + du \otimes 1 = 0$, since du is translation-invariant. This defines our morphism of complexes of sheaves. We check that it gives an isomorphism on the cohomology of the stalks at $\overline{x}$, for any $x \in X$. For this purpose, we filter $[\Omega^*_{\overline{p}}]^T \otimes S(t^*)$ by the subcomplexes of sheaves $\tau_{\leqslant k}[\Omega^*_{\overline{p}}]^T \otimes S(t^*)$, and we filter $(\Omega^*_{\overline{p}})_T$ by the subcomplexes corresponding to the simplicial complexes of sheaves $(\tau_{\leqslant k}\Omega^*_{\overline{p},X\times T^n})_n$. On the subquotients of these filtrations, we have an induced morphism, which on the cohomology of the stalks at $\overline{x}$, gives a map $H^*(\Omega^*_{\overline{p},x}) \otimes S(t^*) \to H^*(\Omega^*_{\overline{p},x}) \otimes H^*(BT)$. This map is $id \otimes Car$, where Car sends $u \in t^*$ to $du \in H^2(BT)$ as before. Car is an isomorphism hence we have a quasi-isomorphism on stalks. It follows that an inclusion of complexes of sheaves $[\Omega^*_{\overline{p}}]^T \otimes S(t^*) \to (\Omega^*_{\overline{p}})_T$ is a quasi-isomorphism. Hence we deduce

Proposition 3.1.3. *$IH^j_{\overline{p},T}(X)$ is equal to the hypercohomology group*
$H^j(\overline{X}, [\Omega^*_{\overline{p}}]^T \otimes S(t^*))$, as an $S(t^*) = H^*(BT)$-module.

If $\overline{\ell}$, $\overline{m}, \overline{p}$ are perversities such that $\overline{\ell} + \overline{m} \leqslant \overline{p}$, the product $IH^*_{\overline{\ell},T}(X) \otimes IH^*_{\overline{m},T}(X) \to IH^j_{\overline{p},T}(X)$ is obtained from the product on the level of complexes of sheaves:

$$[\Omega^*_{\overline{\ell}}]^T \otimes S(t^*) \otimes [\Omega^*_{\overline{m}}]^T \otimes S(t^*) \to [\Omega^*_{\overline{p}}]^T \otimes S(t^*).$$

3.1.4. We will use the Cartan model to investigate the stalk at some $\overline{x} = \pi(x)$, for $x \in X$. For this purpose, we generalize in the obvious way the definition of the Cartan model $[\Omega^*_{\overline{p}}]^T \otimes S(t^*)$ to any abelian compact Lie group T (not necessarily connected). Now, given $x \in X$, we may find a T_x-stable <u>slice</u> S at X, such that $T \overset{T_x}{\times} S$ is a neighborhood of x in X, and with S transverse to all strata. Then X is given the induced stratification, which is T_x-stable. We then have a restriction map $[\Omega^*_{\overline{p},T \overset{T_x}{\times} S}]^T \to [\Omega^*_{\overline{p},S}]^{T_x}$ hence a restriction morphism

$$[\Omega^*_{\overline{p},T \overset{T_x}{\times} S}]^T \otimes S(t^*) \to [\Omega^*_{\overline{p},S}]^{T_x} \otimes S(t^*_x).$$

This is clearly a quasi-isomorphism. Indeed, it is even a filtered quasi-isomorphism, i.e. it induces a quasi-isomorphism

$$(\tau_{\leqslant \ell}[\Omega^*_{\overline{p},T \overset{T_x}{\times} S}]^T) \otimes S(t^*) \to (\tau_{\leqslant \ell}[\Omega^*_{\overline{p},S}])^{T_x} \otimes S(t^*_x).$$

Since we want to look at the cohomology of the stalks at $\overline{x}$, we restrict to the case where S is a "conical" slice. Then, since T_x acts continuously on $\Omega^*_{\overline{p}}(S)$, the cohomology of $[\Omega^*_{\overline{p}}(S)]^{T_x}$ is the same as the space of T_x-invariant elements in $IH^*_{\overline{p}}(S)$, which is the cohomology of the stalk $\Omega^*_{\overline{p},x}$ (up to a shift of dim (S)). Hence we have a spectral sequence (where $E_2^{p,q} = 0$ for p odd)

$$E_2^{p,q} = [H^q(S, \Omega^*_{\overline{p}})]^{T_x} \otimes S^{p/2}(t_x^*) \Rightarrow \underline{H}^{p+q}(\Omega^*_{\overline{p},T})_{\overline{x}}.$$

Since $S^{p/2}(t_x^*) = H^p(BT_x)$, this identifies with the spectral sequence of (2.4.3) (with p-degrees shifted by n), in view of the above filtered quasi-isomorphism of complexes of sheaves. The differential d_2 is specified by the values of $d_2(\omega \otimes 1)$, for ω a closed, T_x-invariant differential form in $\Omega^*_{\overline{p}}(S)$. It is given by $d_2(\omega \otimes 1) = \sum_k i(\xi_k)\omega \otimes u_k$ with the notations of (3.1.1) (so (ξ_k) is a basis of t_x, (u_k) the dual basis of t_x^*).

3.1.5. The <u>co-stalk</u> cohomology, i.e. the hypercohomology with support in $\overline{x}$, is analysed in a similar (but dual) manner. In this case we use the morphism

$$\Gamma_c(S, \Omega^*_{\overline{p}})^{T_x} \otimes S(t_x^*) \to \Gamma_c(T \overset{T_x}{\times} S, \Omega^*_{\overline{p}})^T \otimes S(t^*)$$

obtained as follows; replacing, if necessary, T by a finite covering group (which changes nothing), we may assume that T splits as a product $T = T_x \times N$ of compact Lie groups; then we may identify $T \overset{T_x}{\times} S$ with $N \times S$, in such a way that the action T_x becomes the action of T_x on S, and the action of N is by translation on N; then we map $S(t_x^*)$ to $S(t^*)$, viewing t_x as direct factor of t; and we map $\alpha \in \Gamma_c(S, \Omega^*_{\overline{p}})^{T_x}$ to $\omega \wedge \alpha$ on $N \times S$, where ω is a fixed translation-invariant volume form on N.

So we obtain a spectral sequence

$$E_2^{p,q} = [H_c^q(S, \Omega^*_{\overline{p}})]^{T_x} \otimes S^{p/2}(t_x^*) \Rightarrow H_{\overline{x}}^{p+q}(\Omega^*_{\overline{p},G})$$

where $H_c^p(S, \Omega^*_{\overline{p}})$ may be identified with the co-stalk of $\Omega^*_{\overline{p}}$ at X, up to a shift.

§**3.2** In this section, we investigate Poincaré-Verdier duality for the Cartan-model of equivariant intersection cohomology. So we have a compact abelian Lie group T acting differentibly on a C^∞-pseudo-manifold X. We have a perversity $\overline{p}$ and the complementary perversity $\overline{q}$. The complexes $[\Omega^*_{\overline{p}}]^T \otimes S(t^*)$ and $[\Omega^*_{\overline{q}}]^T \otimes S(t^*)$ are graded differential sheaves of modules over the graded ring $S(t^*)$, where t^* is put in degree 2. In fact, the stalks of either complex are free $S(t^*)$-modules. Hence we may define the Verdier dual complex of $[\Omega^*_{\overline{p}}]^T \otimes S(t^*)$, as the complex of sheaves on $\overline{X}$

$$\overline{U} \mapsto \mathrm{Hom}_{S(t^*)}(\Gamma_c(\overline{U}, [\Omega^*_{\overline{p}}]^T \otimes S(t^*)),\ S(t^*))$$

(i.e. it is not necessary to consider an injective resolution of $S(t^*)$, as in [**9, V,** §**7**]). This is a graded sheaf of modules over $S(t^*)$.

We are going to define a $S(t^*)$-linear morphism of complexes of sheaves from $[\Omega^*_{\overline{p}}]^T \otimes S(t^*)$ to the Verdier dual of $[\Omega^*_{\overline{p}}]^T \otimes S(t^*)$ shifted by n. This amounts to constructing, for each open $\overline{U}$ of $\overline{X}$, a $S(t^*)$-bilinear pairing of complexes $\Gamma(\overline{U}, [\Omega^*_{\overline{q}}]^T \otimes S(t^*)) \otimes \Gamma_c(\overline{U}, [\Omega^*_{\overline{p}}]^T \otimes S(t^*)) \to S(t^*)$. This is defined as follows: $[\alpha \otimes P] \otimes [\beta \otimes Q] \mapsto (\int_U \alpha \wedge \beta).PQ$; since $\alpha \wedge \beta$ belongs to $\Omega^*_{\overline{t}}(U)$, where $\overline{t}$ is the maximal perversity $\overline{t}(c) = c - 2$ (cf. [**22**], [**23**]), its part of maximal degree has to be zero near X_{n-2}, so that $\alpha \wedge \beta$ is in fact a differential form on $U - X_{n-2}$ with compact support, and the integral $\int_U \alpha \wedge \beta$ makes good sense. The fact that this gives a pairing of complexes follows from the following observation:

(i) $\int_U d\alpha \wedge \beta = (-1)^{k+1} \int_U \alpha \wedge d\beta$ for $\alpha \in [\Omega^k_{\overline{q}}]^T$, $\beta \in [\Omega^\ell_{\overline{p}}]^T$

(ii) $\int_U [i(\xi)\alpha] \wedge \beta = (-1)^{k+1} \int_U \alpha \wedge [i(\xi)\beta]$ for $\xi \in t$.

This pairing is obviously compatible with restriction of open sets. Now we want to show that this morphism from $[\Omega^*_{\overline{q}}]^T \otimes S(t^*)$ to the dual of $[\Omega^*_{\overline{p}}]^T \otimes S(t^*)$ is a quasi-isomorphism. We work at a point $\overline{x} = \pi(x)$; the first observation is that we may assume that $T = T_x$, i.e. that x is a fixed point of T. Indeed, with the notations of §3.1.5, taking $U = T \overset{T_x}{\times} S$, $\Gamma(\overline{U}, [\Omega^*_{\overline{q}}]^T \otimes S(t^*))$ is quasi-isomorphic to the tensor product of $\Gamma(N, \Omega^*_N \otimes S(\mathcal{N}^*))$ and of $\Gamma(\overline{S}, [\Omega^*_{\overline{q}}]^{T_x} \otimes S(t^*_x))$ (where $\overline{S} = S/T_x$) and the pairing on the tensor product complex is the tensor product of a pairing

$$\Gamma(N, \Omega^*_N)^N \otimes (\mathcal{N}^*)) \otimes \Gamma(N, \Omega^*_N)^N \otimes S(\mathcal{N}^*)) \to S(\mathcal{N}^*)$$

and of a pairing

$$\Gamma(\overline{S}, [\Omega^*_{\overline{q}}]^{T_x} \otimes S(t^*_x)) \otimes \Gamma_c(\overline{S}_c[\Omega^*_{\overline{p}}]^{T_x} \otimes S(t^*_x)) \to S(t^*_x).$$

We wish to show that it gives a quasi-isomorphism from $\Gamma(N, \Omega_N^*)^N \otimes S(\mathcal{N}^*))$ to its $S(\mathcal{N}^*)$-dual. Since N is isogenous to a torus, we may easily reduce the verification to the case $N = S^1$. The cohomology of this complex is concentrated in degree 0, and is generated there by the global section 1 (identity element of the algebra). The dual complex has cohomology concentrated in degree 0, with generator the $S(\mathcal{N}^*) = \mathbb{C}[u]$-linear map which sends $\alpha \otimes P(u)$ to $(\int_{S^1} \alpha).P(u)$ for $P \in \mathbb{C}[u]$. The cohomology class of this linear map is killed by u, since the linear form $\alpha \otimes P(u) \mapsto (\int_{S^1} \alpha)uP(u)$ is the boundary of the linear form $\alpha \otimes P(u) \mapsto (\int_{S^1} \alpha\omega)P(u)$ where $\omega = du$ is the volume form on S^1. Since we map 1 to a generator of the cohomology of the dual complex, our map is indeed a quasi-isomorphism.

Now we have to show that the other pairing induces a quasi-isomorphism. So we are reduced to the case $T_x = T$. Our pairing maps $\Gamma(\overline{U}, \tau_{\leqslant \ell}[\Omega_{\overline{q}}^*]^T \otimes S(t^*))$, to the dual of $\Gamma_c(\overline{U}, ([\Omega_{\overline{p}}^*]^T / \tau_{\leqslant n-\ell}[\Omega_{\overline{p}}^*]^T) \otimes S(t^*))$, so it suffices to show that it induces a perfect pairing between the cohomology of the subquotients of the $\tau_{\leqslant ?}$ filtrations, which are free $S(t^*)$-modules. Then we get a pairing between $H^\ell(U, \Omega_{\overline{q}}^*)^T \otimes S(t^*)$ and $H_c^{n-\ell}(U, \Omega_{\overline{p}}^*)^T \otimes S(t^*)$. But the pairing between $H^\ell(U, \Omega_{\overline{q}}^*)$ and $H_c^{n-\ell}(U, \Omega_{\overline{p}}^*)$ is perfect, and we still get a perfect pairing on the spaces of T-invariants. The pairing, extended $S(t^*)$-linearly, remains perfect. This concludes the proof, and we may state the

Theorem 3.2.1. *The Verdier-Poincaré dual of the $S(t^*)$-complex of sheaves $[\Omega_{\overline{p}}^*]^T \otimes S(t^*)$ is canonically isomorphic to $[\Omega_{\overline{q}}^*]^T \otimes S(t^*)$, shifted by n.*

Corollary 3.2.2. *If the compact abelian Lie group T acts differentiably on the C^∞-pseudomanifold X, which is compact, then there is a spectral sequence in the category of graded $S(t^*)$-modules*

$$E_2^{i,j} = Ext^i_{S(t^*)}(IH^{n-j}_{\overline{p},T}(X), S(t^*)) \Rightarrow IH^{i+j}_{\overline{q},T}(X)$$

for $(\overline{p}, \overline{q})$ complementary perversities.

This sort of Poincaré duality was first proved by Bredon [**10**] for actions of $\mathbb{Z}/p \cdot \mathbb{Z}$. Let us make this more concrete in case $T = S^1$.

We then have an exact sequence

$$0 \to \operatorname{Ext}^1_{\mathbb{C}[u]}(IH^{n-m-1}_{\overline{p},S^1}(X), \mathbb{C}[u]) \to IH^m_{\overline{q},S^1}(X) \to \operatorname{Hom}_{\mathbb{C}[u]}(IH^{n-m}_{\overline{p},S^1}(X), \mathbb{C}[u]) \to 0$$

In case S^1 acts trivially on $X, IH^*_{\overline{p},S^1}(X) = IH^*_{\overline{p}}(X) \underset{\mathbb{C}}{\otimes} \mathbb{C}[u]$ so the Ext^1 term vanishes and we just have ordinary Poincaré duality for IH^*, tensored with $\mathbb{C}[u]$. On the other hand, if the action is free, then $\overline{X} = X/S^1$ is a stratified pseudomanifold, and $IH^m_{\overline{p},S}(X) = IH^m_{\overline{p}}(X)$, which is a <u>torsion</u> $\mathbb{C}[u]$-module. Then we have

$$IH^m_{\overline{q}}(\overline{X}) = \operatorname{Ext}^1_{\mathbb{C}[u]}(IH^{n-m-1}_{\overline{p}}(\overline{X}), \mathbb{C}[u]).$$

This is also isomorphic to the ordinary dual of $IH^{n-m-1}_{\overline{p}}(\overline{X})$, so that we recover Poincaré duality for the intersection cohomology of the quotient pseudomanifold $\overline{X}$, of dimension $n-1$.

Even for smooth manifolds, the Poincaré duality obtained here (with $S(t^*)$ as coefficient ring) appears new. It suggests an analogous duality for the delocalized equivariant cohomology of [**2**]; such a duality exists, at least for $T = S^1$, as we will show in a future paper.

§4. The localization theorem and a fixed point theorem.

§4.1 Let a compact torus T act on a pseudomanifold X, preserving a topological stratification of X, as in §2.1. Assume X is compact, let X^T be the fixed point set of T in X, fix a perversity $\overline{p}$. There is a natural map of $S(t^*) = H^*(BT)$-modules

$$IH^*_T(X, X - X^T) \xrightarrow{\varphi} IH^*_T(X)$$

where the first group is a relative equivariant intersection cohomology group. As in [**28, Theorem 4.4**], the following <u>localization theorem</u> holds:

Theorem 4.1.1. *Let $\Sigma \subset t$ be $\bigcup_{x \in X - X^T} t_x$, where t_x is the Lie algebra of the stabilizer T_x. Then Σ is a closed subvariety of t (for the Zariski topology), and if A is the localization of $S(t^*)$ with* *respect to the multiplicative subset of functions which are invertible on $t - \Sigma$, induces an isomorphism*

$$IH_T^*(X, X - X^T) \underset{S(t^*)}{\otimes} A \xrightarrow{\sim} IH_T^*(X) \underset{S(t^*)}{\otimes} A.$$

Proof. It suffices to prove that $IH_T^*(X - X^T) \underset{S(t^*)}{\otimes} A$ is zero. In view of the "local to global" spectral sequence for the complex of sheaves $P_T^* = (P_{\overline{p}}^*)_T$ on $\overline{X}$, it suffices to prove that for every element $f \in S(t^*)$ which vanishes on Σ, some power f^N annihilates all stalks of P_T^* at all points of $\overline{X - X^T}$. Now consider the spectral sequence, for $x \in X - X^T$

$$E_2^{p,q} = \underline{H}^p(P_x^{\bullet})^{T_x} \otimes H^q(BT_x) \to H^{p+1}(P_T^{\bullet})_{\overline{x}}$$

considered in (2.4.3). Clearly f annihilates the E_2-term of this spectral sequence; since $E_2^{p,q}$ may be $\neq 0$ only for $0 \leqslant p \leqslant n-2$, the E_∞-term is killed by f^{n-1}, with $n = \dim(X)$.

Remark 4.1.2. *A similar localization theorem holds for the Cartan model of equivariant intersection cohomology, with exactly the same proof.*

Remark 4.1.3. *Using Poincaré duality for equivariant intersection of cohomology, as established in §3.2, we may translate Theorem 4.1.1 into the equivalent statement that the restriction map $IH_T^*(X) \to H^*(X^T, i^*(P_G^{\bullet}))$ becomes an isomorphism upon tensoring with A, i denoting the inclusion $X^T \overset{i}{\hookrightarrow} \overline{X}$. However $H^*(X^T, i^*(P_G^{\bullet}))$ is not the equivariant intersection cohomology of X^T, but that of a suitable T-invariant small neighborhood of X^T in X. For convenience, we state this as*

Corollary 4.1.3. *The restriction morphism*

$$IH_T^*(X) \to \varinjlim_U IH_T^*(U)$$

(where U runs over T-invariant neighborhoods of X^T in X) becomes an isomorphism after tensorization by A.

§4.2. To get fixed point theorems from the Segal-Quillen type localization theorem of §4.1, one needs, as usual, degeneracy of the spectral sequence of §2.3:

$$E_2^{p,q} = H^p(BT) \otimes IH^q(X) \Rightarrow IH_T^{p+q}(X).$$

However, the new and unusual fact is that the analogous "local" (near X^T) spectral sequence need not degenerate. Let us use $IH_T^*(Nbh\ X^T)$ to denote $\varinjlim_U IH_T^*(U)$ as in Corollary 4.1.3. This local spectral sequence is

$$E_2^{p,q} = H^p(BT) \otimes IH^q(NbhX^T) \Rightarrow IH_T^{p+q}(Nbh\ X^T).$$

The following theorem is an obvious consequence of Theorem 4.1.1 (two isomorphic free modules over A have the same rank).

Theorem 4.2.1.

(i) *if the global spectral sequence degenerates, we have* *inequalities:*

$$\sum_i \dim\ IH^{2i}(X) \leqslant \sum_i \dim\ IH^{2i}(Nbh\ X^T)$$
$$\sum_i \dim\ IH^{2i+1}(X) \leqslant \sum_i \dim\ IH^{2i+1}(Nbh\ X^T)$$

(ii) *if the global and the local spectral sequences both degenerate, we have equalities:*

$$\sum_i \dim\ IH^{2i}(X) = \sum_i \dim\ IH^{2i}(Nbh\ X^T)$$
$$\sum_i \dim\ IH^{2i+1}(X) = \sum_i \dim\ IH^{2i+1}(Nbh\ X^T)$$

This may be used to show that $X^T \neq \emptyset$. See [**10**] for a fixed point theorem for elementary abelian p-groups

Now let us give an important case where the global spectral sequence degenerates. We assume that X is a projective algebraic variety (maybe singular) over $\mathbb{C}$; we do not assume that the torus T acts algebraically on X, but we do assume that it preserves some Whitney stratification of X. We take for perversity the middle perversity $\overline{m}(c) = \frac{c-2}{2}$ (cf. [**22**], [**23**]). Let $\eta \in H^2(X)$ be the class of a hyperplane section. Since the intersection complex

$IC^{\bullet}_{\overline{m}}$ is pure by a theorem of Gabber [**18**], see also [**3, Corollaire 5.3.4**]), it follows from a theorem of Deligne that the hard Lefschetz theorem holds for $IH^*(X)$ ([**15**]). In other words, the iterated cup-product by η induces an isomorphism $IH^{n-i}(X) \xrightarrow{\cup\eta^i} IH^{n+i}(X)$ for all i (where $n = \dim_{\mathbb{C}} X$). If η lifts to a class $\widetilde{\eta}$ in $H^2_T(X)$, then $\widetilde{\eta}$ operates on the global spectral sequence and $\widetilde{\eta}^i$ induces an isomorphism

$$E_2^{p,-n-i} \xrightarrow{\cup\widetilde{\eta}^i} E^{p,n+i}$$

This allows us to apply a theorem of Blanchard ([**7**], and Deligne [**14**]) and we obtain

Theorem 4.2.2. *Assume η lifts to a class in $H^2_T(X)$ (for example assume $H^1(X,\mathbb{C}) = 0$). Then the global spectral sequence degenerates, X^T is not empty, and the inequalities (4.2.1)(i) hold.*

This is very analogous to theorems proved in [**8**]. The fact that the fixed point set is non-empty is proved in [**11**] for an elementary abelian p-group, and this implies the case of a torus action considered above. It is harder to obtain degeneracy of the local spectral sequence. Let us denote by $i : X^T \hookrightarrow X$ the inclusion. We now assume that T acts on X by algebraic automorphisms.

Theorem 4.2.3. *As in 4.2.2, assume η lifts to a class in $H^2_T(X)$. Assume also that $i^*\underline{IC}^{\bullet}_X$ is a pure complex on X^T. Then the local spectral sequence also degenerates, hence the equalities (4.2.1)(ii) hold.*

Indeed, since $i^*\underline{IC}^{\bullet}_X$ is a pure complex on X^T, it is the direct sum of $\mathcal{H}^i[-i]$, where $\mathcal{H}^i$ is the i-th perverse cohomology sheaf of $i^*\underline{IC}^{\bullet}_X$ [**3, Théorème 5.4.5**]. Now each $\mathcal{H}^i$ is a pure perverse sheaf, hence the hard Lefschetz theorem holds for $H^*(X^T, \mathcal{H}^i)$ [**16, Théorème 6.2.13**]. The restriction of η to $H^2(Nbh\ X^T)$ obviously has a lift to $H^2_T(Nbh\ X^T)$; so we apply Blanchard's theorem again to each spectral sequence for each $\mathcal{H}^i$. Details are left to the reader.

4.2.4. There are some cases where $i^*\underline{IC}^{\bullet}_X$ is known to be pure:

(a) if X^T is included in the smooth locus of X

(b) for $T = S^1$, if the action of T extends to an algebraic action of $\mathbb{C}^*$ on X, and for each connected component F of X^T, the action of $\mathbb{C}^*$ is either contracting or expanding near F (this follows from a result of Ginsburg [**19, Prop. 15.2**]).

4.2.5. Let X be the flag variety of a complex semi-simple Lie group G, $T \subset G$ a maximal compact torus, w an element of the Weyl group, $\overline{X}_w \subset X$ the corresponding Schubert variety. T operates on $\overline{X}_w$, and $(\overline{X}_w)^T = \bigcup_{\substack{y \in W \\ y \leq W}} \{y\}$. Since $H^1(\overline{X}_w, \mathbb{C}) = 0$, the global spectral sequence degenerates (4.2.2). Now for $y \leqslant w$, $IH^i(Nbhy)$ is zero for i odd; in the local spectral sequence, $E_2^{p,q} = 0$ unless both p and q are even; hence the local spectral sequence degenerates, and the equalities 4.2.1 (ii) hold. Hence we obtain

Proposition:.

(a) $IH^i(\overline{X}_w) = 0$ *for* i *odd*

(b) $\sum_i \dim IH^{2i}(\overline{X}_w) = \bigcup_{\substack{y \in W \\ y \leqslant w}} P_{y,w}(1)$

where $P_{y,w}$ *is the polynomial of Kazhdan-Lusztig* [**26**].

4.2.6 Let $Y \subset \mathbb{P}_n(\mathbb{C})$ be a smooth projective variety, and let $X \subset \mathbb{P}_{n+1}(\mathbb{C})$ be the projective cone over Y (closure in $\mathbb{P}_{n+1}(\mathbb{C})$ of the cone $C(Y) \subset \mathbb{C}^{n+1}$). Then there is an obvious algebraic action of $\mathbb{C}^*$ on X, hence $S^1 \subset \mathbb{C}^*$ acts on X. X^{S^1} is the union of $\{0\}$ and of Y, which is the intersection of X with the hyperplane at infinity in $\mathbb{P}_{n+1}(\mathbb{C})$. The first component satisfies 4.2.4 (b), the second 4.2.4 (a). Furthermore X is simply-connected. Hence by 4.2.3, the equalities 4.2.1 (ii) hold. Let us see how these equalities follow from the hard Lefschetz theorem for the smooth variety Y. Let $P^i(Y)$ be the primitive part of $IH^i(Y)$. Let U be the smooth open set $U = X - \{0\}$ of X. Then we have:

(i) for $i < n+1$, $IH^i(X) = H^i(U)$

(ii) for $i = n+1$, $IH^{n-1}(X) = im[H_c^{n+1}(U) \to H^{n+1}(U)]$

(iii) for $i > n+1$, $IH^i(X) = H_c^i(U)$

(see [**22**]). We have $H^j(U) = H^j(Y)$, $H_c^j(U) = H^{j+2}(Y)$. The map $H^{n-1}(Y) \simeq H_c^{n+1}(U) \to H^{n+1}(U) \simeq H^{n+1}(Y)$ is the cup-product with η, which is bijective. Hence we have

$$\sum_{i\equiv n(\mathrm{mod}\ 2)} \dim\ IH^i(X) = \sum_{i\equiv n(\mathrm{mod}\ 2)} \dim\ H^i(Y) + \dim\ H^n(Y)$$

$$\sum_{i\equiv n+1(\mathrm{mod}\ 2)} \dim\ IH^i(X) = \sum_{i\equiv n(\mathrm{mod}\ 2)} \dim\ H^i(Y) + \dim\ H^{n-1}(Y)$$

The local intersection cohomology groups $IH^i(Nbh\ 0)$ at 0 are deduced from the groups $H^*(C(Y) - \{0\})$. Now $C(Y) - \{0\}$ is a $\mathbb{C}^*$-bundle over Y; using the fact that cup-product with η gives an injection $H^j(Y) \hookrightarrow H^{j+2}(Y)$ for $j \leqslant n-1$, one gets

$$\sum_{i \text{ even}} \dim\ IH^i(Nbh\ 0) = \sum_{\substack{i \text{ even} \\ i\leqslant n}} \dim\ P^i(Y) \quad \text{and}$$

$$\sum_{i \text{ odd}} \dim\ IH^i(bh\ 0) = \sum_{\substack{i \text{ odd} \\ i\leqslant n}} \dim\ P^i(Y) \ .$$

Near the component Y of S^{S^1}, one easily gets:

$$\sum_{i \text{ even}} \dim\ IH^i(Nbh\ Y) = \sum_{i \text{ even}} \dim\ H^i(Y)$$

and similarly for the sum of i odd. Hence the equalities (4.2.1) (ii) amount to the following

$$\dim\ H^n(Y) = \sum_{\substack{i\equiv n \mod 2 \\ i\leqslant n}} \dim\ P^i(Y)$$

$$\dim\ H^{n-1}(Y) = \sum_{\substack{i\equiv n \mod 2 \\ i\leqslant n-1}} \dim\ P^i(Y)$$

which follow from the hard Lefschetz theorem.

This seems to suggest that our fixed point theorem uses the full force of the hard Lefschetz theorem.

References

1. M. F. Atiyah and R. Bott, *The moment map and equivariant cohomology*, Topology **23** (1984), 1–28.
2. P. Baum, J.-L. Brylinski and R. D. MacPherson, *Cohomologie équivariante délocalisée*, C.R.A.S. Paris **300** (1985), 605–608.
3. A. A. Beilinson, J. Berstein and P. Deligne, *Faisceaux pervers*, in "Analyse et Topologie sur les espaces singuliers," Astérisque vol. 100, Soc. Math. Fr., 1982.

4. N. Berline and M. Vergne, *Classes caractéristiques équivariantes*, C.R.A.S. Paris **295** (1982), 539–541.
5. N. Berline and M. Vergne, *Fourier transforms of orbits of the coadjoint representation*, in "Representation theory of reductive Lie groups," Progress in Math. vol. 40, Birkhaäuser Boston, 1983.
6. A. Bialynicki-Birula, *Some theorems on actions of algebraic groups*, Ann. of Math **98** (1973), 480–497.
7. A. Blanchard, *Sur les variétés analytiques complexes*, Ann. Sci. Ec. Norm. sup. **73** (1956), 157–202.
8. A. Borel, "Seminar on transformation groups," Annals of Math. studies vol. 46, Princeton Univ. Press, 1960.
9. A. Borel, "Intersection cohomology," Progress in Math. vol. 50, Birkhäuser Boston, 1984.
10. G. Bredon, *Fixed points of actions on Poincaré duality spaces*, Topology **27** (1988), 459–472.
11. W. A. Browder, *Actions of elementary abelian p-groups*, Topology **27** (1988), 459–472.
12. J. B. Carrell and M. Goresky, *A decomposition theorem for the integral homology of a variety*, Invent. Math. **73** (1983), 367–381.
13. H. Cartan, *La transgression dans un groupe de Lie et dans un fibré principal*, in "Colloque de topologie (espaces fibrés) du C.B.R.M," G. Thore, Liege (1951), 57–71.
14. P. Deligne, *Théorème de Lefschetz et critéres de dégénérescences de suites spectrales*, Publ. Math. I.H.E.S. **35** (1968), 97–126.
15. P. Deligne, *Théorie de Hodge III*, Publ. Math. I.H.E.S. **44** (1974), 5–77.
16. P. Deligne, *La conjecture de Weil II*, Publ. Math. I.H.E.S. **52** (1980), 138–252.
17. E.M. Friedlander, "Etale homotopy of simplicial schemes," Annals of Math. Studies 104, Princeton Univ. Press, 1982.
18. O. Gabber, *Pureté de la cohomologie de MacPherson-Goresky (rédigé par P. Deligne)*, preprint I.H.E.S. 1981.
19. V. Ginsburg, *A proof of the Deligne-Langlands conjecture*, Soviet Math. Dokl. **35** (1987), 304–308.
20. V. Ginsburg, *Symplectic geometry and equivariant cohomology*, preprint Moscow 1985.
21. R. Godement, "Topologie algébrique et théorie des faisceaux," Actual. Sci. Industr., Hermann, 1958.
22. M. Goresky and R. MacPherson, *Intersection homology theory*, Topology **19** (1980), 135–162.
23. M. Goresky and R. D. MacPherson, *Intersection homology II*, Invent. Math. **73** (1983), 77–129.
24. M. Goresky and R. D. MacPherson, "Stratified Morse Theory," Ergeb. der math. 3 Folge, bd. 14, Springer-Verlag, 1988.
25. R. Joshua, *Vanishing of odd-dimensional intersection cohomology*, Math. Zeit. **195** (1987), 239–253.

26. D. Kazhdan and G. Lusztig, *Representations of Coxeter groups and Hecke algebras*, Invent. Math. **53** (1979), 165–184.
27. F. Kirwan, "Cohomology of quotients in symplectic and algebraic geometry," Mathematical Notes, vol. 31, Princeton Univ. Press, 1984.
28. J. Mather, *Notes on topologial stability*, Preprint Harvard Univ.
29. D. Quillen, *The spectrum of an equivariant cohomology ring*, Ann. of Math. **94** (1971), 549–572.

Pennsylvania State University
Department of Mathematics
305 McAllister
University Park, PA 16802

Contemporary Mathematics
Volume **139**, 1992

Some Remarks on Regular Weyl Group Orbits and the Cohomology of Schubert Varieties

JAMES B. CARRELL

1. Introduction

Let G be a semi-simple linear algebraic group over $\mathbb{C}$, B a Borel subgroup of G and G/B the flag variety of G. By a classical result of Schubert calculus, the Schubert varieties $X_w = \overline{BwB/B}$, as w varies over the Weyl group W of G, determine a basis $\{s_w | w \in W\}$ of $H_.(G/B; \mathbf{Z})$. Here s_w is by definition the element of $H_{2\ell(w)}(G/B; \mathbf{Z})$ determined by X_w, where $\ell(w)$ is the length of w.

The purpose of this note is to give an elementary algebraic treatment of the cohomology algebra over $\mathbb{Q}$ of a Schubert variety. This is based on the fact that the cohomology algebra over $\mathbb{Q}$ of the flag variety is isomorphic with the graded algebra formed from the filtered algebra of rational polynomials on the W-orbit $W \cdot t$ of a rational regular element t of the Cartan subalgebra on which W acts. More generally, the cohomology algebra over $\mathbb{Q}$ of a Schubert variety X_w is isomorphic with the corresponding graded algebra for the subvariety $W(w^{-1}) \cdot t$ of $W \cdot t$ consisting of $\{u \cdot t | u \leq w^{-1}\}$, where $<$ is the usual partial order on W determined by B. We show that the ideal of $W(w) \cdot t$ is explicitly determined by a remarkable polynomial Q introduced in [BGG, §3] and the difference operators A_w introduced in [BGG] and also in [D]. Regular W-orbits are explicity mentioned in [BGG] in the context of Schubert polynomials and a theorem of Kostant. The upshot of this note is that there is a precise connection between these topics. In addition, the classical picture of the cohomology algebra $H^{\cdot}(G/B, \mathbb{Q})$ as the coinvariant algebra of W extends to Schubert varieties.

The author wishes to thank S. Kumar for his remarks.

2. Preliminaries

1. Let $\mathbf{h}$ be a Cartan subalgebra of a semi-simple complex Lie algebra $\mathbf{g}$,

---

[1] Partially supported by a grant of the Natural Sciences and Engineering Research Council of Canada.

1980 Mathematics Subject Classification (1985 Revision) 14 M15, 20G10

This paper is in final form and no version of it will be submitted elsewhere.

and let $\Delta \subset \mathbf{h}^*$ denote the corresponding root system. Choose a set of simple roots $\pi = \{\alpha_1, ..., \alpha_\ell\}$. The corresponding simple reflections in the Weyl group W of $(\mathbf{g}, \mathbf{h})$ will be denoted by $r_1, ..., r_\ell$. Recall that the *length* $\ell(w)$ of w is the least number of r_i involved in an expression $w = r_{i_1}...r_{i_j}$ and any such expression of w with $\ell(w)$ factors is called *reduced.* $<$ will denote the usual partial order on W characterized as follows: if $w = s_1...s_k$ is a reduced expression, then $W(w) := \{v \leq w\}$ consists of all $v = s_{i_1}...s_{i_m}$ where $1 \leq i_1 < ... < i_m \leq k$.

For $\alpha \in \Delta$, let $\alpha^\vee \in \mathbf{h}$ be the coroot associated to α and $\mathbf{h}_{\mathbb{Q}} \subset \mathbf{h}$ the $\mathbb{Q}$ span of $\alpha_1^\vee, ..., \alpha_\ell^\vee$. R will denote the algebra Sym $(\mathbf{h}_{\mathbb{Q}}^*)$ of rational polynomials on $\mathbf{h}_{\mathbb{Q}}$ with the natural grading. The action $w \mapsto w \cdot s$ of W on $\mathbf{h}$ stabilizes $\mathbf{h}_{\mathbb{Q}}$ so W acts dually on R: for $f \in R$ *and* $s \in \mathbf{h}_{\mathbb{Q}}$, $w \cdot f(s) := f(w^{-1} \cdot s)$.

2. We now recall the basic operators A_w on R associated to each w in W. A_e is the identity. If $\ell(w) = 1$, say $w = r_i$, then $A_w(f) = (f - w \cdot f)/\alpha_i$. A fundamental result ([BGG], [D]) is that if $r_{i_1}...r_{i_k}$ and $r_{j_1}...r_{j_k}$ are reduced expressions for the same w, then $A_{r_{i_1}}...A_{r_{i_k}} = A_{r_{j_1}}...A_{r_{j_k}}$. Hence A_w is defined as $A_{r_{i_1}}...A_{r_{i_k}}$. Moreover, $A_{r_{i_1}}...A_{r_{i_k}} = 0$ if $r_{i_1}...r_{i_k}$ is not reduced, so $A_v A_w = A_{vw}$ if $\ell(vw) = \ell(v) + \ell(w)$ and $A_v A_w = 0$ otherwise.

We will need an expression for A_w on the set $\mathbf{h}_{\text{reg}}$ consisting of the regular elements in $\mathbf{h}$. Recall that t is called *regular* if $\alpha(t) \neq 0$ for all $\alpha \in \Delta$. Clearly $\mathbf{h}_{\text{reg}}$ is the union of all Weyl chambers in $\mathbf{h}$ or, equivalently, is the set of all t in $\mathbf{h}$ such that the isotropy group $W_t = \{w \in W | w \cdot t = t\}$ is trivial, since W acts freely on the chambers. The following appears in a slightly different form in [D].

LEMMA 2.1. *Let* $w \in W$. *Then there exist regular functions* $a_{u,w}(u \leq w)$ *on* $\mathbf{h}_{\text{reg}}$ *such that if* $g \in R$ *and* $t \in \mathbf{h}_{\text{reg}}$, *then*

$$A_{w^{-1}} g(t) = \sum_{u \leq w} a_{u,w}(t) g(u \cdot t). \tag{2.2}$$

Moreover,

$$a_{w,w}(t) = \prod_{\gamma \in \Delta_+ \cap w\Delta_-} \gamma(w \cdot t)^{-1} \tag{2.3}$$

Consequently, $a_{w,w}$ *is nowhere zero on* $\mathbf{h}_{\text{reg}}$.

Proof. Let w have reduced expression $r_{i_1}...r_{i_k}$. Then $r_{i_k}...r_{i_1}$ is a reduced expression of w^{-1}, so

$$A_{w^{-1}} = \frac{1 - r_{i_k}}{\alpha_{i_k}} \cdots \frac{1 - r_{i_1}}{\alpha_{i_1}}.$$

The first identity follows by expanding and applying the above description of $<$. For the second identity, notice that the coefficient $a_{w,w}(t)$ of $g(w \cdot t)$ is the inverse of

$$\begin{aligned}(-1)^k \alpha_{i_1}(r_{i_2} \cdots r_{i_k}(t))\alpha_{i_2}(r_{i_3} \cdots r_{i_k}(t)) \cdots \alpha_{i_k}(t) \\ = \prod_{1 \leq j \leq k} (r_{i_1} \cdots r_{i_{j-1}}) \cdot \alpha_{i_j}(w \cdot t) \\ = \prod_{\gamma \in \Delta_+ \cap w\Delta_-} \gamma(w \cdot t)\end{aligned}$$

by [B, Ch. 6].

3. The rings $A(W(w) \cdot t)$.

1. Assume $t \in \mathbf{h}_{\mathbb{Q}\mathrm{reg}}$ and $w \in W$. In this section we study the coordinate ring $A(W(w) \cdot t)$ of the finite reduced variety $W(w) \cdot t \subset \mathbf{h}$. By definition, $A(W(w) \cdot t) = R/I(W(w) \cdot t)$, where $I(W(w) \cdot t)$ is the ideal in R of $W(w) \cdot t$. Thus $A(W(w) \cdot t)$ has an increasing filtration $F_k(A(W(w) \cdot t) := j_w(F_k R)$, where $j_w : R \rightarrow A(W(w) \cdot t)$ is the quotient map and $F_k R = \{f \in R | \deg f \leq k\}$. We find a natural basis of each $F_k A(W(w) \cdot t)$ and describe the associated graded algebras $Gr A(W(w) \cdot t)$. These algebras are then identified with certain quotients of the coinvariant algebra $S_W = R/J$ of W, where J is the ideal in R generated by the homogeneous W-invariants. We begin by characterizing $I(W(w) \cdot t)$ in terms of the BGG-operators.

Let $I(w, t) := \{f \in R | A_{v^{-1}} f(t) = 0 \text{ if } v \leq w\}$.

THEOREM 3.1. *For any $w \in W$ and $t \in \mathbf{h}_{\mathbb{Q},\mathrm{reg}}$, $I(w,t) = I(W(w) \cdot t)$.*

Proof. Suppose that $g \in I(W(w) \cdot t)$. By (2.2), $A_{v^{-1}} g(t) = \sum_{u \leq v} a_{u,v}(t) g(u \cdot t)$, so it is immediate that $g \in I(w,t)$. Conversely, if $g \in I(w,t)$, then $A_e g(t) = g(t) = 0$. Now use induction. Suppose that $g(u \cdot t) = 0$ if $u \leq w$ and $\ell(u) < j \leq \ell(w)$. If $v \leq w$ and $\ell(v) = j$, then using (2.2) again gives

$$A_{v^{-1}} g(t) = a_{v,v}(t) g(v \cdot t) + \sum_{u < v} a_{u,v}(t) g(u \cdot t) = 0.$$

By the induction hypothesis, $A_{v^{-1}} g(t) = a_{v,v}(t) g(v \cdot t)$, so $g(v \cdot t) = 0$ since $a_{v,v}$ never vanishes on $\mathbf{h}_{\mathrm{reg}}$. Hence $g \in I(W(w) \cdot t)$, and the proof is finished.

2. We now reproduce the construction in [BGG: §3] of the remarkable polynomial Q of degree $r := \ell(w_0)$ such that $Q(w \cdot t) = \delta_{w,w_0}$. To construct Q, we only need to know that R is a finitely generated R^W-module and that $S_{W,k}$ (the homogeneous component of S_W in degree k) vanishes if $k > r$. The first fact is a general property of linear actions of a finite group ([H, pp.60 - 61]), and

the second is a consequence of Chevalley's theorem on W -invariants ([Ch] or [H, p. 93]).

LEMMA 3.2. *For any $t \in \mathbf{h}_{\mathbb{Q}\mathrm{reg}}$, there exists a Q_t in R of degree r such that $Q_t(w \cdot t) = \delta_{w,w_0}$ for all $w \in W$.*

Proof. Since t is regular, $w \cdot t \neq w_0 \cdot t$ if $w \neq w_0$. Hence there exists a Q' in R such that $Q'(w \cdot t) = \delta_{w,w_0}$. For example, if $w \neq w_0$, let $\ell_w \in \mathbf{h}^*_{\mathbb{Q}}$ have the property that $\ell_w(w_0 \cdot t) = 1$ and $\ell_w(w \cdot t) = 0$. Then take $Q' = \prod_{w \neq w_0} \ell_w$. Next, choose $F_1, \ldots, F_k \in R$ spanning R as an R^W-module such that $\deg F_i \leq r$ for each i. Writing $Q' = \sum a_i F_i$, where each $a_i \in R^W$, we may set $Q_t := \sum a_i(t) F_i$. Note that by lemma 2.1,

$$A_{w_0} Q_t(t) = a_{w_0,w_0}(t) Q_t(w_0 \cdot t) = \Big(\prod_{\alpha \in \Delta_+} \alpha(w_0 \cdot t)\Big)^{-1} \tag{3.3}$$

Hence the degree of Q_t is r.

3. We will now consider various aspects of the ring $A(W \cdot t)$, first obtaining a basis of each $F_k A(W \cdot t) := j(F_k R)$, where $j : R \to A(W \cdot t)$ is the quotient homomorphism. It is useful to note that by (2.2), $A_v(I(W \cdot t)) \subseteq I(W \cdot t)$ for every $v \in W$. Therefore A_v defines a $\mathbb{Q}$-linear endomorphism $\bar{A}_v$ of $A(W \cdot t)$ sending $j(f)$ to $j(A_v f)$. Moreover, if $\ell(v) > k$, then $\bar{A}_v$ annihilates $F_k A(W \cdot t)$. Set $Q_{t,v} := A_{v w_0} Q_t$. Clearly $\deg Q_{t,v} \leq \ell(v)$.

THEOREM 3.4. *The elements $\bar{Q}_v := j(Q_{t,v})$ with $\ell(v) \leq k$ are a basis of $F_k A(W \cdot t)$ for $k = 0, 1, 2, \ldots$.*

Proof. We will first show that the $\bar{Q}_v (v \in W)$ give a basis of $A(W \cdot t)$. Since $W \cdot t$ is reduced and t is regular, $\dim_{\mathbb{Q}} A(W \cdot t) = |W|$. Thus it will suffice to show that the $\bar{Q}_v$ are independent.

LEMMA 3.5. *Suppose $u \neq v$ and $\ell(u) \geq \ell(v)$. Then $A_{u^{-1}} Q_{t,v} = 0$.*

Proof. By definition, $A_{u^{-1}} Q_{t,v} = A_{u^{-1}} A_{v w_0} Q_t = 0$ unless $\ell(u^{-1} w w_0) = \ell(u^{-1}) + \ell(v w_0)$. But $\ell(v w_0) = \ell(w_0) - \ell(v)$ and $\ell(u^{-1}) = \ell(u)$, so the equality implies $\ell(u^{-1} v w_0) \geq \ell(w_0)$. This means $u = v$, contrary to assumption.

Now suppose $\sum c_v \bar{Q}_v = 0$. By (3.3), $\bar{A}_{w_0}(\bar{Q}_{w_0})(t) \neq 0$, so $c_{w_0} = 0$. Make the induction assumption that $c_v = 0$ if $\ell(v) > j$ and let $\ell(u) = j$. By the last lemma,

$$\bar{A}_{u^{-1}}\Big(\sum c_v \bar{Q}_v\Big) = c_u \bar{A}_{u^{-1}}(\bar{Q}_u) = 0.$$

But $A_{u^{-1}} A_{u w_0} = A_{w_0}$, so $c_u = 0$, and hence the $\bar{Q}_v$ are independent. It follows that the $\bar{Q}_v$ are a basis of $A(W \cdot t)$ and consequently $F_k A(W \cdot t) = F_r A(W \cdot t)$ if $k \geq r$. To finish the proof of (3.4), we need

LEMMA 3.6. *Suppose $\sum c_v \bar{Q}_v$ lies in $F_k(A(W \cdot t)$. Then $c_v = 0$ if $\ell(v) > k$.*

Proof. Let u be a maximal length element of W such that $c_u \neq 0$. By the last lemma, $\bar{A}_{u^{-1}}(\sum c_v \bar{Q}_v) = c_u \bar{A}_{u^{-1}}(Q_u)$ which is nonzero. However, if $\ell(u) > k$, then $\bar{A}_{u^{-1}}$ annihilates $F_k A(W \cdot t)$, so $\ell(u) \leq k$ and the lemma is proven.

Lemma 3.6 implies immediately that the $\bar{Q}_v$ with $\ell(v) \leq k$ span $F_k A(W \cdot t)$, which completes the proof of theorem 3.4.

4. We will now show S_W is obtained from that $A(W \cdot t)$ by a natural construction as long as t is regular. Let $S = \oplus_{p \geq 0} S_p$ be a graded $\mathbb{Q}$-algebra and $I \subseteq S$ be an ideal. Then S/I has a natural filtration $F_k(S/I) \subseteq F_{k+1}(S/I)$ satisfying $F_k F_j \subseteq F_{k+j}$. Namely, set $F_k(S/I) = i(\oplus_{p \leq k} S_p)$, where $i : S \to S/I$ is the quotient map. Let

$$Gr(S/I) := F_0 \oplus F_1/F_0 \oplus F_2/F_1 \oplus \dots \tag{3.7}$$

Clearly $Gr(S/I)$ is a graded $\mathbb{Q}$-algebra. On the other hand, I determines a homogeneous ideal grI, namely the ideal generated by the highest terms of the elements of I.

LEMMA 3.8. *Let S be a graded ring and I an ideal in S. Then there is a canonical isomorphism of graded rings*

$$\mu : S/grI \longrightarrow Gr(S/I) \tag{3.9}$$

where $Gr(S/I)$ is the graded ring associated to the natural filtration of S/I as above. μ is determined by the condition that if $f \in S_p$, then $\mu(f \bmod grI)$ is the unique element of $F_p/F_{p-1}(F_p := F_p(S/I))$ determined by f.

For a proof, see [Kr,p.134]. Let $GrA(W \cdot t)$ denote the graded algebra obtained from the degree filtration of $A(W \cdot t)$. Also let $R_t^W = \{f \in R^W | f(t) = 0\}$ and put $J_t = RR_t^W$

THEOREM 3.10. *For any $t \in \mathbf{h}_{\mathbb{Q},\mathrm{reg}}$, $I(W \cdot t) = J_t$ and hence $grI(W \cdot t) = J$. Consequently the natural map*

$$\mu : S_W = R/grI(W \cdot t) \to GrA(W \cdot t)$$

is an isomorphism of graded $\mathbb{Q}$-algebras.

Proof. Clearly $J_t \subseteq I(W \cdot t)$ and $J = grJ_t$, so it suffices to show $I(W \cdot t) \subseteq J_t$. We use induction on k to show $F_k I(W \cdot t) \subseteq F_k J_t$ for all $k \geq 0$. Let $f \in F_k I(W \cdot t)$. Then for any simple reflection s, $A_s f \in F_{k-1} I(W \cdot t) = F_{k-1} J_t$, so $s \cdot f - f \in F_k J_t$. It follows that for all $w \in W, w \cdot f - f \in F_k J_t$. Hence $f^\# - f \in F_k J_t$ where $f^\# = |W|^{-1} \sum_{w \in W} w \cdot f$. Since $f^\# \in F_k J_t, f$ does also and the result is proved.

Thus the ideals $I(W \cdot t), J_t$ and $I(w_0, t) = \{f \in R | A_w f(t) = 0 \text{ for all } w \in W\}$ coincide for all regular t.

COROLLARY 3.11. *The Hilbert series of* $GrA(W \cdot t)$ *is* $\sum_{w \in W} s^{\ell(w)}$. *Consequently,*

$$\sum_{w \in W} s^{\ell(w)} = \prod_{1 \leq i \leq \ell} \frac{(1 - s^{d_i})}{(1 - s)} \tag{3.12}$$

where $d_1, \ldots, d_\ell$ *are the degrees of the fundamental generators of* J.

Proof. The first assertion follows from (3.7). The second uses the theorem and Chevalley's expression for the Hilbert series of S_W.

REMARK. (3.12) was originally proved by Bott using the topology of Lie groups [B]. The first algebraic proof is due to Solomon [S]. Also see [C §9.4].

COROLLARY 3.13. *For any* $k = 0, 1, \ldots, r$, *the set of leading terms* $Q^*_{t,v}$ *of the* $Q_{t,v}$ *(where* $\ell(v) = k$*) projects to a basis* $Q^*_v := Q^*_{t,v}$ *(mod* J*) of* $S_{W,k}$.

5. In this paragraph we will extend the results 3.3, 3.4 and 3.5 to $A(W(w) \cdot t)$. Assume t is an element of $\mathbf{h}_{\mathbb{Q},\mathrm{reg}}$. We begin be finding a nice complement to $I(W \cdot t)$ in $I(W(w) \cdot t)$. Let $i_w : W(w) \cdot t \longrightarrow W \cdot t$ be the inclusion and $i^*_w : A(W \cdot t) \longrightarrow A(W(w) \cdot t)$ the associated comorphism.

LEMMA 3.14. *The kernel of* i^*_w *is the subspace spanned by the* $\bar{Q}_v$ *where* $v \not\leq w$.

Proof. Since the $\bar{Q}_v$ with $v \not\leq w$ span a subspace of $A(W \cdot t)$ having the same dimension as $\ker i^*_w$, it suffices to show that $Q_{t,v}$ lies in $I(W(w) \cdot t)$ if $v \not\leq w$, i.e. that $A_{u^{-1}} A_{vw_0} Q_t(t) = 0$ if $u \leq w$. The only situation in which $A_{u^{-1}} A_{vw_0} Q_t(t) \neq 0$ is when $u^{-1} v w_0 = w_0$, i.e. when $u = v$. However, this cannot occur since $u \leq w$ and $v \not\leq w$.

A consequence of this proof is that for any w in W,

$$I(W(w) \cdot t) = \langle Q_{t,v} | v \not\leq w \rangle + I(W \cdot t). \tag{3.15}$$

THEOREM 3.16. *Let* $w \in W$. *Then* $F_k A(W(w) \cdot t)$ *has a* $\mathbb{Q}$*-basis composed of the* $i^*_w(\bar{Q}_v)$ *with* $v \leq w$ *and* $\ell(v) \leq k$. *The* $\bar{Q}_u$ *with* $u \not\leq w$ *and* $\ell(u) \leq k$ *form a* $\mathbb{Q}$*-basis of* $\ker i^*_w \cap F_k A(W \cdot t)$.

This is an immediate consequence of theorem 3.4 and lemma 3.14. We next give a set of generators of the ideal $grI(W(w) \cdot t)$ The problem is that for arbitrary ideals I, J in a graded ring S, it is not in general true that $gr(I + J) = grI + grJ$.

LEMMA 3.17. *For each* w *in* W *and* t *in* $\mathbf{h}_{\mathbb{Q},\mathrm{reg}}$,

$$grI(W(w) \cdot t) = \langle Q^*_{t,v} | v \not\leq w \rangle + J.$$

Proof. It is clear that $\langle Q^*_{t,v} | v \not\leq w \rangle + J \subseteq grI(W(w) \cdot t)$. Now suppose that $f \in F_k I(W(w) \cdot t)$. Then in $A(W \cdot t)$, $\bar{f} = \sum c_v \bar{Q}_v$ where $c_v = 0$ if $\ell(v) > k$, so

$$f = \sum_{\substack{v \not\leq w \\ \ell(v) \leq k}} c_v Q_{t,v} + g = h + g$$

where $g \in F_kI(W \cdot t)$. We may assume $k \leq r$ and that $c_v \neq 0$ for some v with $\ell(v) = k$. Then h^* and g^* cannot cancel since $A_{v^{-1}}h^* \neq 0$ while $A_{v^{-1}}g^* = 0$. Since $grI(W \cdot t) = J$, it follows that f^* lies in $\langle Q^*_{t,v} | v \not\leq w\rangle + J$, proving the lemma.

Let $S_W(w) := R/grI(W(w) \cdot t)$ and define elements Q^{*w}_v in $S_W(w)$ by $Q^{*w}_v := Q^*_{t,v} \bmod(grI(W(w) \cdot t))$. We have

THEOREM 3.18. *For each $w \in W$ and $t \in \mathbf{h}_{\mathbb{Q},\mathrm{reg}}$,*

(1) $S_W(w) \cong GrA(W(w) \cdot t)$ as graded $\mathbb{Q}$-algebras,

(2) $\dim_{\mathbb{Q}} S_W(w)_k = \#\{v | v \leq w$ and $\ell(v) = k\}$ and a basis is the set of Q^{*w}_v, where $v \leq w$ and $\ell(v) = k$, and

(3) a basis of the kernel of the natural map from $S_{W,k}$ onto $S_W(w)_k$ is the set of Q^*_u with $u \not\leq w$ and $\ell(u) = k$.

Proof. This follows immediately from lemma 3.8 and theorem 3.16.

We have now completed our algebraic treatment of S_W and, more generally, the algebras $S_W(w)$ for $w \in W$. In the next section we will relate these algebras to cohomology of Schubert varieties.

4. Cohomology of Schubert varieties.

1. We now apply the results of §3 to describe the cohomology of Schubert varieties in G/B. We assume that B is the Borel subgroup of G corresponding to the choice of basis π of Δ (§2). The ordering $<$ on W has the property that $X_w := \overline{BwB/B}$ is given by $\cup_{v \leq w} BvB/B$.

Recall the fundamental W-equivariant degree doubling $\mathbb{Q}$-algebra morphism $\beta : R \longrightarrow H^{\cdot}(G/B; \mathbb{Q})$ with kernel J determined by $\beta(\omega) = c_1(L_\omega)$, where ω is a weight and L_ω is the line bundle on G/B canonically associated to ω. β induces an isomorphism of S_W onto $H^{\cdot}(G/B; \mathbb{Q})$ which will also be denoted by β. If $S_{\mathbb{C},W}$ is the covariant algebra defined over $\mathbb{C}$ using $R_{\mathbb{C}} := \mathrm{Sym}(\mathbf{h})$ instead of R and if t is any element of $\mathbf{h}_{\mathrm{reg}}$, then there is a degree doubling $\mathbb{C}$-algebra isomorphism α from $GrA_{\mathbb{C}}(W \cdot t)$ onto $H^{\cdot}(G/B; \mathbb{C})$ making a commutative diagram of $\mathbb{C}$-algebra isomorphisms

$$\begin{array}{ccccc} S_{\mathbb{C},W} & & \xrightarrow{\mu} & & GrA_{\mathbb{C}}(W \cdot t) \\ & \beta \searrow & & \swarrow \alpha & \\ & & H^{\cdot}(G/B; \mathbb{C}) & & \end{array} \tag{4.1}$$

where β and μ are defined analogously to their $\mathbb{Q}$ counterparts and $A_{\mathbb{C}}(W \cdot t)$ denotes the complex coordinate ring of $W \cdot t$, i.e. $R_{\mathbb{C}}/I_{\mathbb{C}}(W \cdot t)$. It follows easily that there also exists a degree doubling $\mathbb{Q}$-algebra isomorphism α from $GrA(W \cdot t)$

onto $H^{\cdot}(G/B;\mathbb{Q})$, provided $t \in \mathbf{h}_{\mathbb{Q},\mathrm{reg}}$, such that $\alpha\mu = \beta$ when $\mathbb{C}$ is replaced by $\mathbb{Q}$. The advantage of working over $\mathbb{C}$ is that the machinery of the theory of vector fields [ACL] also guarantees an isomorphism α_w of $GrA_{\mathbb{C}}(W(w)\cdot t)$ onto $H^{\cdot}(X_{w^{-1}};\mathbb{C})$, as long as t is regular, such that the following diagram of $\mathbb{C}$-algebra morphisms commutes:

$$\begin{array}{ccc} GrA_{\mathbb{C}}(W\cdot t) & \xrightarrow{\alpha} & H^{\cdot}(G/B;\mathbb{C}) \\ i_w^*\downarrow & & \downarrow i_{w^{-1}}^* \\ GrA_{\mathbb{C}}(W(w)\cdot t) & \xrightarrow{\alpha_w} & H^{\cdot}(X_{w^{-1}};\mathbb{C}) \end{array} \tag{4.2}$$

where i_w denotes either inclusion $W(w)\cdot t \subseteq W\cdot t$ or $X_w \subseteq G/B$, and i_w^* denotes either corresponding induced homomorphism.

2. Using theorem 3.18, we immediately obtain an analog of (4.2) for $\mathbb{Q}$. Moreover we get

THEOREM 4.3. *For any $w \in W$, there exists a commutative diagram of $\mathbb{Q}$-algebra morphisms*

$$\begin{array}{ccc} S_W & \xrightarrow{\beta} & H^{\cdot}(G/B;\mathbb{Q}) \\ i_w^*\downarrow & & \downarrow i_{w^{-1}}^* \\ S_W(w) & \xrightarrow{\beta_w} & H^{\cdot}(X_{w^{-1}};\mathbb{Q}) \end{array}$$

where β and β_w are degree doubling isomorphisms and i_w^*, $i_{w^{-1}}^*$ are the restriction (i.e. quotient) maps.

Note that the advantage of replacing S_W by $GrA(W\cdot t)$ is that one thereby obtains a description of the ideal in R containing J giving the kernel of $i_w^*\beta$: namely $grI(W(w)\cdot t)$. Following [BGG], define $P_v \in S_W$ by $P_v := Q_{v^{-1}}^*$. Also define $P_v^w \in S_W(w^{-1})$ as $i_{w^{-1}}^*(P_v)$. From theorem 3.18, we also deduce

COROLLARY 4.4. *The $\beta_{w^{-1}}(P_v^w)$, where $v \le w$, give a $\mathbb{Q}$-basis of $H^{\cdot}(X_w;\mathbb{Q})$. Similarly, the $\beta(P_u)$, where $u \not\le w$, give a $\mathbb{Q}$-basis of the kernel of i_w^*.*

We now come to our final remark. Let $s_w \in H_{2\ell(w)}(G/B;\mathbb{Q})$ denote the class supported by X_w, and let $<,>$ denote the natural pairing between $H^{\cdot}(G/B;\mathbb{Q})$ and $H_{\cdot}(G/B;\mathbb{Q})$.

THEOREM 4.5 ([BGG]). *For each w in W, there exists a nonzero $\lambda_w \in \mathbb{Q}$ so that $<\beta(P_v), s_w> = \lambda_w\delta_{v,w}$. In other words, the basis $\beta(P_w)(w \in W)$ of $H^{\cdot}(G/B;\mathbb{Q})$ is proportional to the dual of the basis $s_w(w \in W)$ of $H_{\cdot}(G/B;\mathbb{Q})$.*

Proof. If $\ell(w) \ne \ell(v)$, then $<\beta(P_v), s_w> = 0$ by definition. Suppose that $\ell(w) = \ell(v)$. Then $<\beta(P_v), s_w> = <\beta(P_v), i_{w*}([X_w])> = <i_w^*(\beta(P_v)), [X_w]>$

$= < \beta_{w^{-1}}(P_v^w), [X_w] > = 0$ unless $v = w$, since if $v \not\leq w$ then $i_{w^{-1}}^*(P_v) = 0$. Here, $[X_w] \in H_{2\ell(w)}(X_w; \mathbb{Q})$ denotes the fundamental class. Since $< \beta(P_v), s_w > \neq 0$ for some $v \in W$, it follows that $< \beta(P_v), s_w > = \lambda_w \delta_{v,w}$ for some nonzero $\lambda_w \in \mathbb{Q}$.

REFERENCES

[ACL] Akyildiz, E., Carrell, J.B., and Lieberman, D.I., *Zeros of holomorphic vector fields on singular spaces and intersection rings of Schubert varieties*, Compositio Math. **57** (1986), 237-248.

[BGG] Bernstein, I.N., Gelfand, I.M. and Gelfand, S.I., *Schubert cells and cohomology of the space G/P*, Russian Math. Surveys, **28** (1973), 1-26.

[B] Bott, R., *An application of Morse theory to the topology of Lie groups*, Bull. Soc. Math. France, **84** (1956), 251 - 282.

[Bo] Bourbaki, N., *Groups et algèbres de Lie VI*, Hermann, Paris (1968).

[C] Carter, R., *Simple groups of Lie type*, John Wiley and Sons, London (1972).

[Ch] Chevalley, C., *Invariants of finite groups generated by reflections*, Amer. Journal of Math. **78** (1955), 778-782.

[D] Demazure, M., *Invariants symmetriques des groupes de Weyl et torsion*, Inv. Math. **29** (1973), 287-301.

[H] Hiller, H., *Geometry of Coxeter groups*, Research Notes in Math. **54**, Pitman Boston London Melbourne (1982).

[K] Kostant, Bertram, *Lie algebra cohomology and generalized Schubert cells*, Annals of Math. **77** (1963), 72-144.

[Kr] Kraft, H.P., *Geometrische Methoden in der Invariantentheorie*, Aspects of Mathematics, Vieweg (1984).

[S] Solomon, L., *Invariants of finite reflection groups*, Nagoya Math. J. **22** (1963), 57 - 64.

DEPARTMENT OF MATHEMATICS, UNIVERSITY OF BRITISH COLUMBIA, VANCOUVER, B.C. V6T 1Z2
Department of Mathematics, University of British Columbia, 121-1984 Mathematics Road, Vancouver, B.C. V6T 1Z2
E-mail: carrell@mtsg.ubc.ca

Contemporary Mathematics
Volume **139**, 1992

Infinitesimal Kazhdan-Lusztig Theories

EDWARD CLINE, BRIAN PARSHALL, AND LEONARD SCOTT

ABSTRACT. Let G be a semisimple, simply connected algebraic group defined and split over $\mathbf{F}_p$, $p > 0$. In this paper, we extend our earlier work on Kazhdan-Lusztig theories for highest weight categories to include the representation theory of G_1T, the pull-back of a maximal split torus T through the Frobenius morphism on G. We reduce the Lusztig conjecture for G to a question involving "baby" Verma modules for G_1T as well as to a question involving the finite Chevalley groups. In relating the Kazhdan-Lusztig theories for G and G_1T, we require no hypothesis that the prime p be very large. Assuming G_1T has a Kazhdan-Lusztig theory, we give a complete calculation of the $\mathrm{Ext}^\bullet$ groups between irreducible G_1T-modules having p-regular highest weights.

In recent work [**CPS5,6**], the authors have attempted to formalize the ingredients necessary for the existence and truth of a Kazhdan-Lusztig conjecture in the context of a general highest weight category $\mathcal{C}$ (as defined in [**CPS4**]). One of the key ideas centers on the notion of an abstract Kazhdan-Lusztig theory. This concept is given in terms of derived category filtrations, but is equivalent to an even-odd vanishing property for Ext^n groups between simple objects and Weyl and "induced" objects. As an important consequence, when $\mathcal{C}$ has a Kazhdan-Lusztig theory, all simple objects are represented in a certain "q-analogue" of the Grothendieck group of $\mathcal{C}$ (which is defined in terms of the derived category). When combined with existence of so-called pre-Hecke operators, these representations of simple objects permit the combinatorial determination of all the Ext^n groups between them. In addition, we introduced the notion of an abstract Kazhdan-Lusztig conjecture. When $\mathcal{C}$ has enough pre-Hecke operators, we established that the validity of this conjecture is equivalent to the assertion that $\mathcal{C}$ has a Kazhdan-Lusztig theory and is also equivalent to weaker conditions

1991 *Mathematics Subject Classification*. Primary 20G15, 20G05.
Key words and phrases. Representations of algebraic groups, Kazhdan-Lusztig theories, highest weight categories.
Research supported in part by NSF Group Project Grant DMS–890–2661.

This paper is in final form and no version of it will be submitted for publication elsewhere.

involving only Ext^1 groups. This work both simplifies the task of proving such a conjecture and demonstrates that the validity of the conjecture enables the above combinatorial determination of Ext^n groups between simple objects.

The above theory applies to at least three important situations in representation theory: semisimple algebraic groups G over fields of positive characteristic, complex semisimple Lie algebras (the category $\mathcal{O}$), and quantum groups at a root of 1. In each case the abstract Kazhdan-Lusztig conjecture is indeed an actual conjecture of Kazhdan-Lusztig or of Lusztig. In particular, we obtained a substantial simplification of Lusztig's famous conjecture **[L1]** for the characters of the irreducible representations of a semisimple algebraic group in positive characteristic. In the category $\mathcal{O}$, where the Kazhdan-Lusztig conjecture is known to be true, we obtained a complete calculation of all Ext^n groups between *any* two simple modules having integral high weights. Also, assuming just announced results of Kazhdan-Lusztig **[KL]** concerning Lusztig's quantum group conjecture **[L4]**, the theory gives a determination for quantum groups of the groups Ext^n between irreducible modules having l-regular high weights.

By and large, however, the above approach applies only to the case in which the underlying weight poset is *finite*, although in many cases in which the weight poset is bounded below, one can reduce to the finite case. Our methods have not readily lent themselves to the case of highest weight categories arising from such group schemes as G_rT, the pull-back through the rth power of the Frobenius of the maximal torus. (Observe that the representation theory of G_1T is closely related to that of the restricted Lie algebra of G.)

In this paper, we take up the question of extending the methods of our previous paper to include, for example, the case of G_1T. Most importantly, we answer the question (left open in the 1990 MSRI conference) as to whether a reduction of the Lusztig conjecture involving Weyl module quotients has an analogue involving much simpler "baby" Verma modules for G_1T. The general machinery to accomplish this is presented in §§1–4, while §5 focuses on G_1T, and (5.10) gives the reduction in question. As another point of interest of this paper, we require no "generic" or "very large prime" hypothesis in relating the Kazhdan-Lusztig theories of G and G_1T. In particular, Lusztig's conjecture for G is implied by the existence of a Kazhdan-Lusztig theory for G_1T for $p \geq h$, and is equivalent to the latter for $p \geq 2h-3$. (Here h denotes the Coxeter number of G.) See Theorem 5.5. At the character formula level, this result is implicit in Kato **[K;Conj. 5.5]**; see Theorem 5.9. A similar equivalence with the "Vogan conjecture" (see **[CPS6;5.10c]**) was proved by Kaneda **[Ka;4.15]**. We take this opportunity to point out that the Janzten region ordinarily used in formulating the Lusztig conjecture does not include all restricted weights unless $p \geq 2h-3$, so that the conjecture of a G_1T theory for for $p \geq h$ might be regarded as more natural.

We are also able to give a reduction of the Lusztig conjecture to a question involving the finite groups $G(q)$ of $\mathbf{F}_q$ rational points. This question is in the spirit of the G_1T reduction above.

Finally, it seems likely that the methods of this paper also apply to the quantum group analogue of G_1T. We hope to pursue this later.

It is a pleasure to thank the Mathematics Institute, University of Warwick for its hospitality while this paper was completed.

1. Preliminaries

Fix a field k, and let $\mathcal{C}$ be a highest weight category over k having weight poset Λ. We will assume the conventions of [**CPS5;§1**]. Thus, we assume that every object in $\mathcal{C}$ has finite length and that the opposite category $\mathcal{C}^{\mathrm{op}}$ is also a highest weight category having the same weight poset Λ. (Recall that Λ is interval finite.) For a weight $\lambda \in \Lambda$, let $L(\lambda)$, $A(\lambda)$, and $V(\lambda)$ denote the corresponding simple, "induced", and "Weyl" objects of $\mathcal{C}$. We sometimes require that $\mathcal{C}$ have a duality D, in the sense of [**CPS4**]. Finally, we often fix a function $\ell : \Lambda \to \mathbf{Z}$, which we call a "length" function on Λ.

This paper is largely devoted to establishing results which can be applied to the following example (for the case $r = 1$).

1.1 EXAMPLE. Let G be a semisimple, simply connected algebraic group defined and split over $\mathbf{F}_p$ for a prime p. Assume $k = \bar{\mathbf{F}}_p$. Let T be a maximal split torus in G, let Φ be the root system of T in G. Fix a positive set Φ^+ of roots and let B^+ (resp., B) be the Borel subgroup containing T which is defined by Φ^+ (resp., $\Phi^- = \Phi \setminus \Phi^+$). We put U equal to the unipotent radical of B. Let $X(T)$ be the group of rational characters on T, and denote the set of dominant integral weights (relative to B^+) by $X(T)^+$. Let $Q(T)$ be the root lattice in $X(T)$. Also, α_0 will always denote the maximal short root in Φ^+, and $\rho = (1/2)\sum_{\alpha\in\Phi^+}\alpha$ the Weyl weight on T. Let

$$C^+ = \{x \in X(T)\otimes\mathbf{R} \mid 0 < (x+\rho,\alpha^\vee) < p, \quad \forall\alpha\in\Phi^+\},$$

and put $C^+_{\mathbf{Z}} = X(T)\cap C^+$. The closure of C^+ is denoted $\bar{C}^+$, and $\bar{C}^+_{\mathbf{Z}} = \bar{C}^+ \cap X(T)$. Recall that a weight $\lambda \in X(T)$ is restricted provided that $1 \leq (\lambda+\rho,\alpha^\vee) \leq p$ for all simple roots α.

Besides the Weyl group $W = N(T)/T$, we will work with the affine Weyl group W_p generated by the affine reflections

$$s_{\alpha,np} : X(T)\otimes\mathbf{R} \to X(T)\otimes\mathbf{R}$$

given by

$$x \mapsto s_{\alpha,np}\cdot x = x - (x+\rho,\alpha^\vee)\alpha + pn\alpha$$

for $\alpha\in\Phi, n\in\mathbf{Z}$. Any $w \in W_p$ can be written as $w = at_{p\xi}$, where $a \in W$ and $t_{p\xi}$ is translation by $p\xi$, $\xi \in Q(T)$. Let $\Sigma \subset W_p$ be the set of reflections in the walls of C^+. Thus, (W_p,Σ) is a Coxeter system. If $s \in \Sigma$, let $F_s \subset \bar{C}^+$ be the associated face of the alcove C^+. A face F of any alcove A is of type s provided that F is W_p-conjugate to F_s.

The set $X(T)$ has two natural poset structures. First, define $\lambda \leq \nu$ provided $\nu - \lambda$ is a sum of positive roots. Second, define $\lambda \uparrow \nu$ provided there is a sequence $\lambda = \lambda_0 < \lambda_1 < \cdots < \lambda_t = \nu$ such that for $0 \leq i < t$, $\lambda_{i+1} = s_{\alpha,pm} \cdot \lambda_i$ for some affine reflection $s_{\alpha,pm}$. There is a similarly defined partial ordering $\uparrow$ on the set of alcoves; see **[J2;§II.6]**.

Let $F : G \to G$ be the Frobenius morphism, and, for an integer $r \geq 1$, let $G_rT = (F^r)^{-1}(T)$, the pull-back of T through F^r. Similarly, we consider $B_rT = (F^r|_B)^{-1}(T)$, the pull-back of T through the restriction $F^r|_B$. Also, $B_r^+T = (F^r|_{B^+})^{-1}(T)$. Let $\mathcal{C}^r$ be the category of rational, finite dimensional modules for the group scheme G_rT. It is well-known (see **[CPS2]** or **[J1,2]**) that the irreducible G_rT-modules are indexed by $X(T) \cong X(B_r^+T)$: for $\lambda \in X(T)$, let $L_r(\lambda)$ be the unique irreducible G_rT-module having B_r^+T-fixed line of weight λ. There is a natural duality $D : \mathcal{C}^r \to (\mathcal{C}^r)^{\mathrm{op}}$ in the sense of **[CPS4]**. For $V \in \mathrm{Ob}(\mathcal{C}^r)$, DV is defined by "twisting" the dual module V^* by first the automorphism on G_rT induced by conjugation by a representative for the long word $w_0 \in W$ and then by a graph automorphism inducing the opposition involution $\iota \equiv -w_0$ on $X(T)$. Also, for $\lambda \in X(B_rT)$, let $A_r(\lambda)$ denote the induced module $\mathrm{ind}_{B_rT}^{G_rT}\lambda$. Then

$$V_r(\lambda) = \mathrm{ind}_{B_r^+T}^{G_rT}(\lambda - 2(p^r - 1)\rho) \cong DA_r(\lambda).$$

It follows from **[PS;§5]** that $\mathcal{C}^r$ is a highest weight category with weight poset $(X(T), <)$, where the induced and Weyl modules are the $A_r(\lambda)$ and $V_r(\lambda)$ just defined.

Assume $p \geq h$, the Coxeter number of G, so that we can fix $\lambda \in C_{\mathbf{Z}}^+$, and consider the poset $(W_p \cdot \lambda, \uparrow)$. Let $\mathcal{C}_\lambda^r$ be the full subcategory of $\mathcal{C}^r$ consisting of objects having composition factors $L_r(\nu)$, $\nu \in W_p \cdot \lambda$. According to **[Do;Cor. 1]** (or see **[J1]**), if $L_r(\tau)$ is a composition factor of $A_r(\sigma)$, then $\tau \uparrow \sigma$. Thus, $A_r(\sigma) \in \mathrm{Ob}(\mathcal{C}_\lambda^r)$ for $\sigma \in W_p \cdot \lambda$. On the other hand, let $I_r(\tau)$ be the injective envelope of $L_r(\tau)$ in the highest weight category $\mathcal{C}^r$. Then $I_r(\tau)$ has an increasing filtration $\{0\} = F_0(\tau) \subset F_1(\tau) \subset \cdots$ with $F_1(\tau) \cong A_r(\tau)$ and, for $i > 1$, $F_i(\tau)/F_{i-1}(\tau) \cong A_r(\sigma_i)$ for $\tau < \sigma_i$. Fix a weight $\sigma = \sigma_i$ for $i > 1$. By Brauer-Humphreys reciprocity (see **[CPS3;Theorem 3.11]**), the number of times $A_r(\sigma)$ occurs as a section in this filtration equals the multiplicity $[V_r(\sigma) : L_r(\tau)]$ of $L_r(\tau)$ as a composition factor of $V_r(\sigma)$. On the other hand, $[V_r(\sigma) : L_r(\tau)] = [A_r(\sigma) : L_r(\tau)]$ because $V_r(\sigma)$ and $A_r(\sigma)$ have the same image in the Grothendieck group of G_rT **[CPS2;1.4.3]**. Therefore, using **[Do;Cor. 1]** again, we conclude that $\tau \uparrow \sigma$. It follows that $I_r(\tau)$ belongs to $\mathcal{C}_\lambda^r$ for $\tau \in W_p \cdot \lambda$. Also, we have shown that $I_r(\tau)/A_r(\tau)$ has a filtration with sections of the form $A_r(\sigma)$ with $\tau \uparrow \sigma$. Dual remarks apply to the Weyl module $V_r(\tau)$ and the projective envelope $P_r(\tau)$ of $L_r(\tau)$.

The above remarks establish that $\mathcal{C}_\lambda^r$ is a highest weight category in the above sense with weight poset $(W_p \cdot \lambda, \uparrow)$.

Returning to the general case, let Γ be an ideal in the weight poset Λ of a highest weight category $\mathcal{C}$, and define $\mathcal{C}[\Gamma]$ to be the full subcategory of $\mathcal{C}$ consisting of all objects having composition factors $L(\nu)$ with $\nu \in \Gamma$. Also, suppose the coideal $\Omega = \Lambda \setminus \Gamma$ is finite. Both $\mathcal{C}[\Gamma]$ and the quotient category $\mathcal{C}(\Omega) = \mathcal{C}/\mathcal{C}[\Gamma]$ are highest weight categories with weight posets Γ and Ω, respectively. Let $i_* \equiv i_{\Gamma *} : \mathcal{C}[\Gamma] \to \mathcal{C}$ (resp., $j^* \equiv j^*_\Omega : \mathcal{C} \to \mathcal{C}(\Omega)$) be the corresponding inclusion functor (resp., quotient functor). Then i_* and j^* are exact functors which carry the simple, induced, and Weyl objects in their respective domains to the corresponding objects in their respective ranges. Furthermore, j^* sends the injective envelope $I(\lambda)$ of $L(\lambda)$ in $\mathcal{C}$ onto the injective envelope of the simple object $j^*L(\lambda)$ in $\mathcal{C}(\Omega)$ for $\lambda \in \Omega$. For details, see [**CPS4;Lemma 1.4**].

In addition, the functor i_* induces a full embedding

(1.2) $$i_* : D^b(\mathcal{C}[\Gamma]) \to D^b(\mathcal{C})$$

of bounded derived categories. This is proved in [**CPS3;Theorem 3.9**] in case Γ is finitely generated. However, since in this paper objects in $\mathcal{C}$ all have finite length, the finitely generated case implies i_* is a full embedding in general. Similarly, a standard truncation argument shows that i_* defines an equivalence of $D^b(\mathcal{C}[\Gamma])$ onto the relative derived category $D^b_{D^b(\mathcal{C}[\Gamma])}(\mathcal{C})$, defined as the full subcategory of $D^b(\mathcal{C})$ consisting of objects with cohomology in $\mathcal{C}[\Gamma]$.

Now assume that $\Omega \subset \tilde{\Omega}$ are finite coideals in Λ. Thus, $\tilde{\Omega} \setminus \Omega$ is an ideal in $\tilde{\Omega}$, so that the full highest weight subcategory $\mathcal{C}[\tilde{\Omega} \setminus \Omega]$ of the highest weight category $\mathcal{C}(\tilde{\Omega})$ is defined as above. By [**CPS3;Theorem 3.9(b)**], we obtain a full recollement diagram

(1.3) $$D^b(\mathcal{C}[\tilde{\Omega} \setminus \Omega]) \underset{\overset{i^!}{\longleftarrow}}{\overset{\overset{i^*}{\longleftarrow}}{\xrightarrow{i_*}}} D^b(\mathcal{C}(\tilde{\Omega})) \underset{\overset{j_*}{\longleftarrow}}{\overset{\overset{j_!}{\longleftarrow}}{\xrightarrow{j^*}}} D^b(\mathcal{C}(\Omega))$$

(in the sense of [**BBD;§3**]). In this diagram i_* is the derived functor of $i_{\tilde{\Omega}\setminus\Omega *}$ and j^* is the derived functor of j^*_Ω defined in the previous paragraph. (Again, it is not necessary to assume that Λ is finitely generated.)

2. Kazhdan-Lusztig theories

We continue to assume that $\mathcal{C}$ is a highest weight category with weight poset Λ as discussed in §1. Suppose there is a fixed length function $\ell : \Lambda \to \mathbf{Z}$ defined on the weight poset of $\mathcal{C}$. Let t be an indeterminate. For an object X in $D^b(\mathcal{C})$, define its (left) Poincaré polynomial in t, t^{-1} to be

$$p_{\nu,X} = p^L_{\nu,X} = \sum_{n=-\infty}^{\infty} \dim \operatorname{Hom}^n_{D^b(\mathcal{C})}(X, A(\nu))t^n$$

for weights $\nu \in \Lambda$. To see that $p_{\nu,X}$ belongs to $\mathbf{Z}[t, t^{-1}]$, its suffices to suppose that $X \cong L(\lambda)$ for some $\lambda \in \Lambda$. Then $p_{\nu,\lambda} \equiv p_{\nu,L(\lambda)}$ is a polynomial in t of degree at most the length of a maximal chain of weights from λ to ν

[CPS3;Lemma 3.8b]. (For X in $\mathcal{C}$, $p_{\nu,X}$ is also a polynomial in t.) Similarly, we define the (right) Poincaré polynomial by

$$p^R_{\nu,X} = \sum_{n=-\infty}^{\infty} \dim \mathrm{Hom}^n_{D^b(\mathcal{C})}(V(\nu), X)t^n$$

for $\nu \in \Lambda$. Again, these are actual Laurent polynomials (and polynomials in t if X lies in $\mathcal{C}$). Also, if $\mathcal{C}$ has a duality D, then $p_{\nu,X} = p^R_{\nu,DX}$ for all X. In particular, $p_{\nu,\lambda} = p^R_{\nu,\lambda}$ for all weights ν, λ. We now make the following definition which is basic to this paper. (Observe, for $X, Y \in \mathrm{Ob}(\mathcal{C})$, that $\mathrm{Hom}^\bullet_{D^b(\mathcal{C})}(X,Y) \cong \mathrm{Ext}^\bullet_{\mathcal{C}}(X,Y)$.)

2.1 Definition. Let $\mathcal{C}$ be a highest weight category with fixed length function $\ell : \Lambda \to \mathbf{Z}$. Then $\mathcal{C}$ is said to have a *Kazhdan-Lusztig theory* (relative to ℓ) provided that

$$\mathrm{Ext}^n_{\mathcal{C}}(L(\lambda), A(\nu)) \neq 0 \Rightarrow \ell(\lambda) - \ell(\nu) \equiv n \pmod 2$$

and

$$\mathrm{Ext}^n_{\mathcal{C}}(V(\nu), L(\lambda)) \neq 0 \Rightarrow \ell(\lambda) - \ell(\nu) \equiv n \pmod 2$$

for all weights $\lambda, \nu \in \Lambda$.

In general, define $\mathcal{E}^L(\mathcal{C},\ell)$ (resp., $\mathcal{E}^R(\mathcal{C},\ell)$) to be the full subcategory of $D^b(\mathcal{C})$ whose objects X satisfy the condition

$$\mathrm{Hom}^n(X, A(\nu)) \neq 0 \ \ (\text{resp., } \mathrm{Hom}^n(V(\nu), X)) \neq 0) \ \Rightarrow n \equiv \ell(\nu) \pmod 2.$$

Put $\hat{\mathcal{E}}^L(\mathcal{C},\ell) = \mathcal{E}^L(\mathcal{C},\ell) \oplus \mathcal{E}^L(\mathcal{C},\ell)[1]$ (resp., $\hat{\mathcal{E}}^R(\mathcal{C},\ell) = \mathcal{E}^R(\mathcal{C},\ell) \oplus \mathcal{E}^R(\mathcal{C},\ell)[1]$). Arguing as in **[CPS6;2.2]**, $\mathcal{C}$ has a Kazhdan-Lusztig theory if and only if each $L(\lambda)$ belongs to $\hat{\mathcal{E}}^L(\mathcal{C},\ell) \bigcap \hat{\mathcal{E}}^R(\mathcal{C},\ell)$.

For a Laurent polynomial $f \equiv f(t) \in \mathbf{Z}[t,t^{-1}]$, write $\bar{f} \equiv \bar{f}(t) = f(t^{-1})$. The *Kazhdan-Lusztig polynomial* associated to weights λ, ν is defined to be

$$P_{\nu,\lambda} = t^{\ell(\lambda)-\ell(\nu)}\bar{p}_{\nu,\lambda}.$$

Under certain conditions, these Kazhdan-Lusztig polynomials are, in fact, actual *polynomials* in t. For example, suppose that the length function ℓ is *compatible* with the poset structure on Λ (i. e., $\nu < \lambda \Rightarrow \ell(\nu) < \ell(\lambda)$). Then the result **[CPS3;Lemma 3.8b]** referred to above immediately implies that, if $\lambda \neq \nu$, $P_{\nu,\lambda}$ is a polynomial in t of degree at most $\ell(\lambda) - \ell(\nu) - 1$. If $\mathcal{C}$ has a Kazhdan-Lusztig theory, the Kazhdan-Lusztig polynomial is, in addition, a polynomial in t^2.

The definition of a Kazhdan-Lusztig theory given in (2.1) is the same as that given in **[CPS5]**, but more general than given in **[CPS6;Defn. 2.1]**. However, the two notions agree when the weight poset Λ is finite **[CPS5;Theorem 2.4]**. Also, in the case of finite Λ, the subcategories $\hat{\mathcal{E}}^L(\mathcal{C},\ell)$ and $\hat{\mathcal{E}}^R(\mathcal{C},\ell)$ defined above coincide with the corresponding subcategories in **[CPS5;§2]**.

In general, we are interested in this paper in the case in which the weight poset is *not* finite. Observe that any poset Λ can be written as a union of finitely generated ideals Γ. Because of the full embedding (1.2), $\mathcal{C}$ has a Kazhdan-Lusztig theory if and only if each $\mathcal{C}[\Gamma]$ has a Kazhdan-Lusztig theory (relative to the length function $\ell|_\Gamma$). For a fixed finitely generated ideal Γ, we can write

$$\Gamma = \bigcup_m \Omega_m \tag{2.2}$$

where the Ω_m are finite subcoideals of Γ. (Here m runs over some index set, which we assume is ordered by the relation: $m \leq n$ if and only if $\Omega_m \subseteq \Omega_n$.) The decomposition (2.2) is possible because Γ is finitely generated and interval finite. For each index m, let let $j_m^* : \mathcal{C}[\Gamma] \to \mathcal{C}(\Omega_m)$ be the quotient functor $j_{\Omega_m}^*$.

The following theorem further reduces the question of when $\mathcal{C}[\Gamma]$ has a Kazhdan-Lusztig theory.

2.3 THEOREM. *Let $\mathcal{C}$ be a highest weight category with finitely generated weight poset Λ. Represent Λ as a union as in (2.2). Suppose that $X \in Ob(D^b(\mathcal{C}))$ has the property, for some fixed index m, that any composition factor $L(\nu)$ of $H^\bullet(X)$ satisfies $\nu \in \Omega_m$.*

(a) *Then, for any index $n \geq m$, we have natural isomorphisms*

$$Hom^\bullet(X, A(\nu)) \cong Hom^\bullet(j_n^* X, j_n^* A(\nu)), \quad Hom^\bullet(V(\nu), X) \cong Hom^\bullet(j_n^* V(\nu), j_n^* X)$$

for all weights $\nu \in \Omega_n$. (Recall that $j_n^* A(\nu)$ is isomorphic to the induced object in the highest weight category $\mathcal{C}(\Omega_n)$ corresponding to $\nu \in \Omega_n$. Similarly, $j_n^* V(\nu)$ is isomorphic to the Weyl object in $\mathcal{C}(\Omega_n)$ corresponding to ν.)

(b) *For any $n \geq m$ and for any $\nu \in \Omega_n$, we have an equality*

$$p_{\nu, j_n^* X} = p_{\nu, X}, \quad p^R_{\nu, j_n^* X} = p^R_{\nu, X}$$

of Poincaré polynomials.

PROOF. We first prove (a). By an easy truncation argument, together with the five lemma, we can reduce to the case in which $X \in \mathrm{Ob}(\mathcal{C})$. Fix $n \geq m$ and let $\nu \in \Omega_n$. Let $I^\bullet$ be a minimal injective resolution of $A(\nu)$ in $\mathcal{C}$. For a non-negative integer t and a weight τ, the assertion that $L(\tau)$ lies in the socle of I^t implies that $\mathrm{Ext}^t_{\mathcal{C}}(L(\tau), A(\nu)) \neq 0$ which, in turn, implies that $\tau \geq \nu$ and t is bounded by the length of a maximal chain of weights from ν to τ by **[CPS3;Lemma 3.8b]**. Since Ω_n is a coideal, the weight τ lies in Ω_n. Since Ω_n is finite, we conclude that $I^\bullet$ is a finite complex. Thus, we can choose an index $N \geq n$ such that any composition factor $L(\tau)$ of any I^t lies in Ω_N. Now we use **[CPS4;Lemma 1.4c]** to conclude that

$$\mathrm{Ext}^\bullet_{\mathcal{C}}(X, A(\nu)) \cong \mathrm{Ext}^\bullet_{\mathcal{C}(\Omega_N)}(j_N^* X, j_N^* A(\nu)). \tag{2.3.1}$$

Consider the recollement diagram (1.3) with $\Omega = \Omega_n$ and $\tilde{\Omega} = \Omega_N$. As noted above, a simple object $L(\tau)$ in the socle of any I^t satisfies $\tau \geq \nu$, and hence $\tau \in \Omega_n$. Thus, I^t is a direct sum of injective hulls $I(\nu)$ of simple objects $L(\nu)$ for $\nu \in \Omega_n$, and $j_N^* I^t$ has this same property. Since $j_N^* I^\bullet$ is an injective resolution in $\mathcal{C}(\Omega_N)$ of $j_N^* A(\nu)$, we conclude that $i^! j_N^* A(\nu) = 0$. (Recall that $i^!$ is the right derived functor of the functor $\mathcal{C}(\tilde{\Omega}) \to \mathcal{C}[\Gamma]$, $\Gamma = \tilde{\Omega} \backslash \Omega$, which assigns to an object M in $\mathcal{C}(\tilde{\Omega})$ the largest subobject in $\mathcal{C}[\Gamma]$.) Therefore, using the distinguished triangle

$$i_* i^! j_N^* A(\nu) \to j_N^* A(\nu) \to j_* j^* j_N^* A(\nu) \to,$$

we conclude that $j_N^* A(\nu) \cong j_* j^* j_N^* A(\nu)$. Hence,

$$\operatorname{Ext}^\bullet_{\mathcal{C}(\Omega_N)}(j_N^* X, j_N^* A(\nu)) \cong \operatorname{Ext}^\bullet_{\mathcal{C}(\Omega_N)}(j_N^* X, j_* j^* j_N^* A(\nu)).$$

Since j^* is left adjoint to j_*, we finally obtain that this latter group is isomorphic to

$$\operatorname{Ext}^\bullet_{\mathcal{C}(\Omega_n)}(j^* j_N^* X, j^* j_N^* A(\nu)) \cong \operatorname{Ext}^\bullet_{\mathcal{C}(\Omega_n)}(j_n^* X, j_n^* A(\nu)).$$

Thus,

$$\operatorname{Hom}^\bullet_{\mathcal{C}}(X, A(\nu)) \cong \operatorname{Hom}^\bullet_{\mathcal{C}}(j_n^* X, j_n^* A(\nu)),$$

by (2.3.1).

A dual argument establishes that

$$\operatorname{Hom}^\bullet(V(\nu), X) \cong \operatorname{Hom}^\bullet(j_n^* V(\nu), j_n^* X),$$

completing the proof of (a).

Finally, (b) follows from (a), together with the definitions of the Poincaré polynomials. □

The following result is an immediate consequence of the theorem, together with Definition 2.1.

2.4 Corollary. *Let $\mathcal{C}$ be a highest weight category with finitely generated weight poset Λ expressed as the union (2.2) of its finite subcoideals. Then $\mathcal{C}$ has a Kazhdan-Lusztig theory relative to a fixed length function $\ell : \Lambda \to \mathbf{Z}$ if and only if each of the quotient highest weight categories $\mathcal{C}(\Omega_m)$ has a Kazhdan-Lusztig theory relative to the length function $\ell_m = \ell|_{\Omega_m}$.* □

3. Cohomology calculations

Recall that for a highest weight category $\mathcal{C}$, together with a length function $\ell : \Lambda \to \mathbf{Z}$ defined on its weight poset, we have the general notion of its left and right Grothendieck groups $K_0^L(\mathcal{C}, \ell)$ and $K_0^R(\mathcal{C}, \ell)$. These are discussed in **[CPS5;§2]**, where it is proved that $K_0^L(\mathcal{C}, \ell)$ (resp., $K_0^R(\mathcal{C}, \ell)$) is a free module over the ring $\mathbf{Z}[t, t^{-1}]$ of Laurent polynomials in an indeterminate t with basis $\{[V(\lambda) \mid \lambda \in \Lambda\}$ (resp., $\{[A(\lambda)] \mid \lambda \in \Lambda\}$.

In what follows, it will be convenient to make use of two subgroups of the ordinary Grothendieck group $K_0(\mathcal{C})$. Thus, define $K_0(\mathcal{C})_V$ (resp., $K_0(\mathcal{C})_A$) to be the free abelian subgroup of $K_0(\mathcal{C})$ with basis $\{[V(\nu)] \mid \nu \in \Lambda\}$ (resp., $\{[A(\nu)] \mid \nu \in$

$\Lambda\}$). Observe that, when Λ is a finite set, we have equalities $K_0(\mathcal{C}) = K_0(\mathcal{C})_V = K_0(\mathcal{C})_A$. Also, there is a natural pairing

$$\langle\ ,\ \rangle : K_0(\mathcal{C})_V \times K_0(\mathcal{C})_A \to \mathbf{Z} \tag{3.1}$$

obtained by putting $\langle[V(\nu)], [A(\tau)]\rangle = \delta_{\nu,\tau}$.

Assume that we are given a highest weight category $\mathcal{C}$ with weight poset Λ. Define the *complete left Grothendieck group* $\hat{K}_0^L(\mathcal{C})$ of $\mathcal{C}$ to consist of formal expressions

$$x = \sum_{\nu\in\Gamma,\, n\in\mathbf{Z}} a_{n,\nu}[V(\nu)]t^n, \ a_{n,\nu} \in \mathbf{Z} \tag{3.2}$$

for some finitely generated ideal Γ (depending on x), which can be rewritten

$$x = \sum_{\nu} p_\nu[V(\nu)] = \sum_{n=-\infty}^{k} f_n t^n \quad (\text{some } k \in \mathbf{Z})$$

with each $p_\nu \in \mathbf{Z}[t, t^{-1}]$ and with each $f_n \in K_0(\mathcal{C})_V$. Thus, in (3.2), the $a_{n,\nu} = 0$ for all sufficiently large n (independently of ν), while, for a fixed ν, $a_{n,\nu} = 0$ for $n << 0$ and, for a fixed n, $a_{n,\nu} = 0$ for all but finitely many ν. Also, regarding $a_{n,\nu}$ as defined for all $\nu \in \Lambda$, we have that $a_{n,\nu} = 0$ for ν outside some finitely generated ideal.

The dual notion of the *complete right Grothendieck group* $\hat{K}_0^R(\mathcal{C})$ consists of formal expressions

$$y = \sum_{\nu\in\Gamma,\, n\in\mathbf{Z}} b_{n,\nu}[A(\nu)]t^n, \ b_{n,\nu} \in \mathbf{Z} \tag{3.3}$$

for some finitely generated ideal Γ, and which can be rewritten

$$y = \sum_{\nu} q_\nu[A(\nu)] = \sum_{n=k}^{\infty} g_n t^n \quad (\text{some } k \in \mathbf{Z})$$

with each $q_\nu \in \mathbf{Z}[t, t^{-1}]$ and each $g_n \in K_0(\mathcal{C})_A$.

Suppose that $\mathcal{C}$ has a duality D. Then D defines an isomorphism

$$[D] : \hat{K}_0^L(\mathcal{C}) \to \hat{K}_0^R(\mathcal{C}) \tag{3.4}$$

satisfying

$$[D](x) = \sum a_{n,\nu}[A(\nu)]t^{-n}$$

for x as in (3.2).

Let $\mathbf{Z}[[t]]_t$ be the ring of formal Laurent power series in t. We extend the pairing (3.1) to give a pairing

$$\langle\ ,\ \rangle : \hat{K}_0^L(\mathcal{C}) \times \hat{K}_0^R(\mathcal{C}) \to \mathbf{Z}[[t]]_t \tag{3.5}$$

defined by

$$\langle x, y\rangle = \sum_{n,m} \langle f_n, g_m\rangle t^{m-n}$$

for x (resp., y) as in (3.2) (resp., (3.3)). Both $\hat{K}_0^L(\mathcal{C})$ and $\hat{K}_0^R(\mathcal{C})$ are naturally $\mathbf{Z}[t,t^{-1}]$-modules, and the pairing (3.5) is $\mathbf{Z}[t,t^{-1}]$-sesquilinear in the sense that

$$\langle fx, y\rangle = \langle x, \bar{f}y\rangle$$

for $f \in \mathbf{Z}[t,t^{-1}]$.

We will need the following lemma.

3.6 LEMMA. *Let $\mathcal{C}$ be a highest weight category with weight poset Λ. Let $X \in Ob(D^b(\mathcal{C}))$.*

(a) *For an integer $n \geq 0$, the groups*

$$Hom^n(X, A(\nu)) \text{ and } Hom^n(V(\nu), X)$$

are nonzero for at most finitely many weights $\nu \in \Lambda$.

(b) *Let Γ be the ideal generated by weights γ such that $L(\gamma)$ is a composition factor of $H^\bullet(X)$. Then the Poincaré polynomials $p_{\nu,X}$ and $p^R_{\nu,X}$ vanish for all $\nu \notin \Gamma$.*

PROOF. We can clearly assume in both (a) and (b) that $X \in \mathrm{Ob}(\mathcal{C})$.

We first prove (a). If $P_\bullet$ is a projective resolution of X, then

$$\mathrm{Ext}^n(X, A(\nu)) \neq 0 \Rightarrow \mathrm{Hom}(P_n, A(\nu)) \neq 0$$

which in turn implies that $L(\nu)$ is a composition factor of P_n. Since P_n has finite length, P_n has only finitely many composition factors. A dual argument establishes the assertion concerning the $\mathrm{Ext}^n(V(\nu), X)$. Part (a) follows.

Finally, part (b) is an immediate consequence of **[CPS3;Lemma 3.8b]**, together with its dual version. □

Because of the above lemma, an object X in $D^b(\mathcal{C})$ defines elements

$$(3.7)\qquad [X]_L^\infty = \sum_\nu \bar{p}_{\nu,X}[V(\nu)] \in \hat{K}_0^L(\mathcal{C}),\ [X]_R^\infty = \sum_\nu p^R_{\nu,X}[A(\nu)] \in \hat{K}_0^R(\mathcal{C}).$$

Thus, $[V(\nu)]_L^\infty = [V(\nu)] \in \hat{K}_0^L(\mathcal{C})$ and $[A(\nu)]_R^\infty = [A(\nu)] \in \hat{K}_0^L(\mathcal{C})$. We have the following basic result:

3.8 THEOREM. *Suppose that $\mathcal{C}$ is a highest weight category with fixed length function $\ell : \Lambda \to \mathbf{Z}$. Let $X \in Ob(\hat{\mathcal{E}}^L(\mathcal{C},\ell))$ (resp., $Y \in Ob(\hat{\mathcal{E}}^R(\mathcal{C},\ell))$). Then we have*

$$(3.8.1)\qquad \langle [X]_L^\infty, [Y]_R^\infty\rangle = \sum_{i=-\infty}^{\infty} \dim Hom^i_{D^b(\mathcal{C})}(X,Y)t^i.$$

PROOF. For some finitely generated ideal Γ, we have that $X, Y \in \mathrm{Ob}(D^b_{\mathcal{C}[\Gamma]}(\mathcal{C}))$, so that, by (1.2), we can without loss of generality, assume that Λ is finitely generated. For a given i, the coefficient c_i of t^i in $\langle [X]_L^\infty, [Y]_R^\infty\rangle$ is, by definition, the coefficient of t^i in the expression

$$(3.8.2)\qquad \sum_\tau p^L_{\tau,X} p^R_{\tau,Y}.$$

In a representation (2.2) of Λ as a union of its finite coideals, let m be an index sufficiently large so that: (a) all the weights τ which contribute to a coefficient of t^i in (3.8.2) belong to Ω_m, and (b) if $L(\sigma)$ is a composition factor of $H^\bullet(X)$ or $H^\bullet(Y)$, then $\sigma \in \Omega_m$. For the highest weight category $\mathcal{C}(\Omega_m)$ the theorem follows from [**CPS5;Prop. 2.3**]. Hence, by Theorem 2.3a, we have that

$$c_i = \dim \operatorname{Hom}^i(j_m^* X, j_m^* Y).$$

However, by [**CPS4;Lemma 1.4c**], for sufficiently large m we have

$$\operatorname{Hom}^n(j_m^* X, j_m^* Y) \cong \operatorname{Hom}^n(X, Y). \quad \square$$

As a consequence of the above theorem, we obtain the following result.

3.9 COROLLARY. *Suppose that $\mathcal{C}$ has a Kazhdan-Lusztig theory relative to a length function ℓ on its weight poset. Then for any $\lambda, \nu \in \Lambda$, dim $Ext^i(L(\lambda), L(\nu))$ is the coefficient of t^i in the power series*

$$\sum_\tau p_{\tau,\lambda} p^R_{\tau,\nu}. \quad \square$$

We now define Euler characteristics

$$\hat{\chi}^L : \hat{K}_0^L(\mathcal{C}) \to \hat{K}_0(\mathcal{C}), \ \hat{\chi}^R : \hat{K}_0^R(\mathcal{C}) \to \hat{K}_0(\mathcal{C})$$

into the complete Grothendieck group $\hat{K}_0(\mathcal{C})$ of $\mathcal{C}$ (see §6) by setting

(3.10) $$\hat{\chi}^L(x) = \sum_n (-1)^n f_n \text{ and } \hat{\chi}^R(y) = \sum_n (-1)^n g_n$$

for x as in (3.2) and y as in (3.3).

3.11 PROPOSITION. *For an object X in $\mathcal{C}$, we have*

(3.11.1) $$\hat{\chi}^L([X]_L^\infty) = \iota([X])$$

where $\iota : K_0(\mathcal{C}) \to \hat{K}_0(\mathcal{C})$ is the natural inclusion of the ordinary Grothendieck group into the complete Grothendieck group. Dually, we have

(3.11.2) $$\hat{\chi}^R([X]_R^\infty) = \iota([X]).$$

PROOF. Replacing the weight poset of $\mathcal{C}$ be a sufficiently large finitely generated ideal Γ and $\mathcal{C}$ by $\mathcal{C}[\Gamma]$, we can assume at the start that $\mathcal{C}$ has finitely generated weight poset Λ. For each index n in (2.2), let

$$\delta_n : \hat{K}_0(\mathcal{C}) \to K_0(\mathcal{C}(\Omega_n))$$

be the natural projection homomorphism obtained from the isomorphism $\hat{K}_0(\mathcal{C}) \cong \varprojlim K_0(\mathcal{C}(\Omega_n))$. The quotient functor $j_n^* : \mathcal{C} \to \mathcal{C}(\Omega_n)$ induces a homomorphism $[j_n^*] : K_0(\mathcal{C}) \to K_0(\mathcal{C}(\Omega_n))$ of Grothendieck groups. Then $\delta_n \circ \iota = [j_n^*]$ for any index n. To show that (3.10.1) is valid, it is therefore enough to prove that

$$\delta_n \hat{\chi}^L([X]_L^\infty) = [j_n^*]([X])$$

for all sufficiently large n. However,

$$\delta_n \hat{\chi}^L([X]_L^\infty) = \sum_{\nu\in\Omega_n} p_{\nu,X}(-1)[j_n^* V(\nu)] = \sum_{\nu\in\Omega_n} p_{\nu,j_n^* X}(-1)[j_n^* V(\nu)]$$

by Theorem 2.3b, once n is sufficiently large that $H^\bullet(X)$ has composition factors $L(\lambda)$ with $\lambda \in \Omega_n$. By **[CPS5;Prop. 3.2]**, this latter expression equals $[j_n^*]([X])$, as desired.

A dual argument now establishes the assertion for the right Euler characteristic. □

3.12 EXAMPLE. We return to the situation of Example 1.1 for $r = 1$. Assume that $p \geq h$, the Coxeter number of G. Fix a p-regular weight $\lambda \in C_{\mathbf{Z}}^+$. The highest weight category $\mathcal{C}_\lambda^1$ has weight poset $(W_p \cdot \lambda, \uparrow)$.

For $n = 0, 1, \cdots$, let $\wp_n$ denote the corresponding generalized Kostant partition function on $X(T)$; thus, $\wp_n(\tau)$ is the number of ways that the weight τ can be written as a sum of exactly n positive roots. Consider the formal T-character

$$\sigma = \prod_{\alpha>0}(1 - e^{-p\alpha}t^{-2})^{-1} = \sum_{n\geq 0,\theta} \wp_n(\theta)e^{-p\theta}t^{-2n}. \tag{3.12.1}$$

For any weight $\nu \in W_p \cdot \lambda$, we define

$$[V_1(\nu)] \cdot \sigma = \sum_{n\geq 0,\theta} \wp_n(\theta)[V_1(\nu - p\theta)]t^{-2n} \in \hat{K}_0^L(\mathcal{C}_\lambda^1).$$

Similarly,

$$[A_1(\nu)] \cdot \bar{\sigma} = \sum_{n\geq 0,\theta} \wp_n(\theta)[A_1(\nu - p\theta)]t^{2n} \in \hat{K}_0^R(\mathcal{C}_\lambda^1).$$

Clearly, for $x = \sum p_\nu[V_1(\nu)]$ (resp., $y = \sum q_\nu[A_1(\nu)]$) as in (3.2) (resp., (3.3)) for $\mathcal{C}_\lambda^1$, $x \cdot \sigma = \sum p_\nu[V_1(\nu)] \cdot \sigma$ (resp., $y \cdot \bar{\sigma} = \sum q_\nu[A_1(\nu)] \cdot \bar{\sigma}$) is a well-defined element in $\hat{K}_0^L(\mathcal{C}_\lambda^1)$ (resp., $\hat{K}_0^R(\mathcal{C}_\lambda^1)$.)

Similarly, for $\xi \in X(T)$, let $\lambda' \in C_{\mathbf{Z}}^+$ be so that $p\xi + \lambda \in W_p \cdot \lambda'$. Then we can define $x \cdot e^{p\xi} = \sum p_\nu[V_1(\nu + p\xi)] \in \hat{K}_0^L(\mathcal{C}_{\lambda'}^1)$ (resp., $y \cdot e^{p\xi} = \sum q_\nu[A_1(\nu + p\xi)] \in \hat{K}_0^R(\mathcal{C}_{\lambda'}^1)$). Observe that $p\xi + W_p \cdot \lambda \subset W_p \cdot \lambda'$ since W_p is normalized by translations by $p\xi$. We then have the following result. (Compare **[Ka;Lemma 4.6]**.)

3.12.2 THEOREM. *Assume the above notation, and let $\xi \in X(T)$. Assume that $\lambda' \in C_{\mathbf{Z}}^+$ satisfies $p\xi + \lambda \in W_p \cdot \lambda'$. Assume also that $p > h$. We have*

$$[L_1(p\xi + \lambda)]_L^\infty = (\sum_{w\in W} t^{-\ell(w)}[V_1(p\xi + w \cdot \lambda)]) \cdot \sigma \in \hat{K}_0^L(\mathcal{C}_{\lambda'}^1). \tag{3.12.2.1}$$

Also,

$$[L_1(p\xi + \lambda)]_R^\infty = (\sum_{w\in W} t^{\ell(w)}[A_1(p\xi + w \cdot 0)]) \cdot \bar{\sigma} \in \hat{K}_0^R(\mathcal{C}_{\lambda'}^1). \tag{3.12.2.2}$$

PROOF. Using **[J2;p. 355]**, we can assume that $\lambda = 0$. Clearly,

$$p_{\tau,p\xi} = p_{\tau - p\xi,0}$$

for all $\tau \in X(T)$. Hence,

$$[L_1(p\xi)]_L^\infty = [L_1(0)]_L^\infty \cdot e^{p\xi},$$

so to prove (3.12.2.1), it suffices to consider the special case in which $\xi = 0$.

Observe that

$$p_{\nu,0}^L = \sum_n \dim \operatorname{Ext}_{\mathcal{C}}^n(L_1(0), A_1(\nu))t^n = \sum_n \dim H^n(B_1T, \nu)t^n,$$

since the exactness of the induction functor $\operatorname{ind}_{B_1T}^{G_1T}$ (see [**CPS1;§4**]) implies that

$$\operatorname{Ext}_{G_1T}^\bullet(L_1(0), A_1(\nu)) \cong \operatorname{Ext}_{B_1T}^\bullet(L_1(0), \nu) \cong H^\bullet(B_1T, \nu).$$

If $\mathfrak{n}$ denotes the Lie algebra of U, the Frobenius twist $\mathfrak{n}^{*(1)}$ of the dual T-module has character $\sum_{\alpha\in\Phi^+} e^{p\alpha}$. We give $\mathfrak{n}^{*(1)}$ homological degree 2 and form the symmetric algebra $S^\bullet(\mathfrak{n}^{*(1)})$. Since $p > h$, it is known (see [**FP;Cor. 2.6**]) that

$$H^\bullet(U_1, k) \cong S^\bullet(\mathfrak{n}^{*(1)}) \otimes H^\bullet(\mathfrak{n}, k), \tag{3.12.2.3}$$

an isomorphism of rational T-modules, where the Koszul cohomology of the Lie algebra $\mathfrak{n}$ is described as a rational T-module by its character:

$$\operatorname{ch} H^n(\mathfrak{n}, k) \cong \sum_{w\in W, \ell(w)=n} e^{-w\cdot 0}. \tag{3.12.2.4}$$

(For another development of $H^\bullet(U_1, k)$, see [**AJ**].) Also,

$$H^\bullet(B_1T, \nu) \cong (H^\bullet(U_1, k) \otimes \nu)^T \cong H^\bullet(U_1, k)_{-\nu},$$

the $-\nu$-weight space in $H^\bullet(U_1, k)$, while (3.11.2.3) and (3.11.2.4) imply that the formal T-character of $H^n(U_1, k)$ is given by the expression

$$\sum_{2m+\ell(w)=n, \theta} \wp_m(\theta) e^{(-w\cdot 0)+p\theta}.$$

Thus,

$$[L_1(0)]_L^\infty = \sum_\nu \bar{p}_{\nu,0}[V_1(\nu)] = (\sum_{w\in W} t^{-\ell(w)}[V_1(w\cdot 0)]) \cdot \sigma,$$

as required for (3.12.2.1). Finally, (3.12.2.2) now follows from the identity

$$D(x \cdot \sigma) = D(x) \cdot \bar{\sigma}$$

for $x \in \hat{K}_0^L(\mathcal{C}_0^1)$. □

Recall that Lusztig [**L2**] has defined a function $d(A, B)$ on pairs A, B of alcoves: let $\mathcal{S}(A, B)$ be the set of hyperplanes separating A and B. If $H \in \mathcal{S}(A, B)$, let $\epsilon(H) = 1$ (resp., -1) if A lies on the negative (resp., positive) side of H. (Observe that H is defined by an equation of the form $(x + \rho, \beta^\vee) = pn$ for some positive root β, so that H has a positive side defined by $(x+\rho,\beta^\vee) > np$ and a negative side defined by $(x+\rho,\beta^\vee) < np$.) Then

$$d(A,B) = \sum_{H\in\mathcal{S}(A,B)} \epsilon(H). \tag{3.12.3a}$$

We use this function to define, for any $\lambda \in C^+_{\mathbf{Z}}$, a "generic" length function $\ell_{\text{gen}} \equiv \ell$ on $W_p\cdot\lambda$. Namely, if $\tau = z\cdot\lambda \in W_p\cdot\lambda$, we put

$$\ell(\tau) = d(C^+, z\cdot C^+). \tag{3.12.3b}$$

We shall require some elementary properties of this function. For these, it is often convenient to work with the extended affine Weyl group $\tilde{W}_p$, generated by reflections $s_{\alpha,pr}$ and translations $t_{p\xi}$ for $\xi \in X(T)$. Thus, $\tilde{W}_p$ contains W_p as a normal subgroup, and $\tilde{W}_p = W_p\cdot N$, where N is the stabilizer in $\tilde{W}_p$ of the alcove C^+. The Bruhat length function ℓ on W_p extends to a length function on $\tilde{W}_p$ by putting $\ell(zn) = \ell(z)$ for $n\in N$, $z\in W_p$.

For any $\theta \in X(T)$, define

$$\text{ht}(\theta) = (1/2)\sum_{\alpha>0}(\theta,\alpha^\vee). \tag{3.12.4}$$

In general, the identification $w\mapsto w\cdot C^+$ between W_p and the set of alcoves does not satisfy the identity $d(C^+, w\cdot C^+) = \ell(w)$. However, we have the following result. (Compare [**K**].)

3.12.5 LEMMA. *Let $z\in\tilde{W}_p$ be expressed as $z = t_{p\theta}a$ for $\theta\in X(T)$ and $a\in W$. Then*

$$d(C^+, z\cdot C^+) = -\ell(a) + 2ht(\theta). \tag{3.12.5.1}$$

If $z\cdot C^+$ is dominant (i. e., it is contained in $X(T)^+ - \rho$), then

$$\ell(z) \equiv d(C^+, z\cdot C^+) = -\ell(a) + 2ht(\theta). \tag{3.12.5.2}$$

PROOF. First, assume that $z\cdot C^+$ is dominant. Then the equality $\ell(z) = d(C^+, z\cdot C^+)$ is well-known, since, in this case, $d(C^+, z\cdot C^+)$ is the number of alcoves separating C^+ and $z\cdot C^+$. A direct calculation shows that $\theta\in X(T)^+$. Also, z is a distinguished *right* coset representative for W in $\tilde{W}_p$ in the sense that $\ell(bz) = \ell(b) + \ell(z)$ for all $b\in W$. (See [**CPS6;4.8**], for example. The discussion there easily carries over to $\tilde{W}_p$.) In particular,

$$\ell(z) + \ell(a) = \ell(z) + \ell(a^{-1}) = \ell(t_{pa^{-1}\theta}).$$

Since $\theta\in X(T)^+$, [**IM;Prop. 1.23**] implies that

$$\ell(t_{pa^{-1}\theta}) = \sum_{\alpha>0}|(a^{-1}\theta,\alpha^\vee)| = \sum_{\alpha>0}(\theta,\alpha^\vee) = 2\text{ht}(\theta).$$

Hence,

$$\ell(z) = 2\text{ht}(\theta) - \ell(a),$$

establishing (3.12.5.2).

To see (3.12.5.1), observe first that $d(C^+, z \cdot C^+) = d(t_{p\xi} \cdot C^+, (t_{p\xi} z) \cdot C^+)$, for all $\xi \in X(T)$. Hence, using the formula

$$d(A,B) + d(B,D) + d(D,A) = 0 \tag{3.12.5.3}$$

for alcoves A, B, D (see **[L2;1.4.1]**), we easily reduce (3.12.5.1) to the special case in which $z \cdot C^+$ is dominant. Now apply (3.12.5.2). □

We conclude this section with the following result, using the length function (3.12.3). It is a corollary of (the proof of) (3.12.2).

3.12.6 PROPOSITION. *Assume $p > h$. Let $\xi \in X(T)$ and $\lambda \in C^+_{\mathbf{Z}}$. Choose $\lambda' \in C^+_{\mathbf{Z}}$ so that $p\xi + \lambda \in W_p \cdot \lambda'$. Then $L_1(p\xi+\lambda)$ belongs to $\mathcal{E}^L(\mathcal{C}^1_{\lambda'}, \ell) \bigcap \mathcal{E}^R(\mathcal{C}^1_{\lambda'}, \ell)$.*

PROOF. As before, we can assume $\lambda = 0$. Since $L_1(p\xi) \cong L_1(0) \otimes p\xi$ and $A_1(\nu) \cong A_1(\nu - p\xi) \otimes p\xi$, we can easily reduce to the special case in which $\xi = 0$ (and hence $\lambda' = 0$). By the proof of (3.12.2), $\mathrm{Hom}^n(L_1(0), A_1(\nu)) \neq 0$ implies that $\nu = t_{p\theta} a \cdot 0$ for some $\theta \in Q(T)$ and some $a \in W$ with $\ell(a) \equiv n \pmod 2$. Thus, (3.12.3) implies that $L_1(0)$ lies in $\mathcal{E}^L(\mathcal{C}^1_0, \ell)$. By duality, it also lies in $\mathcal{E}^R(\mathcal{C}^1_0, \ell)$. □

4. Hecke operators

In this section, we adapt the notion **[CPS6]** of an abstract Hecke operator on a highest weight category to the present context. We begin with the following combinatorial definition, a variation on **[CPS6;§4]**.

4.1 DEFINITION. Let Λ be a poset having a length function $\ell : \Lambda \to \mathbf{Z}$ and having a nonempty subset Λ_0 of *initial elements*.

(a) By a *weak reflection* on Λ we mean a pair (Λ_s, s) consisting of a nonempty ideal $\Lambda_s \subset \Lambda$ and a function $s : \Lambda_s \to \Lambda_s$ such that $s^2 = \mathrm{id}$ and for any $\lambda \in \Lambda_s$ either $\lambda s \leq \lambda$, or $\lambda s > \lambda$; in the latter case we assume $\ell(\lambda s) = \ell(\lambda) + 1$. (For economy of notation, we will often denote a weak reflection (Λ_s, s) by the associated mapping s. In most of the new applications we have in mind for this paper, it turns out, unlike the situation in **[CPS6]**, that $\Lambda_s = \Lambda$ for a weak reflection s.)

(b) Suppose that S is a set of weak reflections on Λ. For $\lambda, \nu \in \Lambda$, we write $\nu \uparrow_S \lambda$ (or just $\nu \uparrow \lambda$ if S is clear from context) provided there exists a chain
$$\nu = \nu_0 < \nu_1 < \cdots < \nu_n = \lambda$$
in Λ such that for $0 \leq i < n$ there is a weak reflection $s_i \in S$ satisfying $\nu_i \in \Lambda_{s_i}$ and $\nu_i s_i = \nu_{i+1}$.

(c) Let S be a set of weak reflections on Λ. We say that Λ is *weakly S-connected* provided given any $\lambda \in \Lambda$, there exists an initial element $\nu \in \Lambda_0$ such that $\nu \uparrow \lambda$.

We emphasize that in the definition below the length function $\ell : \Lambda \to \mathbf{Z}$ is arbitrary, except that it must satisfy the condition that $\ell(\lambda s) = \ell(\lambda) + 1$ whenever $\lambda < \lambda s \in \Lambda_s$ for some weak reflection. The following definition directly generalizes **[CPS6;Defn. 4.2]** in the sense that we use the more general notion of $\hat{\mathcal{E}}^L(\mathcal{C}, \ell)$.

4.2 DEFINITION. (**[CPS6;Defn. 4.2]**) Let $\mathcal{C}$ be a highest weight category with duality $D : \mathcal{C} \to \mathcal{C}$ and with a length function $\ell : \Lambda \to \mathbf{Z}$. Consider the following data (i)–(ii):

(i) A weak reflection (Λ_s, s) on Λ.

(ii) Let $\mathfrak{G} \equiv \mathfrak{G}(\Lambda_s) \subset \mathrm{Ob}(\mathcal{C})$ be the set consisting of all objects isomorphic to some $V(\lambda)$, $A(\lambda)$, or $L(\lambda)$ for $\lambda \in \Lambda_s$. We assume that we are given a mapping

$$\beta_s \equiv \beta : \mathfrak{G} \to \mathrm{Ob}(D^b_{\mathcal{C}[\Lambda_s]}(\mathcal{C})),$$

where $D^b_{\mathcal{C}[\Lambda_s]}(\mathcal{C})$ denotes the relative derived category in $D^b(\mathcal{C})$ associated to $\mathcal{C}[\Lambda_s]$.

Then we say that β_s is a *pre-Hecke operator* (of type s) provided the following conditions (a)–(c) are satisfied:

(a) $\beta_s D(X) \cong D\beta_s(X)$ for any $X \in \mathfrak{G}$.

(b) Let $\lambda \in \Lambda_s$. If $\lambda < \lambda s$, there is a distinguished triangle

$$V(\lambda s) \to \beta_s V(\lambda) \to V(\lambda)[1] \to$$

(dually, there is a distinguished triangle

$$A(\lambda)[-1] \to \beta_s A(\lambda) \to A(\lambda s) \to;)$$

if $\lambda > \lambda s$, there is a distinguished triangle

$$V(\lambda)[-1] \to \beta_s V(\lambda) \to V(\lambda s) \to$$

(dually, there is a distinguished triangle

$$A(\lambda s) \to \beta_s A(\lambda) \to A(\lambda)[1] \to);$$

and, finally, if $\lambda s = \lambda$, then

$$\beta_s V(\lambda) \cong V(\lambda)[1] \oplus V(\lambda)[-1]$$

(dually, $\beta_s A(\lambda) \cong A(\lambda)[-1] \oplus A(\lambda)[-1]$).

(c) Let $\lambda \in \Lambda_s$ satisfy $\lambda s > \lambda$. If $L(\lambda) \in \mathrm{Ob}(\hat{\mathcal{E}}^L(\mathcal{C}, \ell))$, then the *adjoint condition*

$$\mathrm{Hom}^n(\beta_s L(\lambda), A(\nu)) \cong \mathrm{Hom}^n(L(\lambda), \beta_s A(\nu))$$

holds for all weights $\nu \in \Lambda_s$ and all integers n.

Finally, if β_s is a pre-Hecke operator, then we say β_s is a *Hecke operator* if, in addition, β_s satisfies the following *decomposition property:*

(d) Let $\lambda \in \Lambda_s$ be such that $\lambda < \lambda s$ and $L(\lambda) \in \mathrm{Ob}(\hat{\mathcal{E}}^L(\mathcal{C}, \ell))$. Then $\beta_s L(\lambda)$ is a direct sum of objects $L(\nu)$ with $\nu \in \Lambda_s$.

All the observations of **[CPS6;Remark 4.3]** apply in the present context, in view of the discussion concerning (1.2). If $L(\lambda)$ belongs to $\hat{\mathcal{E}}^L(\mathcal{C}, \ell)$ (as defined following (2.1)), then $\beta_s L(\lambda)$ is isomorphic to an object in $\mathcal{C}$ for any pre-Hecke operator β_s with $\lambda < \lambda s \in \Lambda_s$. In addition, if $L(\lambda)$ belongs to $\mathcal{E}^L(\mathcal{C}, \ell)$ (resp., $\mathcal{E}^L(\mathcal{C}, \ell)[1]$), then $\beta_s L(\lambda)$ belongs to $\mathcal{E}^L(\mathcal{C}, \ell)[1]$ (resp., $\mathcal{E}^L(\mathcal{C}, \ell)$). Also, let S_{pre} (resp., S_{Hck}) denote the set of all weak reflections s associated to a pre-Hecke (resp., Hecke) operator β_s of type s.

For weights $\nu < \lambda$, put

$$\mu(\nu, \lambda) = \dim_k \mathrm{Ext}^1(L(\lambda), A(\nu)).$$

Thus, $\mu(\nu, \lambda)$ is the coefficient of t in the Poincaré polynomial $p_{\nu,\lambda}$. The statement and the proof of the following result concerning the explicit decomposition of $\beta_s L(\lambda)$ is essentially the same as that of **[CPS6;Prop. 4.4]**.

(4.3) PROPOSITION. *Consider a highest weight category $\mathcal{C}$ as in Definition 4.2. Let s be a weak reflection, and let β_s be a pre-Hecke operator of type s. Let $\lambda \in \Lambda_s$ be a weight such that $\lambda < \lambda s$, $L(\lambda) \in Ob(\hat{\mathcal{E}}^L(\mathcal{C}, \ell))$, and $\beta_s L(\lambda)$ is completely reducible. Then we have*

$$\beta_s L(\lambda) \cong L(\lambda s) \oplus \bigoplus_{\nu s \leq \nu < \lambda} \mu(\nu, \lambda) L(\nu).$$

(In particular, this decomposition holds if β_s is a Hecke operator and λ is any weight such that $L(\lambda) \in Ob(\hat{\mathcal{E}}^L(\mathcal{C}, \ell))$ and $\lambda < \lambda s \in \Lambda_s$.) Finally, $L(\lambda s)[-\ell(\lambda s)]$ lies in $\mathcal{E}^L(\mathcal{C}, \ell)$. □

4.3.1 REMARK. In the above result, if we replace the condition that $\beta_s L(\lambda)$ is completely reducible by the condition that $L(\lambda s)$ is a direct summand of $\beta_s L(\lambda)$, then the final assertion of the proposition is still valid. This is a corollary of the proof.

Let β_s be a pre-Hecke operator of type s on $\mathcal{C}$. As in **[CPS6;Remark 4.5]**, β_s defines a natural $\mathbf{Z}[t, t^{-1}]$-linear operator $[\beta_s]$ on the complete left Grothendieck group $\hat{K}^L(\mathcal{C}[\Lambda_s])$. More precisely, for x as in (3.2), we put

$$[\beta_s](x) = \sum a_{n,\nu} [\beta_s][V(\nu)] t^n,$$

where

$$\text{(4.4a)} \qquad [\beta_s][V(\nu)] = \begin{cases} t^{-1}[V(\nu)] + [V(\nu s)], & \nu s > \nu \\ t[V(\nu)] + [V(\nu s)], & \nu s < \nu \\ (t + t^{-1})[V(\nu)], & \nu s = \nu. \end{cases}$$

Clearly, $[\beta_s](x)$ defines an element in $\hat{K}_0^L(\mathcal{C}[\Lambda_s]) \subset \hat{K}_0^L(\mathcal{C})$. Also, suppose that $\lambda \in \Lambda_s$, $\lambda < \lambda s$, and $L(\lambda) \in \mathrm{Ob}(\mathcal{E}^L(\mathcal{C}, \ell))$. Then, as in [**CPS6;Remark 4.5**], we have

$$[\beta_s][L(\lambda)]_L^\infty = [\beta_s L(\lambda)]_L^\infty. \tag{4.4b}$$

These formulas enable recursive determination of $[L(\lambda)]_L^\infty$, provided (4.3) holds and one is able to determine formulas for the terms in (4.3) with smaller weights inductively. The latter step is always possible in the situation of [**CPS6**], where Λ is finite and Λ_0 consists of minimal elements, and we formulated the resulting character formula as an abstract Kazhdan-Lusztig conjecture. The step in question can also be carried out for G_1T, where one has translation by elements in $pX(T)$ available. The program is carried out in §5; see Theorem 5.7 and its proof. If one where to formulate a Kazhdan-Lusztig conjecture for G_1T, it would amount to establishing the formula (5.7.2).

We conclude this section with the following analogue of [**CPS6;Theorem 5.5**]. Its proof is similar to that given in [**CPS6**].

4.5 THEOREM. *Let $\mathcal{C}$ be a highest weight category with duality D and length function $\ell : \Lambda \to \mathbf{Z}$. Let $S \subset S_{\text{pre}}$ be such that Λ is weakly S-connected with respect to some fixed set Λ_0 of initial elements. Assume each $L(\lambda)$, $\lambda \in \Lambda_0$, lies in $\hat{\mathcal{E}}^L(\mathcal{C}, \ell)$. For each $s \in S$, fix a pre-Hecke operator β_s of type s. The following are equivalent:*

(a) *For each $s \in S$, the pre-Hecke operator β_s is a Hecke operator.*

(b) *For each $s \in S$ and weight $\tau \in \Lambda_s$ such that $\tau s > \tau$, the simple module $L(\tau s)$ is a direct summand of $\beta_s L(\tau)$.*

(c) *For each $s \in S$ and weight $\tau \in \Lambda_s$ such that $\tau s > \tau$, we have*

$$Hom(\beta_s L(\tau), L(\tau s)) \neq 0.$$

(d) *$\mathcal{C}$ has a Kazhdan-Lusztig theory.*

(e) *Let $\theta, \nu \in \Lambda$. If $\ell(\theta) \equiv \ell(\nu)$ (mod 2), then*

$$Ext^1(V(\theta), L(\nu)) = 0.$$

PROOF. Assume that β_s is a Hecke operator, for each $s \in S$. Proposition 4.3 shows that, for each weight $\lambda \in \Lambda_s$ such that $\lambda < \lambda s$ and $L(\lambda) \in \mathrm{Ob}(\hat{\mathcal{E}}^L(\mathcal{C}, \ell))$, $\beta_s L(\lambda)$ is completely reducible and contains $L(\lambda s)$ as a direct summand. Moreover, $L(\lambda s)$ also lies in $\mathrm{Ob}(\hat{\mathcal{E}}^L(\mathcal{C}, \ell))$.

If τ is any weight, there is a chain

$$\tau_0 < \tau_1 < \cdots < \tau_n = \tau \tag{4.5.1}$$

with $\tau_0 \in \Lambda_0$ and $\tau_{i+1} = \tau_i s_i$ for some $s_i \in S$ and $\tau_i \in \Lambda_{s_i}$ $(0 \leq i < n)$. Using induction on n, and the hypothesis that $L(\tau_0) \in \mathrm{Ob}(\hat{\mathcal{E}}^L(\mathcal{C}, \ell))$, it follows the same is true for any weight τ. Thus, (a) implies (b).

Assume that (b) holds. By induction on the length of the chain (4.5.1), we again see that $L(\tau)[-\ell(\tau)] \in \mathrm{Ob}(\mathcal{E})^L(\mathcal{C},\ell))$ for each weight τ. Because $\mathcal{C}$ has a duality D, it follows that $L(\tau)[-\ell(\tau)]$ also belongs to $\mathcal{E}^R(\mathcal{C},\ell)$, and (d) follows.

The implications (b) $\Rightarrow$ (c) and (d) $\Rightarrow$ (e) are obvious.

Assume that condition (e) holds. Let $s \in S$ and let $\xi \in \Lambda_s$ with $\xi < \xi s$ and $L(\xi) \in \mathrm{Ob}(\hat{\mathcal{E}}^L(\mathcal{C},\ell))$. Let $X = \beta_s L(\xi)$. Then either X lies in $\mathcal{E}^L(\mathcal{C},\ell)$ or in $\mathcal{E}^L(\mathcal{C},\ell)[1]$. Hence, if

$$\mathrm{Hom}(X, A(\lambda)) \neq 0 \neq \mathrm{Hom}(X, A(\nu)),$$

it follows that $\ell(\lambda) \equiv \ell(\nu) \pmod 2$. It follows that the hypothesis of (e) verifies that of [**CPS5;Theorem 4.1**]. Hence, X is completely reducible. Thus, (e) $\Rightarrow$ (a).

Finally, assume that condition (c) holds. Let $\tau \in \Lambda_s$ with $\tau < \tau s$ be given. Using yet another induction argument, we can assume that $L(\tau)$ belongs to $\hat{\mathcal{E}}^L(\mathcal{C},\ell)$. By [**CPS6;Lemma 5.4**], $L(\tau s)$ has multiplicity one as a composition factor of $\beta_s L(\tau) \in \mathrm{Ob}(\mathcal{C})$. Using the duality operator D and the hypothesis (c), we see that $L(\tau s)$ appears in both the head and the socle of $\beta_s L(\tau)$. Thus, $L(\tau s)$ is a direct summand of $\beta_s L(\tau)$ and $L(\tau s) \in \mathrm{Ob}(\hat{\mathcal{E}}^L(\mathcal{C},\ell))$. Thus, (b) holds. □

5. Representation theory of G_1T

We consider the case of a semisimple, simply connected algebraic group G as in Example 1.1. Thus, $p \geq h$, the Coxeter number of G. Fix a p-regular weight $\lambda \in C^+_{\mathbf{Z}}$. We work with the case $r = 1$, and consider the highest weight category $\mathcal{C}^1_\lambda$ (having weight poset $(W_p \cdot \lambda, \uparrow)$). We use the length function $\ell : W_p \cdot \lambda \to \mathbf{Z}$ defined in (3.12.3). Given any other p-regular weight $\lambda' \in C^+_{\mathbf{Z}}$, the highest weight categories $\mathcal{C}^1_\lambda$ and $\mathcal{C}^1_{\lambda'}$ are equivalent [**J2;p. 335**]. In fact, the equivalence can be chosen so as to map $L_1(w\cdot\lambda)$, $A_1(w\cdot\lambda)$, and $V_1(w\cdot\lambda)$ to $L_1(w\cdot\lambda')$, $A_1(w\cdot\lambda')$, and $V_1(w\cdot\lambda')$, respectively. Thus, for simplicity, we usually state our results for the category $\mathcal{C}^1_0$ obtained by taking $\lambda = 0$, and we use without mention the fact that there is a corresponding result for an arbitrary $\mathcal{C}^1_\lambda$, $\lambda \in C^+_{\mathbf{Z}}$. Also, in what follows, $\mathcal{C}$ denotes the category of finite dimensional rational G-modules. If $\tau \in X(T)^+$, $\mathcal{C}_\tau$ denotes the full subcategory of $\mathcal{C}$ consisting of objects having composition factors $L(\nu)$ for $\nu \in W_p \cdot \tau$.

Fix $s \in \Sigma$ and $\nu_s \in \bar{C}^+_{\mathbf{Z}}$ on the corresponding s-face of C^+. As defined in [**J2;§II.9**], there are corresponding translation functors $T_0^{\nu_s}$ and $T^0_{\nu_s}$, which are adjoint to each other. As in [**CPS6;§4.8**], one can construct an associated "Jantzen reflection" functor

$$\Theta_s = T^0_{\nu_s} \circ T_0^{\nu_s} : \mathcal{C}^1_0 \to \mathcal{C}^1_0.$$

Thus, Θ_s is an exact functor commuting with duality, which is (left and right) adjoint to itself. It follows, for any $M \in \mathrm{Ob}(\mathcal{C}^1_0)$, that we have morphisms

$$\text{(5.1)} \qquad M \xrightarrow{\delta(M)} \Theta_s M \xrightarrow{\epsilon(M)} M.$$

defined using the adjunction morphisms associated to the corresponding translation functors.

Each $s \in \Sigma$ defines a weak reflection on $W_p \cdot 0$ in which $(W_p \cdot 0)_s = W_p \cdot 0$. Namely, for $z \cdot 0 \in W_p \cdot 0$ and $s \in \Sigma$, put $(z \cdot 0)s = zs \cdot 0$. (By **[L2;1.4]**, or the formula stated in (3.12.5), we verify that s satisfies (4.1a).) Then we have the following result.

5.2 THEOREM. *Let $s \in \Sigma$.*

(a) *For $M \in \mathfrak{G} \equiv \mathfrak{G}((W_p \cdot 0))$ (see (4.2)) we have $\epsilon(M) \circ \delta(M) = 0$. Therefore, we define $\beta_s : \mathfrak{G} \to Ob(D^b(\mathcal{C}_0^1))$ by putting, for $M \in \mathfrak{G}$, $\beta_s M$ equal to the sequence $M \to \Theta_s M \to M$ defined in (5.1), regarded as a complex concentrated in degrees $-1, 0, 1$. Then β_s is a pre-Hecke operator.*

(b) *Let $\tau \in W_p \cdot 0$ be restricted. Assume $\tau < \tau s$. Let β_s^G be the pre-Hecke operator of type s defined for G-modules in* **[CPS6;Theorem 4.8.3]**. *Then we have*

$$\beta_s L_1(\tau) \cong \beta_s^G L(\tau)|_{G_1 T}.$$

PROOF. The proof of (a) resembles that of **[CPS6;Theorem 4.8.3]** and is omitted. To prove (b), let $\delta \in W_p \cdot \nu_s$ lie on the s-face of the alcove containing τ. Since τ is restricted and $\tau < \tau s$, we have that δ is also restricted, and hence $L_1(\delta) \cong L(\delta)|_{G_1 T}$. Both $\beta_s L_1(\delta)$ and $\beta_s^G L(\delta)$ are constructed by first forming the tensor product Z of $L(\delta)$ with a certain irreducible G-module, and then projecting Z to the block containing irreducible G- or $G_1 Z$- modules with high weights in $W_p \cdot 0$. For any dominant weight σ, $L(\sigma)|_{G_1 T}$ is a completely reducible $G_1 T$-module with summands of the form $L_1(\xi)$ for $\xi \in W_p \cdot \sigma$ (in fact, $\xi = \sigma - p\theta$ for $\theta \in Q(T)$). It follows that the $G_1 T$-module $\Theta_s L(\tau) \cong T_{\nu_s}^0 L_1(\delta)$ is obtained by restricting to $G_1 T$ the corresponding G-module. Similarly, the complex (5.1) with $M = L_1(\tau)$ is obtained by restricting the corresponding complex for G to $G_1 T$. The assertion in (b) is therefore clear. (Compare **[J2;pp. 334–335]**.) □

Let Γ_J be the set of dominant weights τ which satisfy the Jantzen condition

$$(\tau + \rho, \alpha_0^\vee) \leq p(p - h + 2). \tag{5.3}$$

Recall the classical Lusztig conjecture **[L1;§3, Problem IV]**. It describes a character formula for irreducible rational G-modules $L(\gamma)$ with high weight $\gamma \in \Gamma_J(0) = \Gamma_J \cap W_p \cdot 0$. Explicitly, the conjecture states that given $w \cdot 0 \in \Gamma_J(0)$, we have

$$\operatorname{ch} L(w \cdot 0) = \sum_y (-1)^{\ell(y) - \ell(w)} P_{yw_0, ww_0}(-1) \operatorname{ch} V(y \cdot 0), \tag{5.4}$$

where the sum is over all elements $y \in W_p$ such that $yw_0 \leq ww_0$ and $y \cdot 0 \in X(T)^+$. As usual, $w_0 \in W$ denotes the long word, and the P_{yw_0, ww_0} are the Kazhdan-Lusztig polynomials associated to the Coxeter group W_p. (Our notation differs somewhat from the original formulation in that, because we view our polynomials as polynomials in t, rather than $q = t^2$, we evaluate them at $t = -1$.)

We emphasize that, for any given $w \in W_p$ and $\xi \in C^+_{\mathbf{Z}}$ with $w \cdot 0 \in \Gamma_J(0)$, the above formula for $\operatorname{ch} L(w \cdot 0)$ is completely equivalent to a corresponding formula for $\operatorname{ch} L(w \cdot \xi)$:

$$\operatorname{ch} L(w \cdot \xi) = \sum_y (-1)^{\ell(y)-\ell(w)} P_{yw_0,ww_0}(-1) \operatorname{ch} V(y \cdot \xi). \tag{5.4.1}$$

This formulation is stated in **[J2;§II.7.20]**. (Similar formulas would hold as well for singular weights, though we do not need this.)

We now can prove the following result.

5.5 THEOREM. *Let G be a semisimple, simply connected algebraic group as in (1.1).*

(a) *Assume $p \geq 2h - 3$. If the Lusztig conjecture (5.4) is true for G, the highest weight category $\mathcal{C}^1_0$ has a Kazhdan-Lusztig theory.*

(b) *Conversely, assume that $p \geq h$. Then, if $\mathcal{C}^1_0$ has a Kazhdan-Lusztig theory, the Lusztig conjecture (5.4) is true for G.*

PROOF. We first prove (a). Assume that (5.4) holds for all $w \cdot 0 \in \Gamma_J(0)$ and that $p \geq 2h - 3$. A direct calculation shows that any restricted weight lies in Γ_J. To show $\mathcal{C}^1_0$ has a Kazhdan-Lusztig theory, we claim it suffices to prove that each irreducible G_1T-module $L_1(\omega)$ with ω a restricted weight in $W_p \cdot 0$ belongs to $\hat{\mathcal{E}}^L(\mathcal{C}^1_0, \ell)$. If this condition holds, then the equivalences $\mathcal{C}^1_\lambda \cong \mathcal{C}^1_0$ $(\lambda \in C^+_{\mathbf{Z}})$ imply that, for any restricted weight $\nu \in W_p \cdot \lambda$, we have $L_1(\nu) \in \operatorname{Ob}(\hat{\mathcal{E}}(\mathcal{C}^1_\lambda, \ell))$. Let $\tau \in W_p \cdot 0$ and write $\tau = \tau_0 + p\tau_1$, where τ_0 is restricted (and p-regular). If

$$\operatorname{Ext}^n(L_1(\tau), A_1(\nu)) \neq 0,$$

then

$$\operatorname{Ext}^n(L_1(\tau_0), A_1(\nu - p\tau_1)) \cong \operatorname{Ext}^n(L_1(\tau), A_1(\nu)) \neq 0,$$

so $n \equiv \ell(\tau_0) - \ell(\nu - p\tau_1) \pmod 2$, by assumption. But $\ell(\tau_0) - \ell(\nu - p\tau_1) = \ell(\tau_0) + 2\operatorname{ht}(\tau_1) - \ell(\nu) = \ell(\tau) - \ell(\nu)$, using (3.12.5). Hence, $L_1(\tau) \in \operatorname{Ob}(\hat{\mathcal{E}}(\mathcal{C}^1_0, \ell))$. Finally, applying the duality functor $D : \mathcal{C}^1_0 \to \mathcal{C}^1_0$, we see that $L_1(\tau)$ also belongs to $\hat{\mathcal{E}}^R(\mathcal{C}^1_0, \ell)$. This establishes the claim.

The special case in which $G = \mathrm{SL}_2$ and $p = 2$ is readily handled directly. Hence, we can assume that $p > h$. By (3.12.6), $L_1(0)$ belongs to $\mathcal{E}^L(\mathcal{C}^1_0, \ell)$. Any other restricted weight $\nu = w \cdot 0$ is connected to the weight 0 by a series

$$0 = \nu_0 \uparrow \nu_1 = \nu_0 s_0 \uparrow \cdots \uparrow \nu_k = \nu_{k-1} s_{k-1} = \nu$$

of restricted weights with $s_0, \cdots, s_{k-1} \in \Sigma$. We use the pre-Hecke operators β_{s_i} defined in (5.2). Thus, by (5.2b), $\beta_{s_i} L_1(\nu_i)$ is the restriction to G_1T of its G-analogue $\beta^G_{s_i} L(\nu_i)$. By **[CPS6;Theorems 5.9, 5.5]**, the G-modules $\beta^G_{s_i} L(\nu_i)$ are completely reducible. Hence, the G_1T-modules $\beta_{s_i} L_1(\nu_i)$ are completely reducible. By repeated application of the adjoint condition (4.2c), it follows that each $\beta_{s_i} L_1(\nu_i)$ and hence each $L_1(\nu_i)$ belong to $\hat{\mathcal{E}}^L(\mathcal{C}_0^1, \ell)$. In particular, $L_1(\nu)$ belongs to $\hat{\mathcal{E}}^L(\mathcal{C}_0^1, \ell)$, as desired. This completes the proof of (a).

To prove (b), assume that $\mathcal{C}_0^1$ has a Kazhdan-Lusztig theory; thus, $\mathcal{C}_\xi^1$ has a Kazhdan-Lusztig theory for $\xi \in C_{\mathbf{Z}}^+$. By a result of Kato **[K;5.4]**, if the character formula (5.4.1) holds for a given irreducible G-module $L(\nu)$ with ν a p-regular restricted weight, it holds for all irreducible G-modules $L(\nu + p\mu)$ with μ dominant and $\nu + p\mu \in \Gamma_J$. As a temporary notation, let $\preceq$ denote the partial ordering on weights in which $\omega \preceq \omega'$ iff $\omega' - \omega$ is a linear combination of positive roots with nonnegative rational coefficients. Let Γ_J^{reg} denote the set of p-regular weights in Γ_J, and consider Δ an ideal in Γ_J^{reg} with respect to the order $\preceq$ (which implies Δ is an ideal with respect to the order $\uparrow$). Then the result of Kato's just cited implies that, if the character formula (5.4.1) holds for irreducible modules whose high weights are in Δ but restricted, then it holds for all irreducible modules whose high weights are in Δ. In particular, to establish (b), it is enough to prove that (5.4.1) holds for all restricted $\nu \in \Gamma_J^{\text{reg}}$.

Choose a restricted weight ν which is minimal with respect to $\preceq$ for which (5.4.1) is not necessarily known to hold. Then ν is not minimal with respect to $\uparrow$ (in $W_p \cdot \nu$), so that $\nu = \delta s$ for some restricted weight $\delta < \nu$ and some $s \in \Sigma$. Let Δ be the ideal of all weights in Γ_J^{reg} strictly less than ν with respect to $\preceq$. Thus, (5.4.1) is valid for all restricted weights in Δ, and thus for all weights in Δ. To show that $\operatorname{ch} L(\nu)$ satisfies (5.4.1), it suffices, as in the proof of **[CPS6;Theorem 5.3]**, to show the G-module $M = \beta_s^G L(\delta)$ is completely reducible (see also **[A]**). By **[J2;II.7.19(3)]**, the G-composition factors of M have high weights in the ideal $\Delta \cup \{\nu\}$. It follows in particular that, if $L(\zeta)$ is a G-composition factor of M, $\zeta \in \Gamma_J^{\text{reg}}$. By **[K;Lemma 5.3]**, $L(\zeta) \cong L(\zeta_0) \otimes L(p\zeta_1)$, where ζ_0 is restricted and $\zeta_1 \in \bar{C}_{\mathbf{Z}}^+$. The module $M|_{G_1T}$ and hence the module $M|_{G_1}$ are completely reducible by (4.5) and (5.2b). An application of the Hochschild-Serre sequence **[J2;§II.10.16]** now shows M is completely reducible. This completes the proof. □

An alternate proof for the second half of the theorem can be based on **[L2;5.2]**, together with the recursions **[L2;p. 143]** for the polynomials $Q_{A,C}$. That is, one can compare the character formulas given by the two possible Kazhdan-Lusztig conjectures and check that they agree. (See the proof of the theorem below).

Also, Kato **[K;Defn. 3.3]**, based on ideas in **[L2]**, has developed the notion of *generic* Kazhdan-Lusztig polynomials $\hat{P}_{A,C}$. We record next a formula in $\hat{K}_0^L(\mathcal{C}_0^1)$ in terms of these polynomials for the irreducible objects $L_1(\tau)$ when $\mathcal{C}_0^1$ has a Kazhdan-Lusztig theory.

Kato **[K]** makes use of an antiautomorphism, which we denote by $w \mapsto \bar{w}$, of the extended affine Weyl group $\tilde{W}_p$ It is defined by setting

(5.6) $$\bar{w} = t_{p\xi} a^{-1} \text{ for } w = a t_{p\xi},\ a \in W,\ \xi \in X(t).$$

Kato uses this anti-automorphism to define a right action of $\tilde{W}_p$ on $\mathbf{E} = X(T) \otimes \mathbf{R}$

in terms of the usual left action. Namely, $x \cdot w$ ($x \in \mathbf{E}$, $w \in \tilde{W}_p$) for Kato is $\bar{w} \cdot x$ in the notation of this paper. He also defines a left action of $\tilde{W}_p$ on the set of alcoves by defining, for $y \in W_p$,

$$w(C^- \cdot y) = C^- \cdot wy,$$

where $C^- = w_0 \cdot C^+$ (in our notation!). This system of notation is compatible with that of Lusztig [**L2**]. We also remark that the partial order $\leq$ on the set of alcoves defined in [**L2;1.5**] coincides with the $\uparrow$ partial ordering used in this paper. This follows easily from the definitions, together with [**J2;Cor. II.6.10.**].

Let $* = -w_0 : X(T) \to X(T)$ be the opposition involution. It defines an automorphism $w \mapsto w^*$ on $\tilde{W}_p$ given by $w^* = (w_0 a w_0)t_{p\xi^*}$ for w as in (5.6). One easily calculates that $w_0 \bar{w} w_0 = w^{*-1}$ and that $\bar{w}_0 = w_0$. It follows that the alcove denoted wC^+ in the notation of [**L2**] and [**K**] is the alcove $w^{*-1} \cdot C^+$ in our notation:

$$wC^+ = ww_0(C^- \cdot 1) = C^- \cdot ww_0 = \overline{ww_0} \cdot C^- = \bar{w}_0\bar{w} \cdot (w_0 \cdot C^+) = w^{*-1} \cdot C^+.$$

The reader is cautioned that this identification is only at the level of alcove labeling, and *not* of actions.

In keeping with our convention regarding Kazhdan-Lusztig polynomials, we regard the polynomials $Q_{A,B}$ and $\hat{P}_{A,C}$ of Lusztig and Kato as polynomials in t, rather than $q = t^2$. That is, $\hat{P}_{A,C}(t)$ for us means $\hat{P}_{A,C}(t^2) = \hat{P}_{A,C}(q)$ in [**K**]. We also write

$$\hat{P}_{A,C}(t) = P_{\tau,\omega}(t)$$

for alcoves A, C containing τ and ω, respectively.

5.7 THEOREM. *Let G be as above with $p > h$. Assume that $\mathcal{C}_0^1$ has a Kazhdan-Lusztig theory. Fix $\omega \in W_p \cdot 0$. In the complete left Grothendieck group $\hat{K}_0^L(\mathcal{C}_0^1)$, we have*

$$[L_1(\omega)]_L^\infty = \sum_{\tau \uparrow \omega} t^{\ell(\tau)-\ell(\omega)} \hat{P}_{\tau,\omega}(t)[V_1(\tau)]. \tag{5.7.1}$$

Thus,

$$ch\, L_1(\omega) = \sum_{\tau \uparrow \omega} (-1)^{\ell(\tau)-\ell(\omega)} \hat{P}_{\tau,\omega}(-1)\, ch\, V_1(\tau). \tag{5.7.2}$$

(In these formulas, we are using the length function ℓ on $W_p \cdot 0$ defined in (3.12.3).) Finally, when $\omega = w \cdot 0$ is restricted, we have

$$ch\, L_1(\omega) = ch\, L(\omega) = \sum_y (-1)^{\ell(y)-\ell(w)} P_{yw_0,ww_0}(-1)\, ch\, V(y \cdot 0), \tag{5.7.3}$$

where the sum is over all elements $y \in W_p$ such that $yw_0 \leq ww_0$ and $y \cdot 0$ is dominant, in agreement with (5.4).

PROOF. Following [**K;§3**], the generic polynomial $\hat{P}_{A,C}$ may be defined in terms of the polynomials $Q_{A,C}$ of [**L2**] as follows. If D_C denotes the formal linear combination

$$D_C = \sum_{A \leq C} Q_{A,C}(t) A,$$

then define

$$E_C = \prod_{\alpha > 0} (1 - t^{-2}\theta_{-\alpha})^{-1} D_C.$$

In this expression, $\theta_{-\alpha}$ (α a positive root) is an operator on a certain module $\hat{\mathcal{M}}$ for the Hecke algebra $\mathcal{H}$ of W_p. These operators commute with the action of $\mathcal{H}$ on $\hat{\mathcal{M}}$ and they satisfy

$$\theta_{-\alpha} e_\xi = t^{2\mathrm{ht}(\alpha)} e_{\xi - p\alpha}.$$

In this expression, $\xi \in pX(T) - \rho$ is a special point and $e_\xi = \sum_{\xi \in \bar{A}} A$, the sum of all alcoves containing ξ in its closure. Then the generic polynomials are determined by expressing E_C as a formal linear combination of alcoves (see [**K;Theorem 3.5**])

(5.7.4) $$E_C = \sum_{A \leq C} \hat{P}_{A,C}(t) A.$$

For $y \in W_p$, let T_y be the standard basis element of $\mathcal{H}$. Let $w \in W_p$ be so that $C = (w^*)^{-1} \cdot C^+$ lies in the restricted region. Then Lusztig [**L2;5.2**] proves

(5.7.5) $$D_C = \sum_{y \leq ww_0,\ \ell(yw_0) = \ell(y) + \ell(w_0)} P_{yw_0, ww_0} T_y e_{-\rho}.$$

Observe that the condition that $\ell(yw_0) = \ell(y) + \ell(w_0)$ in (5.7.5) is equivalent to the condition that $y^{*-1} \cdot C^+$ be dominant. An analogous version holds at special points other than $-\rho$, thus describing all D_C. We have $Q_{A,B} = Q_{A+p\xi, B+p\xi}$ for all $\xi \in X(T)$ by [**L2;Cor. 7.4**]. In addition, it follows from [**L2;Prop. 8.7**] that $t_{p\xi} \cdot (\theta_{-\alpha} D_C) = \theta_{-\alpha}(t_{p\xi} \cdot D_C)$ for all alcoves C and weights ξ. (Here $t_{p\xi}$ is regarded as an operator on $\hat{\mathcal{M}}$, translating each alcove in any formal linear combination by $p\xi$.) It follows from the definition of E_C given above that $\hat{P}_{A,C} = \hat{P}_{A+p\xi, C+p\xi}$ for any weight ξ and any pair of alcoves A, C.

For any weight $\lambda \in C^+_{\mathbf{Z}}$, we define a homomorphism $\mathfrak{f}_\lambda \equiv \mathfrak{f} : \hat{K}^L_0(\mathcal{C}^0_1) \to \hat{\mathcal{M}}$ by putting, for $x \in \hat{K}^L_0(\mathcal{C}^0_1)$ as in (3.2),

$$\mathfrak{f}(x) = \mathfrak{f}\Big(\sum_{\nu \in \Gamma, n \in \mathbf{Z}} a_{n,\nu} [V_1(\nu)] t^n\Big)\Big) = \sum_{\nu \in \Gamma, n \in \mathbf{Z}} a_{n,\nu} A_\nu t^{n - \ell(\nu)}.$$

In this expression, A_ν denotes the alcove containing the weight ν. Clearly, in the notation of §3.12, $\mathfrak{f}$ commutes (up to a power of t) with translation by elements $p\xi \in X(T)$, in the sense that

(5.7.6a) $$\mathfrak{f}(x \cdot e^{p\xi}) = (t_{p\xi} \cdot \mathfrak{f}(x)) t^{-2\mathrm{ht}(\xi)}$$

for x as in (3.2). Also, the elements $\tilde{E}_{A_\nu} \equiv t^{-\ell(\nu)} E_{A_\nu}$ obey a similar rule:

(5.7.6b) $$\tilde{E}_{A_{\nu + p\xi}} = (t_{p\xi} \cdot \tilde{E}_{A_\nu}) t^{-2\mathrm{ht}(\xi)}$$

Observe that

$$D_{C^+} = \mathfrak{f}(\sum_{z\in W} t^{-\ell(z)}[V_1(z\cdot 0)]).$$

Hence,

(5.7.7) $$E_{C^+} = \mathfrak{f}((\sum_{z\in W} t^{-\ell(z)}[V_1(z\cdot 0)])\cdot\sigma) = \mathfrak{f}([L_1(0)]_L^\infty),$$

by (3.12.2) and a direct calculation.

Assuming $\mathcal{C}_0^1$ has a Kazhdan-Lusztig theory, we can apply (4.5) and (5.2) to conclude that $\mathcal{C}_0^1$ has enough Hecke operators β_s, $s\in\Sigma$, in the sense that $\Lambda = X(T)$ is weakly S_{Hck}-connected if we take the set of initial elements Λ_0 to be the set $pX(T)-\rho$ of special points in Λ.

We claim now that

(5.7.8) $$\mathfrak{f}([L_1(\omega)]_L^\infty) = \tilde{E}_{A_\omega}$$

for all p-regular weights ω. By (5.7.6a,b), the validity of this formula for a given weight ω is equivalent to its validity for any translated weight $\omega+p\xi$, $\xi\in X(T)$. In particular, we may assume that ω is restricted. We prove (5.7.8) by induction on $\ell(\omega) = d(C^+, A_\omega)$. It is true when $\ell(\omega)=0$ (i. e., $\omega\in C_{\mathbf{Z}}^+$) by (5.7.7). Put $C = A_\omega$ and choose a sequence

$$C^+ = C_0, C_1, \cdots, C_n = A_\omega$$

of alcoves lying in the restricted region such that $C_i \uparrow C_{i+1}$ and C_i is adjacent to C_{i+1}, $0\le i<n$. Put $C=C_n=A_\nu$ and $C'=C_{n-1}$. Let the common face of C and C' be of type s. By **[L2;10.7]**,

(5.7.9) $$D_C = (T_s+1)D_{C'} - \sum_A \mu(A,C')t^{d(A,C')+1}D_A,$$

where the sum is over all alcoves A such that $A\uparrow C'$, $d(A,C')$ odd, and $sA\uparrow A$. Here $\mu(A,C')$ is defined **[L2;8.1]**, for $d(A,C')$ odd and $A\uparrow C'$, to be the coefficient of $t^{d(A,C')-1}$ in the polynomial $Q_{A,C'}$. This power of t is the highest possible power of t occurring in $Q_{A,C'}$ by **[L2;Theorem 5.2]**. Using **[L2;Prop. 8.7]**, we easily obtain that

$$\hat{P}_{A,C'} = \sum_{n,\theta} \wp_n(\theta)t^{-2n+2\mathrm{ht}(\theta)}Q_{A,C'-p\theta}.$$

Note that

$$\deg Q_{A,C'-p\theta} \le d(A,C'-p\theta)-1 = d(A,C')-2\mathrm{ht}(\theta)-1.$$

It follows that $\mu(A,C')$ is the coefficient of $t^{d(A,C')-1}$ in $\hat{P}_{A,C'}$. Also, Lusztig points out in **[L2;10.7]** (see also **[L2;Cor. 8.9, 8.10]**) that the alcoves A appearing in the sum (5.7.9) are all translates by elements of $pX(T)$ of restricted alcoves smaller than C' in the $\uparrow$ partial ordering. Applying the operator $\prod_{\alpha>0}(1-$

$t^{-2}\theta_{-\alpha})^{-1}$ to both sides of (5.7.9), we obtain a similar equation

$$E_C = (T_s + 1)E_{C'} - \sum_A \mu(A, C')t^{d(A,C')+1}E_A, \tag{5.7.10}$$

summed over the same set of alcoves A as in (5.7.9). Equivalently,

$$\tilde{E}_C = t^{-1}(T_s + 1)\tilde{E}_{C'}) - \sum_\nu \mu(A_\nu, A_{\omega'})\tilde{E}_{A_\nu}.$$

In this sum, ν runs over weights W_p-conjugate to ω such that $A = A_\nu$ satisfies the conditions in (5.7.9). Also, $\omega' \in W_p \cdot \omega$ is such that $A_{\omega'} = C'$. Consider the operator $[\beta_s]$ on $\hat{K}_0^L(\mathcal{C}, \ell)$, defined in (4.4a). An easy calculation using **[L2;1.6]** shows that

$$\mathfrak{f}([\beta_s]x) = t^{-1}(T_s + 1)\mathfrak{f}(x)$$

for all $x \in \hat{K}_0^L(\mathcal{C}, \ell)$. Hence, applying (4.3), (5.7.6a,b), and induction, we obtain the desired formula (5.7.8). The desired expression (5.7.1) follows immediately from this, together with the definition of $\mathfrak{f}$.

The character formula (5.7.2) follows from Proposition 3.11. Finally, the asserted character formula (5.7.3) has already been proved in Theorem 5.5 when $p \geq 2h - 3$. However, for $p > h$, it follows easily from (5.7.1) and (5.7.5), using the elementary fact that $\mathrm{ch}V_1(\nu) \cdot \sigma(-1)$ is the character of the Verma module (for the associated complex semisimple Lie algebra) of high weight ν. □

We mention that Kaneda **[Ka;Prop. 4.12]** asserts (without proof) a similar calculation of the Poincaré polynomial for $\mathcal{C}_0^1$ in terms of the generic Kazhdan-Lusztig polynomials under the hypothesis of a "generic" Vogan conjecture.

By (3.7) and (5.7), the validity of the Lusztig conjecture would give a complete calculation of the groups

$$\mathrm{Ext}^\bullet_{G_1T}(L_1(\tau), L_1(\nu))$$

for all irreducible G_1T-modules having high weights which are p-regular. We summarize this in the following result, which results immediately from Corollary 3.9 applied to a subcategory $\mathcal{C}_0^1[\Gamma]$ for a finitely generated ideal Γ containing the relevant dominant weights.

5.8 THEOREM. *Let G be as in (1.1) with $p > h$. Assume that the category $\mathcal{C}_0^1$ has a Kazhdan-Lusztig theory. Consider p-regular weights ν, τ which lie in the same orbit of W_p. Then for any non-negative integer i, we have*

$$\sum_0^\infty \dim \mathrm{Ext}^i_{G_1T}(L_1(\nu), L_1(\tau))t^i = \langle [L_1(\tau)]_L^\infty, [L_1(\nu)]_R^\infty\rangle = \sum_\omega p_{\omega,\tau}p_{\omega,\nu}, \tag{5.8.1}$$

where $p_{\nu,\omega} = t^{\ell(\omega)-\ell(\nu)}\bar{\tilde{P}}_{\tau,\omega}$ in terms of the generic Kazhdan-Lusztig polynomials. In particular, $\mathrm{Ext}^i_{G_1T}(L_1(\tau), L_1(\nu)) = 0$ unless $i \equiv \ell(\tau) - \ell(\nu)$ $(mod 2)$. □

As in **[CPS5]**, the existence of a Kazhdan-Lusztig theory is equivalent, for $p > h$, to the validity of the character formula (5.7.3). This character formula has been conjectured by Kato **[K]** to hold for $p \geq h$, though this does not include all restricted weights when p is small, as remarked in the introduction. Thus, Kato's conjecture is actually stronger than that of Lusztig.

5.9 THEOREM. *Assume that $p > h$. Then C_0^1 has a Kazhdan- Lusztig theory if and only if the character formula (5.7.3) holds for all restricted weights $\omega = w \cdot 0$, $w \in W_p$.*

PROOF. This follows from the arguments in [**CPS6;Theorem 5.3**]. □

Next, exactly as in [**CPS5;Cor. 5.4**] we have the following further equivalent formulations for G_1T of the Lusztig conjecture for G. (More precisely, (5.9a) states a condition on C_0^1 which implies the Lusztig conjecture by (5.5), and is equivalent to it for $p \geq 2h - 3$.) Using Theorem 4.5, the proof is essentially identical to that given in [**CPS5**], and it is omitted.

5.10 THEOREM. *Let G be as in (1.1) with $p \geq h$. The following statements are equivalent.*

(a) *C_0^1 has a Kazhdan-Lusztig theory.*
(b) *If $\tau, \nu \in W_p \cdot 0$ are mirror images of each other in adjacent p-alcoves, then $Ext^1(L_1(\tau), L_1(\nu)) \neq 0$.*
(c) *Let $\tau < \tau s$, where $\tau \in W_p \cdot 0$, $s \in \Sigma$. Then $V_1(\tau s)$ has a quotient with exactly two composition factors $L_1(\tau s)$ and $L(\tau)$.*
(d) *Let $\tau, \nu \in W_p \cdot 0$ satisfy $\ell(\tau) \equiv \ell(\nu)$ (mod 2). Then*
$$Ext^1(V_1(\tau), L_1(\nu)) = 0.$$
(e) *If $\tau, \nu \in W_p \cdot 0$ satisfy $\ell(\tau) \equiv \ell(\nu)$ (mod2). Then $A_1(\nu)/L_1(\nu)$ contains no B_1T-fixed line of weight τ.* □

Parts (b) and (c) are perhaps the most intriguing. They are close to issues raised by S. Doty and J. Sullivan (in lectures and verbal communication to the authors) regarding reductions of Ext calculations for simple modules. In this spirit, we point out that (b) is equivalent to the following assertion:

(. *b′) For any two weights $\tau, \nu \in W \cdot 0$ which are mirror images of each other in adjacent p-alcoves, we have $Ext^1(L_1(\tau), L_1(\nu)) \neq 0$.*

To see this equivalence, let τ, ν be as in (b), and let F be the common face of the alcoves containing these two weights. We assert that the closure $\bar{F}$ contains a special point v, in the sense that for any root α, one of the hyperplanes $H_{\alpha,rp}$ defining the reflection $s_{\alpha,rp}$ passes through v. To verify this, we can assume that the root system Φ is irreducible. The claim is clear if Φ has only one root length, since then all the roots are W-conjugate. So, assume that Φ has two root lengths, and let C be an alcove containing F as a face. In the *extended* affine Weyl group $\tilde{W}_p$, let N_C be the stabilizer of C. Thus, $\tilde{W}_p$ is the semidirect product of N_C and the normal subgroup W_p. In particular, $N_C \cap W = N_C \cap W_p = \{1\}$. Then, if v_0 is a special point lying in $\bar{C}$, it is obvious that only the identity element of N_C stabilizes v_0. Since N_C permutes the special points lying in $\bar{C}$ and since $\bar{F}$

contains all the vertices of $\bar{C}$ except one, it follows that $\bar{F}$ contains a special point v, as claimed. For some weight $\xi \in X(T)$, $v+p\xi = -\rho$, so putting $\tau' = \tau + p\xi$, we see that τ', ν' lie in adjacent alcoves in $W \cdot C^+$. Let τ'', ν'' be the corresponding weights in $W \cdot 0$ which lie in these alcoves. Then

$$\mathrm{Ext}^1(L_1(\tau), L_1(\nu)) \cong \mathrm{Ext}^1(L_1(\tau'), L_1(\nu')) \cong \mathrm{Ext}^1(L_1(\tau''), L_1(\nu'')),$$

using the equivalences $\mathcal{C}_0^1 \cong \mathcal{C}_\lambda^1$, $\lambda \in C_{\mathbf{Z}}^+$. Thus, (b) is implied by (b′). It follows that (b) and (b′) are equivalent.

There is a similar alternative formulation of part (c) in the theorem. We leave details to the reader.

Translating by an appropriate weight in $pX(T)$, we see, in (b)–(e), one of the weights can be assumed to be restricted. In fact, in (b) one can assume both weights are restricted, cf. the argument (c) $\Rightarrow$ (b) $\Rightarrow$ (d) in Theorem 4.5.

Observe that statement (b′) is entirely analogous to a corresponding equivalence of the Kazhdan-Lusztig conjecture (now a theorem) for the category $\mathcal{O}$ for a complex semisimple Lie algebra. In fact, as shown in [**CPS5**], that conjecture is equivalent to the assertion that $\mathrm{Ext}^1_{\mathcal{O}}(L(\tau), L(\nu)) \neq 0$ for any two weights $\tau, \nu \in W \cdot 0$ which lie in adjacent chambers. We mention that certain $\mathrm{Ext}^1_{G_1T}$ calculations between simple modules are given by Kaneda [**Ka**].

Finally, we can also give a reduction to finite groups.

5.11 THEOREM. *Let G be as in (1.1) with $p \geq h$. For $q = p^r$, let $G(q)$ denote the subgroup of $\mathbf{F}_q$-rational points of G. Then the following condition implies the Lusztig conjecture (5.4) for G:*

If $\tau, \nu \in W_p \cdot 0$ are restricted weights in adjacent p-alcoves, then

(5.11.1) $$Ext^1_{G(q)}(L(\nu), L(\tau)) \neq 0$$

for q sufficiently large.

PROOF. Assume the condition holds, so that [**CPSK;Theorem 7.1**] implies that

(5.11.2) $$\mathrm{Ext}^1_G(L(\nu), L(\tau)) \neq 0$$

for any such τ, ν. Let $\tau = \nu s > \nu$, say, where $s \in \Sigma$. As in the proof of (5.5b), it suffices by induction to show that $\beta_s^G L(\nu)$ is completely reducible. By (5.11.2), the proof of [**CPS6;Theorem 5.9**] shows that

$$\mathrm{Hom}_G(\beta_s^G L(\nu), L(\tau)) \neq 0.$$

The proof of [**CPS6;Theorem 5.5**] (i. e., that (6.5c) implies (6.5b) there) shows that $L(\tau)$ is a direct summand of $\beta_s^G L(\nu)$. The complete reducibility of $\beta_s^G L(\nu)$ follows from [**CPS5;Theorem 4.1**], since condition (e) of [**CPS6;Theorem 5.5**] holds inductively for weights in $W_p \cdot 0$ which are less than or equal to ν. $\square$

6. Appendix: Complete Grothendieck Groups

Let $\mathcal{C}$ be a highest weight category with weight poset Λ, as above. Thus, $\mathcal{C}$ is the union of Serre subcategories $\mathcal{C}[\Gamma]$, Γ ranging over the finitely generated ideals of Λ, and each $\mathcal{C}[\Gamma]$ is a highest weight category with weight poset Γ. In turn, each $\mathcal{C}[\Gamma]$ has quotient highest weight categories $\mathcal{C}[\Gamma](\Omega) = \mathcal{C}[\Gamma]/\mathcal{C}[\Gamma\backslash\Omega]$ with weight poset Ω, where Ω ranges over the finite coideals of Γ.

We define the *complete Grothendieck group* $\hat{K}_0(\mathcal{C})$ by

$$\hat{K}_0(\mathcal{C}) = \varinjlim_{\Gamma} \varprojlim_{\Omega\subseteq\Gamma} K_0(\mathcal{C}[\Gamma](\Omega)) \cong \varinjlim_{\Gamma} \hat{K}_0(\mathcal{C}[\Gamma]),$$

where Γ ranges over the finitely generated ideals of Λ, and Ω ranges over the finite coideals of Γ. Obviously, there is a natural map

$$\iota \equiv \iota_{\mathcal{C}} : K_0(\mathcal{C}) \to \hat{K}_0(\mathcal{C})$$

from the ordinary to the complete Grothendieck group, at least for the case $\Lambda = \Gamma$ finitely generated. However, each object X of $\mathcal{C}$ belongs to some $\mathcal{C}[\Gamma]$, so ι is defined in general. The object in $\hat{K}_0(\mathcal{C})$ corresponding to X will be denoted simply by $[X]$, the same notation commonly used in $K_0(\mathcal{C})$, when no confusion can arise. (It will be obvious in a moment that the map ι above is an injection.) We assert that the general element of $\hat{K}_0(\mathcal{C})$ may be uniquely expressed as a formal sum

$$\sum_{\nu} c_\nu[L(\nu)]$$

where ν ranges over Λ, and the c_ν are integers, zero for ν outside some finitely generated ideal of Λ. (As usual $L(\nu)$ is the irreducible object indexed by ν.) The above expression is compatible in an obvious way with the maps $\hat{K}_0(\mathcal{C}[\Gamma]) \to \hat{K}_0(\mathcal{C})$ and $\hat{K}_0(\mathcal{C}[\Gamma]) \to K_0(\mathcal{C}[\Gamma](\Omega))$.

We briefly give the proof of this assertion. Because of the general assumption of highest weight category theory that Λ is interval-finite, each finitely generated ideal Γ is the union of its finite coideals Ω. Observe the following properties of the highest weight categories $\mathcal{C}[\Gamma](\Omega)$ and quotient maps $\mathcal{C}[\Gamma] \to \mathcal{C}[\Gamma](\Omega)$: Each object of $\mathcal{C}[\Gamma](\Omega)$ has finite length, and $\mathcal{C}[\Gamma](\Omega)$ has, up to isomorphism, only finitely many simple objects, the images of the $L(\nu)$ for $\nu \in \Omega$. These images are all nonzero and nonisomorphic. The images of the other simple objects of $\mathcal{C}[\Gamma]$ are zero.

Since $K_0(\mathcal{C}[\Gamma](\Omega))$ can be regarded a product of copies of $\mathbf{Z}$, one copy for each isomorphism class of simple objects in $\mathcal{C}[\Gamma](\Omega)$, we conclude that

$$\hat{K}_0(\mathcal{C}[\Gamma]) \cong \varprojlim K_0(\mathcal{C}[\Gamma](\Omega)) \cong \prod_{\Gamma} \mathbf{Z},$$

an infinite product of copies of $\mathbf{Z}$, indexed by the set Γ. Regarding such a product as a formal sum gives the description above of $\hat{K}_0(\mathcal{C})$ in case $\Lambda = \Gamma$, and the general case follows easily.

References

A. H. Andersen, *An inversion formula for the Kazhdan– Lusztig polynomials for affine Weyl groups*, Adv. in Math. **60** (1986), 125–153.

AJ. H. Andersen and J. Jantzen, *Cohomology of induced representations for algebraic groups*, Math. Ann. **269** (1984), 487–525.

BBD. A. Beilinson, J. Bernstein, and P. Deligne, *Analyse et topologie sur les espaces singulares*, Astérisque **100** (1982).

CPS1. E. Cline, B. Parshall and L. Scott, *Cohomology, hyperalgebras, and representations*, J. Algebra **63** (1980), 98–123.

CPS2. E. Cline, B. Parshall and L. Scott, *Injective modules for infinitesimal algebraic groups, I*, J. London Math. Soc. **31** (1985), 277–291.

CPS3. E. Cline, B. Parshall and L. Scott, *Finite dimensional algebras and highest weight categories*, J. reine angew. Math. **391** (1988), 85–99.

CPS4. E. Cline, B. Parshall and L. Scott, *Duality in highest weight categories*, Contemp. Math **82** (1989), 7–22.

CPS5. E. Cline, B. Parshall and L. Scott, *Abstract Kazhdan-Lusztig theories*, to appear.

CPS6. E. Cline, B. Parshall and L. Scott, *Simulating perverse sheaves in modular representation theory*, to appear.

CPSK.E. Cline, B. Parshall, L. Scott, and W. van der Kallen, *Rational and generic cohomology*, Invent. math. **39** (1977), 143–163.

Do. S. Doty, *The strong linkage principle*, Amer. J. Math. **111** ((1989)), 135–141.

FP. E. Friedlander and B. Parshall, *Cohomology of infinitesimal and discrete groups*, Math. Ann. **273** (1986), 353–374.

IM. N. Iwahori and M. Matsumoto, *On some Bruhat decomposition and the structures of the Hecke rings of p-adic Chevalley groups*, Publ. Math. I.H.E.S. **25** (1965), 5–48.

J1. J. Jantzen, *Darstellungen halbeinfacher Gruppen u. ihrer Frobenius-Kerne*, J. reine Angew. Math. **317** (1980), 157–199.

J2. J. Jantzen, *Representation theory of algebraic groups*, Academic Press, 1987.

Ka. M. Kaneda, *Extensions of modules for infinitesimal algebraic groups*, J. Algebra **122** (1989), 188–210.

K. S. I. Kato, *On the Kazhdan–Lusztig polynomials for affine Weyl groups*, Adv. in Math. **55** (1985), 103–130.

KL. D. Kazhdan and G. Lusztig, *Affine Lie algebras and quantum groups*, International Math. Res. Notices (Duke Math. J.) **2** (1991), 21–29.

L1. G. Lusztig, *Some problems on the representation theory of finite Chevalley groups*, Proc. Symposia Pure Math. **37** (1980), 313–317.

L2. G. Lusztig, *Hecke algebras and Jantzen's generic decomposition patterns*, Adv. in Math. **37** (1980), 121–164.

L3. G. Lusztig, *On quantum groups*, J. Algebra **131** (1990), 466-475.

PS.B. Parshall and L. Scott, *Derived categories, quasi-hereditary algebras, and algebraic groups*, Carlton U. Math. Notes **3** (1988), 1-104.

Department of Mathematics, University of Oklahoma, Norman, OK 73019

Department of Mathematics, University of Virginia, Charlottesville, VA 22902

Contemporary Mathematics
Volume **139**, 1992

Harish-Chandra modules for semisimple Lie groups with one conjugacy class of Cartan subgroups

DAVID H. COLLINGWOOD AND RONALD S. IRVING

1. Introduction

1.1 Description of results. In this paper we study Harish-Chandra modules for a connected noncompact semisimple Lie group G with one conjugacy class of Cartan subgroups. Our goal is to show that the standard principal series modules for G behave exactly like generalized Verma modules in a category $O_{\mathfrak{p}}$ of highest weight modules associated to a complex Lie algebra $\mathfrak{g}$ and a parabolic subalgebra $\mathfrak{p}$; moreover, this analogous behavior is to be understood as resulting from the existence of Harish-Chandra modules for G which play the same role as projective modules for $O_{\mathfrak{p}}$. In addition, we will raise some questions, and offer some negative answers which show why better results cannot be obtained.

Let K be a maximal compact subgroup of G and let $\mathfrak{g}$ be the complexified Lie algebra of G. Fix once and for all a finite dimensional representation F of G. We denote by $\mathcal{HC}_\infty$ the category of Harish-Chandra modules with the same generalized infinitesimal character as F. There exists a finite set $\mathcal{D}$ in one-to-one correspondence with the set of isomorphism classes of irreducible Harish-Chandra modules in $\mathcal{HC}_\infty$ and to each $\delta \in \mathcal{D}$ we can associate a *standard module* $\pi(\delta)$ in $\mathcal{HC}_\infty$ (a principal series representation) which admits a unique irreducible quotient $\bar{\pi}(\delta)$. In this sense, the standard modules are a plausible substitute for generalized Verma modules in $\mathcal{O}_{\mathfrak{p}}$. With this in mind, let us denote $\pi(\delta)$ by $M(\delta)$ and $\bar{\pi}(\delta)$ by $L(\delta)$.

The set $\mathcal{D}$ has a partial order $\leq$ with respect to which $(M(\delta) : L(\delta)) = 1$ and $(M(\delta) : L(\gamma)) > 0$ only if $\gamma \leq \delta$. (Throughout the paper, given a module X and a simple module L, we will denote by $(X : L)$ the multiplicity of L as a composition factor of X.) Moreover, there is a unique maximal element $\delta_{\max}$

1991 *Mathematics Subject Classification*. Primary 22E46, 22E47; Secondary 17B10.

This paper is in final form and no version of it will be submitted for publication elsewhere in $\mathcal{D}$ and $L(\delta_{\max})$ is F. Call a simple module $L(\gamma)$ (as well as the parameter γ) *socular* if $L(\gamma)$ occurs as a summand of the socle of some standard module $M(\delta)$. (Recall that the socle of a module X is the largest semisimple submodule soc X of X.) Let $\mathcal{X}$ denote the collection of socular elements of $\mathcal{D}$. Recall the notion of GK dimension $d(X)$ of a module X and the right cell equivalence relation $\sim_R$ on $\mathcal{D}$; these are reviewed in §2.6. Recall also for a module X the notions of its socle filtration and radical filtration [**17**]. The *Loewy length* of X, denoted $\ell\ell(X)$, is the common length of these filtrations; we call X *rigid* if the two filtrations coincide. We can now state our first main result. It depends on hypotheses described in more detail later.

THEOREM A. *Let G be a connected noncompact semisimple Lie group with one conjugacy class of Cartan subgroups. Assume that all $\delta \in \mathcal{D}$ are of $\mathcal{O}$-type.*

(i) *The following are equivalent:*
  (a) *$L(\delta)$ is socular;*
  (b) *$d(L(\delta)) = d(L(\delta_{\min}))$;*
  (c) *$\delta \sim_R \delta_{\min}$.*

(ii) *Let δ be socular. Then there is a unique maximal element $\delta^{\#}$ in the set $\{\gamma \in \mathcal{D} : (M(\gamma) : L(\delta)) > 0\}$. The module $M(\delta^{\#})$ has simple socle $L(\delta)$, contains $L(\delta)$ with multiplicity 1, is rigid, and satisfies $\ell\ell M(\delta^{\#}) = \ell\ell M(\delta_{\max})$.*

(iii) *Given $\gamma \in \mathcal{D}$, if $\gamma \neq \delta^{\#}$ for any $\delta \in \mathcal{D}$, then $\ell\ell M(\gamma) < \ell\ell M(\delta_{\max})$.*

The theorem yields a characterization of socular simple modules and reveals the existence of a distinguished collection of standard modules, those of maximal Loewy length, in bijective correspondence with the socular simple modules. Such a result is proved for socular simple modules and generalized Verma modules in the category $\mathcal{O}_{\mathfrak{p}}$ in [**16**]. The key to the proof for $\mathcal{O}_{\mathfrak{p}}$ is the existence of projective indecomposable modules, the projective covers in $\mathcal{O}_{\mathfrak{p}}$ of the simple modules. The key idea of this paper is that an analogous family of modules exists in the Harish-Chandra setting. They will no longer be projective, or even canonical, but will satisfy many of the properties of projective indecomposable modules in $\mathcal{O}_{\mathfrak{p}}$.

In order to state our second main result, we need some more notions. We will say a Harish-Chandra module X has a *standard flag* if it has a finite filtration whose successive subquotients are standard modules. One can always re-arrange such a filtration so that all the subquotients isomorphic to a given $M(\varepsilon)$ occur in succession. The standard flag is of $\mathcal{O}$-type if, under such a re-arrangement, no two subquotients $M(\varepsilon)$ extend each other non-trivially. Call a Harish-Chandra module P a *pseudo-projective cover* of a simple module $L(\gamma)$ if (i) P has simple cap $L(\delta)$; (ii) P has a standard flag of $\mathcal{O}$-type; and (iii) P satisfies BGG reciprocity: $(P : M(\delta)) = (M(\delta) : L(\gamma))$, where $(P : M(\delta))$ is the number of occurrences of $M(\delta)$ as subquotient in a standard flag for P. Let $\mathbb{D}$ be the duality on $\mathcal{HC}_{\infty}$, which preserves simple modules; call X *self-dual* if $\mathbb{D}X \cong X$.

THEOREM B. *Let G be a connected noncompact semisimple Lie group with one conjugacy class of Cartan subgroups. Assume that all $\delta \in \mathcal{D}$ are of $\mathcal{O}$-type.*

(i) *For each $\gamma \in \mathcal{D}$ there exists a pseudo-projective cover $P(\gamma)$ of $L(\gamma)$, occurring as a direct summand of $\theta M(\delta_{\max})$ for a suitable composition θ of translation functors.*
(ii) *$P(\gamma)$ is self-dual if and only if γ is socular, in which case $P(\gamma)$ has Loewy length $2\,\ell\ell M(\delta_{\max}) - 1$.*

It is a straightforward matter to deduce Theorem A from Theorem B, in parallel to the analogous deduction in the category $\mathcal{O}_{\mathfrak{p}}$. The simple groups G with one conjugacy class of Cartan subgroups, to which Theorems A and B can apply, have as covering groups either: (i) a complex group, (ii) $Spin(2n+1,1)$, (iii) $SL(n,\mathbb{H})$, or (iv) $E_{6(-26)}$. The $\mathcal{O}$-type hypothesis of the Theorems is satisfied in the first three cases, and in the low rank examples of the fourth case. We fully expect it to be true in general, but have not yet verified it in complete generality.

The integer $\ell\ell M(\delta_{\max})$ can be calculated on a case-by-case basis. One obtains the following values for it, or rather for $t_G = \ell\ell M(\delta_{\max}) - 1$: (i) if G is complex, then $t_G = \sqrt{\ell(w_0)}$, where w_0 is the longest element of the Weyl group; (ii) if $G = Spin(2n+1,1)$, then $t_G = 1$; (iii) if $G = SL(n,\mathbb{H})$, then $t_G = n(n-1)/2$; (iv) if $G = E_{6(-26)}$, then $t_G = 3$. The calculation in case (iii) is the only one which is not straightforward; it can be done using derived functors to relate Harish-Chandra modules for $SL(n,\mathbb{C})$ to Harish-Chandra modules for $SL(n,\mathbb{H})$.

Two aspects of the proofs are worth noting. First, parts of it require working in a filtered version of $\mathcal{HC}_\infty$, with translation functors on the filtered category. In the process, we actually obtain a filtered version of Theorem B. Second, as with $\mathcal{O}_{\mathfrak{p}}$, a crucial step is the result that $M(\delta_{\max})$ has simple socle L and $M(\delta_{\max})/L$ has lower GK dimension than $M(\delta_{\max})$. This involves the calculation of the *multiplicity* (or Bernstein degree) of $M(\delta_{\max})$. The analogous computation in $\mathcal{O}_{\mathfrak{p}}$ is essentially trivial, but here we rely on the Jacquet functor and the Osborne conjecture to transfer our calculation from $\mathcal{HC}_\infty$ to $\mathcal{O}_{\mathfrak{p}}$.

Let us describe the layout of the paper. In §1.2 we explain why one cannot directly mimic in $\mathcal{HC}_\infty$ the arguments of $\mathcal{O}_{\mathfrak{p}}$. Examples show that some of the basic behavior of $\mathcal{O}_{\mathfrak{p}}$ fails to occur in $\mathcal{HC}_\infty$, necessitating the more complicated considerations of this paper. In section 2 we will review the relevant background on Harish-Chandra modules and obtain the just mentioned result on $M(\delta_{\max})$. In section 3 we discuss the filtered version of $\mathcal{HC}_{\max}$, translation functors, Grothendieck groups and the corresponding Hecke modules, and formulate our hypothesis. Section 4 contains the construction of the modules $P(\gamma)$ and an analysis of their radical filtrations. Section 5 treats the self-duality of pseudo-projective modules and the consequences for standard modules, including the proof of Theorem A.

1.2 Description of subtleties.

We wish to discuss two subtleties in dealing with $\mathcal{HC}_\infty$ that determine the approach of this paper: the lack of a good subcategory of $\mathcal{HC}_\infty$ closed under translation, and the difficulty of decomposing the modules arising from $M(\delta_{\max})$ under translation. We begin with the categorical problem.

The pseudo-projective modules $P(\gamma)$ constructed in Theorem B are not canonical; for each γ more than one $P(\gamma)$ may exist with the given properties. Moreover, they are not projective in any obvious subcategory of Harish-Chandra modules. A more satisfying result would be that a full subcategory $\widetilde{\mathcal{HC}}$ of $\mathcal{HC}_\infty$ exists, containing the simple and standard modules, with projective covers $P(\gamma)$ for each $L(\gamma)$ that satisfy the properties of Theorem B. One may further wish that $\widetilde{\mathcal{HC}}$ be closed under all wall-crossing translation functors. In fact, though unstated in Theorem B, a crucial property of the $P(\gamma)$'s is that for suitable translation functors θ_s one can decompose $\theta_s P(\gamma)$ as a direct sum of $P(s \times \gamma)$ and other $P(\varepsilon)$'s. (Here $s \times \gamma$ is a distinguished element of $\mathcal{D}$ below but adjacent to γ in the ordering, and the decomposition allows us to proceed inductively.) Thus it is natural to ask for this to be the case for all $P(\gamma)$'s in a hypothetical $\widetilde{\mathcal{HC}}$ and all θ_s's.

The fundamental fact underlying the approach of this paper is that no such $\widetilde{\mathcal{HC}}$ exists! One already discovers this in the simplest possible case, $G = SL(2, \mathbb{C})$, as we explain in a moment. For complex groups G in general, a substitute for $\widetilde{\mathcal{HC}}$ exists. In this case, the complexified Lie algebra of G has the form $\mathfrak{g} \times \mathfrak{g}$, where $\mathfrak{g}$ is the real Lie algebra of G (with complex structure). A given infinitesimal character χ of $\mathfrak{g} \times \mathfrak{g}$ may be written as $\chi_1 \otimes \chi_2$. Take $\mathcal{HC}'$ to be the full subcategory of $\mathcal{HC}_\infty$ consisting of those modules with generalized infinitesimal character χ such that $1 \otimes \mathcal{Z}(\mathfrak{g})$ acts with *genuine* infinitesimal character χ_2. Then Bernstein and Gelfand prove in **[1]** that $\mathcal{HC}'$ is equivalent to the category $\mathcal{O}_{\chi_1}$. (See also work of Enright **[9]** and Joseph **[23]**.) Moreover, translation functors for $\mathfrak{g} \times 0$ commute with the equivalence and standard modules correspond to Verma modules. Thus $\mathcal{HC}'$ has the desired properties, up to closure under half of the translation functors. It turns out that this is enough for our purposes. One could adopt as our $P(\gamma)$'s the objects in $\mathcal{HC}'$ corresponding to the projective indecomposable modules of $\mathcal{O}$. However, direct sums of these are not in fact closed under all translation functors, and the choice clearly isn't canonical, for by symmetry one may choose an analogous set on the other side.

Let us turn to the case of $SL(2, \mathbb{C})$ to exhibit the non-canonical nature of the $P(\gamma)$'s more explicitly. There are two parameters in $\mathcal{D}$, which we may call $\gamma = \delta_{\max}$ and $\delta = \delta_{\min}$. The standard module $M(\gamma)$ has length two, with cap $L(\gamma)$ and socle $L(\delta)$. The Weyl group is generated by two commuting simple reflections, s and t, which yield the two translation functors θ_s and θ_t. Both functors kill $L(\gamma)$ and send $L(\delta)$ to a non-split extension of $M(\gamma)$ by $L(\delta)$. Call these two extensions X_s and X_t. If they were isomorphic, then the desired category $\widetilde{\mathcal{HC}}$ would exist. However a simple argument, using adjointness of translation functors to and off a wall and using the singular categories at the s, t, and s-t walls, shows that $\dim Hom_{\mathcal{HC}}(X_s, X_t) = 1$. Thus X_s and X_t are not isomorphic. Each of them may be chosen as our $P(\delta)$, along with $M(\gamma)$ as $P(\gamma)$, in which case we obtain the two projective modules of a category of the form $\mathcal{HC}'$. But neither is to be preferred to the other.

In addition to non-canonicality, neither choice of $\mathcal{HC}'$ is closed under translation. To see this, consider the structure of $Y = \theta_s\theta_t L(\delta)$. It is an extension of X_t by itself (and also of X_s by itself, since θ_s and θ_t commute). An adjointness argument shows that Y, which is self-dual, has simple socle and simple top $L(\delta)$. Thus it clearly cannot lie in $\mathcal{HC}'$. The rather complicated structure of Y is worth examination. It has both X_s and X_t as submodules and factor modules. Using the description by Gelfand, Graev, and Ponomarev of the category $\mathcal{HC}_\infty$ in terms of a quiver and relations [**11**], it is possible to determine what the structure of Y is; it may be summarized by the following diagram:

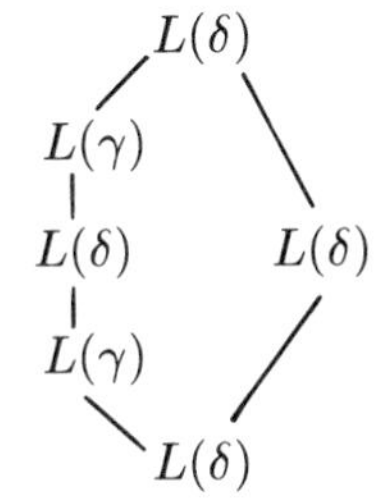

Given the impossibility of finding a subcategory $\widetilde{\mathcal{HC}}$ with the desired properties, one may still wish that an $\widetilde{\mathcal{HC}}$ exists in which a suitable choice of the $P(\gamma)$'s may be realized as projective modules. In particular, if such a subcategory exists, it will be a BGG category in the sense of [**21**]. Thus it will have finite global dimension. In contrast, various other natural subcategories of $\mathcal{HC}_\infty$, such as the subcategory of modules with genuine infinitesimal character, have infinite global dimension. If such a category $\widetilde{\mathcal{HC}}$ exists, it would be a natural setting in which to calculate the spaces $\mathrm{Ext}^i\big(M(\gamma), L(\delta)\big)$. In particular, one might expect that their dimensions are the coefficients of the Lusztig-Vogan polynomials associated to G. This is true for G complex, using $\mathcal{HC}'$, and is also true if G is one of the groups $Spin(2n+1,1)$, with a suitable choice of $\widetilde{\mathcal{HC}}$.

We now turn to the second subtlety, the analysis of translation on $M(\delta_{\max})$. In $\mathcal{O}_\mathfrak{p}$, the generalized Verma module $M_\mathcal{O}(w_{\max})$ associated to the maximal element $w_{\max}$ of $W^\mathfrak{p}$ is projective. (See §2.3 for notation in $\mathcal{O}_\mathfrak{p}$.) Thus, the indecomposable summands of modules of the form $\theta M_\mathcal{O}(w_{\max})$, for θ a composition of translation functors, are projective, and one obtains all projective indecomposable modules $P_\mathcal{O}(w)$ in this manner. The module $P_\mathcal{O}(w)$ is the projective cover of $L_\mathcal{O}(w)$, and satisfies all the properties of a pseudo-projective cover.

One may wish that the desired pseudo-projective modules in $\mathcal{HC}_\infty$ arise as summands of $\theta M(\delta_{\max})$ for suitable θ. This is in fact how we proceed, but great care is required because in general $\theta M(\delta_{\max})$ does not decompose in $\mathcal{HC}_\infty$ as pleasantly as $\theta M_\mathcal{O}(w_{\max})$ does in $\mathcal{O}_\mathfrak{p}$. One already sees this in the $SL(2,\mathbb{C})$ example above: the module $\theta_s\theta_t M(\delta_{\max})$ has a standard flag, but not of $\mathcal{O}$-type.

By analogy with $\mathcal{O}_\mathfrak{p}$-behavior, one would have expected it to decompose as a direct sum of two pseudo-projective covers, but it fails to because X_s and X_t are not isomorphic.

The failure just described is not really to the point. To obtain a pseudo-projective cover for $L(\delta)$, one would expect to choose a composition $\theta = \theta_{s_1} \circ \ldots \circ \theta_{s_k}$ so that the elements $s_1, \ldots, s_k$ yield a path downwards within $\mathcal{D}$ of minimal length from $\delta_{\max}$ to δ. For then $\theta M(\delta_{\max})$ has a standard flag with $M(\delta)$ occurring exactly once as a subquotient, and the indecomposable summand $P(\delta, \theta)$ of $\theta M(\delta_{\max})$ with $M(\delta)$ occurring is a natural choice for a pseudo-projective cover of $L(\delta)$. The composition $\theta_s\theta_t$ of the $SL(2, \mathbb{C})$ example is not of minimal length for any δ.

Let us call θ *minimal* (for δ) if it is of the above form. We now describe an example of a minimal θ such that $P(\delta, \theta)$ fails to be a pseudo-projective cover of $L(\delta)$: it does not have a simple cap, its standard flag is not of $\mathcal{O}$-type, and it does not satisfy BGG reciprocity. Let $G = SL(3, \mathbb{H})$ and let $\mathcal{D}$ be the set $\{\delta_1, \ldots, \delta_{15}\}$, indexed as in Figure 3. The desired θ is $\theta_1\theta_3\theta_4\theta_2$. To see what goes wrong, observe first that $\theta_1\theta_3 M(\delta_2)$ behaves analogously to the module $\theta_s\theta_t M(\delta_{\max})$ of $SL(2, \mathbb{C})$. It is indecomposable with simple cap $L(\delta_4)$ and with a standard flag having as subquotients two copies of $M(\delta_4)$ extending each other non-trivially and two copies of $M(\delta_2)$ extending each other non-trivially. A standard calculation shows that $\theta M(\delta_{\max})$ has as cap, a direct sum of $L(\delta_{10})$, and $L(\delta_4)$. It has as part of its standard flag a submodule Z with simple cap $L(\delta_{10})$ and with standard flag factors $\{M(\delta_i) : i = 2, 4, 5, 7, 8, 10\}$ once each, and $\theta M(\delta_{\max})/Z$ is an extension of $M(\delta_2)$ by $M(\delta_4)$. If Z were to split as a summand, it would be the desired $P(\delta_{10}, \theta)$ and all would be well. However, the analysis of $\theta_1\theta_3 M(\delta_2)$, which occurs as a submodule of $\theta M(\delta_{\max})$, shows that Z does not split. Rather, $\theta M(\delta_{\max})$ is indecomposable and is itself $P(\delta_{10}, \theta)$. (In contrast, if one chooses $\theta' = \theta_3\theta_4\theta_3\theta_2$, one arrives at a similar set-up in which Z does split, becoming the desired pseudo-projective cover of $L(\delta_{10})$.)

The above example shows that the construction of a pseudo-projective cover for $L(\delta)$ will depend on a careful choice of minimal θ for δ. Another description of what has gone wrong in the example gives a clue for choosing the minimal θ. Associated to $\mathcal{HC}_\infty$ and $\mathcal{O}_\mathfrak{p}$ are Hecke modules $\mathcal{M}$ and $\mathcal{M}_\mathcal{O}$, with $\mathbb{Z}[q^{1/2}, q^{-1/2}]$-bases $\{p(\delta) : \delta \in \mathcal{D}\}$ and $\{p_\mathcal{O}(w) : w \in W^\mathfrak{p}\}$. The action of elements θ_s of the Hecke algebra $\mathcal{H}$ on the two bases can be explicitly described. In the case of $\mathcal{O}_\mathfrak{p}$, the formula for $\theta_s p_\mathcal{O}(w)$ corresponds to the decomposition of $\theta_s P_\mathcal{O}(w)$ into indecomposable projective modules. One might analogously expect that pseudo-projective covers exist in $\mathcal{HC}_\infty$ for which the formula for $\theta_s p(\delta)$ corresponds to the decomposition of $\theta_s P(\delta, \theta)$ into indecomposable pseudo-projective modules. The example above exhibits the failure of this phenomenon if θ is not chosen carefully. One finds that $\theta_3\theta_4\theta_2 p(\delta_{\max}) = p(\delta_8)$ and $\theta_1 p(\delta_8) = p(\delta_{10}) + p(\delta_4)$. However, as we saw, $\theta_1 P(\delta_8, \theta_3\theta_4\theta_2)$ is indecomposable.

As a clue to avoiding this behavior, recall for a given group G that there is a functor $\mathcal{L}$ from a suitable $\mathcal{O}_{\mathfrak{p}}$ to $\mathcal{HC}_\infty$ inducing a Hecke module homomorphism $\mathbb{L}$ from $\mathcal{M}_{\mathcal{O}}$ to $\mathcal{M}$. Suppose w and δ are chosen so that $s_1, \ldots, s_k$ is simultaneously a minimal path from $w_{\max}$ to w and from $\delta_{\max}$ to δ. Then $\mathcal{L}(M_{\mathcal{O}}(w)) = M(\delta)$ and $\mathbb{L}(p_{\mathcal{O}}(w)) = p(\delta) + \Sigma_{\varepsilon\in\mathcal{D},\varepsilon>\delta} k_\varepsilon p(\varepsilon)$ for non-negative integers k_ε. Let $\theta = \theta_{s_1} \circ \ldots \circ \theta_{s_k}$. Since $\mathcal{L}$ commutes with translation, $\theta M(\delta_{\max}) = \mathcal{L}(\theta M(w_{\max}))$ and $P(\delta, \theta)$ will be an indecomposable summand of $\mathcal{L}(P_{\mathcal{O}}(w))$. In our bad example, we have $\mathbb{L}(p_{\mathcal{O}}(w)) = p(\delta_{10}) + p(\delta_4)$ and $\mathcal{L}(P_{\mathcal{O}}(w)) = P(\delta_{10}, \theta)$. However if we choose $\theta' = \theta_3\theta_4\theta_3\theta_2$ and the corresponding w', then we obtain $\mathbb{L}(p_{\mathcal{O}}(w')) = p(\delta_{10})$ and the module $\mathcal{L}(P_{\mathcal{O}}(w')) = P(\delta_{10}, \theta')$ is a pseudo-projective cover of $L(\delta_{10})$. Thus we may use the behavior of $\mathbb{L}$ on $p_{\mathcal{O}}(w)$'s as a clue to finding a good minimal θ. This point of view motivates the approach taken in the paper, and the examples illustrate the necessity of an approach that takes into account careful choices of paths from $\delta_{\max}$ to δ.

Both authors were supported by grants from the National Science Foundation and the National Security Agency, most recently under NSA grants MDA904-90-H-4041 and MDA904-90-H-1005 respectively.

2. Preliminaries

2.1 Harish-Chandra modules.

Throughout the paper, G will denote a connected semisimple Lie group with one conjugacy class of Cartan subgroups. Let θ denote an involutive automorphism of G whose fixed point set K is a maximal compact subgroup of G. We have an Iwasawa decomposition $G = KAN$ and a compatible minimal parabolic subgroup P with Langlands decomposition $P = MAN$; P will be connected in our setting and determines a set of positive restricted roots Σ^+. Denote the real Lie algebra of G by $\mathfrak{g}_0$ and its complexification by $\mathfrak{g}$, with the analogous convention for other Lie groups. Let H be a θ-stable Cartan subgroup. It decomposes as $H = TA$, where $T = (K \cap H)$.

A $(\mathfrak{g}, K)$-module is a module V over the universal enveloping algebra $\mathcal{U}(\mathfrak{g})$ together with an action of K such that: (a) if $v \in V$, then $K.v$ spans a finite dimensional space on which the K action is continuous; (b) the differential of the K action coincides with the $\mathfrak{k}$-action. A $(\mathfrak{g}, K)$-module V is said to be *finitely generated* (resp. of *finite length*) if it is as a $\mathcal{U}(\mathfrak{g})$-module. Let $K^\wedge$ denote the set of equivalence classes of continuous irreducible (finite-dimensional) representations of K. Given a finitely generated $(\mathfrak{g}, K)$-module V and $\sigma \in K^\wedge$, we set $V(\sigma)$ equal to the span of all K-invariant finite dimensional subspaces of V isomorphic to σ; then $V = \oplus_{\sigma\in K^\wedge} V(\sigma)$. We say V is *admissible* if $\dim_{\mathbb{C}} V(\sigma)$ is finite, for all $\sigma \in K^\wedge$.

We fix, once and for all, a finite dimensional representation F of G with infinitesimal character χ. Recall that a module X has *generalized infinitesimal character* χ if there is a positive integer i such that $(z - \chi(z))^i.x = 0$ for every $z \in \mathcal{Z}(\mathfrak{g})$ and $x \in X$; this condition is equivalent to requiring that $(\mathrm{Ker}\chi)^i$ annihilate X. Let $\mathcal{HC}_\infty$ denote the category of finitely generated admissible

$(\mathfrak{g}, K)$-modules with generalized infinitesimal character χ. Denote by $\mathcal{HC}_1$ the category of finitely generated $(\mathfrak{g}, K)$-modules with genuine infinitesimal character χ. Then $\mathcal{HC}_1$ is a full abelian subcategory of $\mathcal{HC}_\infty$. We refer to objects of $\mathcal{HC}_\infty$ as *Harish-Chandra modules.*

2.2 Langlands classification.

By our hypothesis that G has a single conjugacy class of Cartan subgroups, H is connected. This means that the Langlands classification of irreducible representations in $\mathcal{HC}_\infty$ reduces to the parametrization of certain positive systems of roots of $\mathfrak{h}$ in $\mathfrak{g}$. To describe this more precisely, fix an Iwasawa Borel subalgebra $\mathfrak{b} \subseteq \mathfrak{p}$ with Levi decomposition $\mathfrak{b} = \mathfrak{h} \oplus \mathfrak{u}$. Let Φ denote the roots of $\mathfrak{h}$ in $\mathfrak{g}$ and Φ^+ the system of positive roots determined by $\mathfrak{b}$. Set $\Phi_\mathfrak{p}$ equal to the roots of $\mathfrak{m} \oplus \mathfrak{a}$ (which are just the roots in Φ which vanish on $\mathfrak{a}$) and $\Phi_\mathfrak{p}^+ = \Phi^+ \cap \Phi_\mathfrak{p}$. We have the Weyl group W (resp. $W_\mathfrak{p}$) of Φ (resp.$\Phi_\mathfrak{p}$) and S the set of generating simple reflections for W. Let $\leq$ be the Bruhat order of W, with the convention that e is the unique minimal element. Let w_0 (resp. $w_\mathfrak{p}$) be the longest element of W (resp. $W_\mathfrak{p}$). The Φ^+-dominant integral highest weight of our fixed finite dimensional representation F of G is $\lambda_\chi - \rho$ for some element $\lambda_\chi \in \mathfrak{h}^*$. Let $W^\mathfrak{p} = \{w \in W : w^{-1}\Phi_\mathfrak{p}^+ \subseteq \Phi^+\}$ and identify $W^\mathfrak{p}$ with the set of minimal length coset representatives for $W_\mathfrak{p}\backslash W$.

Let $W(A) =$ (Normalizer of A in M)/(Centralizer of A in M). We wish to define an action of $W(A)$ on the set $W^\mathfrak{p}$. To each $w \in W^\mathfrak{p}$ we may associate a finite dimensional vector space $E(w)$ and an irreducible representation $(E(w), \zeta_w)$ of MA on $E(w)$ having highest weight $w_\mathfrak{p} w w_0 \lambda_\chi - \rho$; recall that MA is connected. Given $z \in W(A)$, we define a new finite dimensional irreducible representation of MA, denoted $(E(w), z.\zeta_w)$, by $[z.\zeta_w](x) = \zeta_w(z^{-1}xz)$ for $x \in MA$. (Notice in particular that the dimensions of the two representations are the same.) A moment's reflection shows that $(E(w), z.\zeta_w) = (E(y), \zeta_y)$ for a unique $y \in W^\mathfrak{p}$. In this way we have defined an action of the group $W(A)$ on the set $W^\mathfrak{p}$. Let $\mathcal{D}$ be the set of $W(A)$ orbits on $W^\mathfrak{p}$.

It is now possible to describe the Langlands classification in the form we shall need. First, given $w \in W^\mathfrak{p}$, view $E(w)$ as a P-module by letting N act trivially. Let

$$\mathrm{Ind}_P(E(w)) = \{\mathrm{smooth} f : G \to E(w) : f(pg) = [\zeta_w \oplus e^{\rho(A)}](p)f(g), p \in P\},$$

where $\rho(A) = \rho|_\mathfrak{a}$. Then form the principal series representation $I_p(W)$, which is the subspace of K-finite vectors in $\mathrm{Ind}_P(E(w))$. This is a $(\mathfrak{g}, K)$-module and lies in in $\mathcal{HC}_1$.

Given an $I_P(w)$, we may decompose the inducing representation ζ_w on MA as $\delta_w \otimes \exp(\nu_w)$, where δ_w is an irreducible finite dimensional representation of M and $\exp(\nu_w)$ is a character on A. If $(\nu_w, \alpha) \geq 0$ for all $\alpha \in \Sigma^+$, then we say that $I_P(w)$ is in *Langlands quotient position.* If $\delta \in \mathcal{D}$, then there will exist a Weyl group element w_δ in the orbit δ such that $I_P(w_\delta)$ is in Langlands quotient position; we refer to such a w_δ as a *good* representative of δ. We denote $I_P(w_\delta)$

by $M(\delta)$ and refer to this as a *standard module* for G. A priori, since the element w_δ in the orbit δ need not be unique, this definition may seem to depend upon the choice of w_δ in δ. However, given two good representatives w_δ and w_γ in the orbit δ, it follows from Langlands' classification that $I_P(w_\delta) \cong I_P(w_\gamma)$. We can now state the Langlands classification (and Miličić's refinement); see [**31**, §**5**].

THEOREM 2.2.1. *(i)(Miličić) If $\delta \in \mathcal{D}$, then the standard module $M(\delta)$ has a unique irreducible quotient, denoted $L(\delta)$.*

(ii)(Langlands) If V is an irreducible $(\mathfrak{g}, K)$-module in $\mathcal{HC}_\infty$, then there exists a unique $\delta \in \mathcal{D}$ such that $V \cong L(\delta)$.

REMARK 2.2.2. We will make use of an exact contravariant duality functor $\mathbb{D}$ on $\mathcal{HC}_\infty$, which is discussed for instance in [**28**]. With respect to this duality, the modules $\mathbb{D}M(\delta)$ are the opposite standard modules: the ones having $L(\delta)$ as a unique irreducible submodule.

2.3 The Category $\mathcal{O}_\mathfrak{p}$.

We denote by $\mathcal{O}_\mathfrak{p}$ the category of finitely generated $\mathcal{U}(\mathfrak{g})$-modules which are locally $\mathfrak{p}$-finite, $\mathfrak{m}$-semisimple and have a fixed regular integral generalized infinitesimal character. If $w \in W^\mathfrak{p}$, then define $M_\mathcal{O}(w)$ to be the algebraically induced module

$$\mathcal{U}(\mathfrak{g}) \otimes_{\mathcal{U}(\mathfrak{p})} E(w). \tag{2.3.1}$$

The modules $M_\mathcal{O}(w)$ lie in the category $\mathcal{O}_\mathfrak{p}$, as do their unique irreducible quotient modules, denoted $L_\mathcal{O}(w)$. Finally, the category $\mathcal{O}_P$ admits enough projectives and we let $P_\mathcal{O}(w)$ be the indecomposable projective cover of $L_\mathcal{O}(w)$.

2.4 The Bruhat order and translation functors.

Given a simple root $\alpha \in \Phi^+$with associated simple reflection s, we follow [**28**] to form the translation functors ψ_s to the α-wall and ϕ_s away form the α-wall on the category $\mathcal{HC}_\infty$. Let $\theta_s = \phi_s \circ \psi_s$, the functor of translation across the α-wall. All three functors are exact and covariant. Also, ϕ_s and ψ_s are left and right adjoint to each other; this gives rise to natural transformations σ_1 from the identity functor to θ_s and σ_2 from θ_s to the identity functor.

We wish to place some additional structure on the set $\mathcal{D}$. Specifically we need a partial ordering on $\mathcal{D}$ and an action of the set S on $\mathcal{D}$. Let us now show how to do this, following [**30**]. To describe the required data on $\mathcal{D}$ requires the introduction of some more definitions. Given an orbit $\delta \in \mathcal{D}$, set $\Phi^+(\delta) = \{\alpha \in \Phi : (\alpha, w_\mathfrak{p} w_\delta w_0 \lambda_\chi) > 0\}$; i.e., $\Phi^+(\delta)$ is the system of positive roots determined by the highest weight (plus ρ) of the finite dimensional representation $E(w_\delta)$ from which $I_P(w_\delta)$ is induced. Because H is θ-stable, we know that Φ is invariant under the action of θ and this allows us to make the following definition.

DEFINITION 2.4.1. Let α be a simple root for the positive system $\Phi^+(\delta)$. If $\theta\alpha = \alpha$, then we say α is a *compact root.* If $\theta\alpha \neq \pm\alpha$, then we say that α is a *complex root.*

We can speak in the obvious way of *compact reflections* and *complex reflections* about the associated simple roots of $\Phi^+(\delta)$.

Fix $\delta \in \mathcal{D}$ and define

$$\tau(\delta) = \{\alpha \in \Phi^+(\delta) : \alpha \text{ is } \Phi^+(\delta)\text{-simple and either (a) } \alpha \text{ is a compact root or (b) } \alpha \text{ is a complex root and } \theta\alpha \notin \Phi^+(\delta).\}$$

For every $\delta \in \mathcal{D}$, construct an automorphism $\phi(\delta) : \mathfrak{g} \to \mathfrak{g}$ which preserves $\mathfrak{h}$ and carries the positive system $\Phi^+(\delta)$ to our Iwasawa positive system Φ^+. We may now transport the notions of compact and complex $\Phi^+(\delta)$-roots to Φ^+: $\alpha \in \Phi^+$ is compact (resp. complex) if $\alpha = \phi(\beta)$ for some $\Phi^+(\delta)$-simple compact (resp. complex) root β. We define

$$\tau_S(\delta) = \{s_\beta \in S : \ \beta = \phi(\delta)(\alpha), \text{for some } \alpha \in \tau(\delta)\}. \tag{2.4.2}$$

As above, we may speak of compact and complex simple reflections in S for δ.

To define the partial order on $\mathcal{D}$ we will use a notion of length for elements in $\mathcal{D}$ and the so-called *cross action.* If $\delta \in \mathcal{D}$, then set

$$\ell(\delta) = \max\{\ell(y) : y \in W^{\mathfrak{p}} \text{ and } y \text{ is in the orbit } \delta\}. \tag{2.4.3}$$

Next, if $s = s_\beta$ is an element of S and $\delta \in \mathcal{D}$, we define $s \times \delta$ as follows:
(2.4.4)

(i) if s is compact, then $s \times \delta = \delta$;

(ii) if s is complex, then $s \times \delta$ is the unique element in $\mathcal{D}$ satisfying

$$\Phi^+(s \times \delta) = s_{\phi^{-1}(\delta)(\beta)}\Phi^+(\delta).$$

The action of S on $\mathcal{D}$ defined in (2.4.4) is called the *cross action.*

LEMMA 2.4.5 ([**30**,§6]). *(i) Let α be a $\Phi^+(\delta)$-simple complex root and put $\beta = \phi(\alpha)$. If $\theta\alpha \notin \Phi^+(\delta)$, then $\alpha \in \tau(\delta)$, $s_\beta \in \tau_S(\delta)$ and $\ell(s_\beta \times \delta) = \ell(\delta) - 1$.*

(ii) Let α be a $\Phi^+(\delta)$-simple complex root and put $\beta = \phi(\alpha)$. If $\theta\alpha \in \Phi^+(\delta)$, then $\alpha \notin \tau(\delta)$, $s_\beta \notin \tau_S(\delta)$ and $\ell(s_\beta \times \delta) = \ell(\delta) + 1$.

DEFINITION 2.4.6. We introduce a relation $\xrightarrow{s}$ on $\mathcal{D}$ by the rule: $\delta \xrightarrow{s} \gamma$ if $\gamma = s \times \delta$ for some reflection $s \notin \tau_S(\delta)$. The *Bruhat order* on $\mathcal{D}$ is the smallest partial order $<$ with the following property: Fix $\gamma, \gamma' \in \mathcal{D}$ and $s \in S$ with $\gamma' \xrightarrow{s} \gamma$. Suppose that either

(i) $\delta \leq \gamma'$, or

(ii) there is a $\delta' \in \mathcal{D}$ with $\delta' \xrightarrow{s} \delta$ and $\delta' < \gamma'$.

Then $\delta < \gamma$.

REMARK. If $\delta \in \mathcal{D}$ and $s \in S$ then (2.3.4) and (2.3.5) lead to three possibilities for $s \times \delta$: (a) $s \times \delta > \delta$ if $\ell(s \times \delta) > \ell(\delta)$, (b) $s \times \delta < \delta$ if $\ell(s \times \delta) < \ell(\delta)$, and (c) $s \times \delta = \delta$. These parallel the three possibilities for $s \times w$ in a Bruhat poset $W^{\mathfrak{p}}$.

It is for this reason that $\mathcal{D}$ in our cases resembles a Bruhat poset. The above discussion shows that $s \in \tau_S(\delta)$ if and only if case (b) or case (c) holds.

LEMMA 2.4.7 ([**30**, (5.10)]). *Let γ and δ be elements of $\mathcal{D}$ such that $\big(M(\delta) : L(\gamma)\big) > 0$. Then $\gamma \leq \delta$.*

LEMMA 2.4.8 ([**6**, (3.35)]). *(i) If $ws \notin W^{\mathfrak{p}}$, then $\theta_s I_P(w) = \theta_s I_P(ws) = 0$.*

(ii) If $ws \in W^{\mathfrak{p}}$ and $ws < w$, then $\theta_s I_p(w) = \theta_s I_P(ws)$ and there is a nonsplit short exact sequence $0 \to I_P(w) \to \theta_s I_P(w) \to I_P(ws) \to 0$ in $\mathcal{HC}_\infty$.

(iii) If $ws \in W^{\mathfrak{p}}$ and $ws > w$, then $\theta_s I_P(w) = \theta_s I_P(ws)$ and there is a nonsplit short exact sequence $0 \to I_P(ws) \to \theta_s I_P(w) \to I_P(w) \to 0$ in $\mathcal{HC}_\infty$.

(iv) The homomorphisms in (ii) and (iii) are the natural maps arising from the natural transformations σ_1 and σ_2.

COROLLARY 2.4.9. *(i) If $s \times \delta = \delta$, then $\theta_s M(\delta) = 0$.*

(ii) If $s \times \delta < \delta$, then there exists a nonsplit short exact sequence $0 \to M(\delta) \to \theta_s M(\delta) \to M(s \times \delta) \to 0$ in $\mathcal{HC}_\infty$.

(iii) If $s \times \delta > \delta$, then there exists a nonsplit short exact sequence $0 \to M(s \times \delta) \to \theta_s M(\delta) \to M(\delta) \to 0$ in $\mathcal{HC}_\infty$.

Let us also describe the behavior of θ_s on the irreducible modules.

THEOREM 2.4.10. *Let $\delta \in \mathcal{D}$ and $s \in S$.*

(i) If $s \times \delta < \delta$ or $s \times \delta = \delta$, then $\theta_s L(\delta) = 0$.

(ii) If $s \times \delta > \delta$, then $\theta_s L(\delta)$ has simple cap and simple socle isomorphic to $L(\delta)$.

(iii) Let $U_s L(\delta) = rad\theta_s L(\delta)/soc\theta_s L(\delta)$. Then $\big(U_s L(\delta) : L(s \times \delta)\big) = 1$ and every other composition factor of $U_s L(\delta)$ has the form $L(\gamma)$ with $\gamma < \delta$ and $s \in \tau_S(\gamma)$.

(iv)(Vogan's conjecture) $U_s L(\delta)$ is semisimple.

PROOF. Proofs of (i)-(iii) can be found in [**28**] and [**29**]. Part (iv) is equivalent to the Kazhdan-Lusztig conjecture for $\mathcal{HC}_\infty$ which is proved in [**30**]. ♦

2.5 Preliminaries on extensions of standard modules.

We prove in this section two results about standard modules which will be needed later.

PROPOSITION 2.5.1. *Let $\delta \in \mathcal{D}$. Let X be a Harish-Chandra module with simple cap $L(\delta)$ such that every composition factor $L(\gamma)$ of $radX$ satisfies $\ell(\gamma) \leq \ell(\delta)$ and $\gamma \neq \delta$. Then (i) X is a homomorphic image of $M(\delta)$; (ii) Every composition factor $L(\gamma)$ of $radX$ satisfies $\ell(\gamma) < \ell(\delta)$.*

PROOF. We define $P^- = \theta P = MAN^-$; i.e., P^- is the opposite minimal parabolic subgroup to P. We recall the Frobenius reciprocity result [**31**,(3.8.2)]:

$$Hom_{\mathfrak{g},K}\big(\mathbb{D}X, I_{P^-}(w_\delta)\big) = Hom_{MA}\big(H_0(\mathfrak{n}^-, \mathbb{D}X), E(w_\delta)\big). \tag{2.5.2}$$

Assume for the moment that we can show

$$\text{(2.5.3)} \qquad Hom_{\mathfrak{m}\oplus\mathfrak{a}}\big(H_0(\mathfrak{n}^-, \mathbb{D}X), E(w_\delta)\big) \neq 0.$$

Then (2.5.2) will produce a non-zero map $T : \mathbb{D}X \to I_{P^-}(w_\delta)$. By duality, we get a non-zero map $\mathbb{D}T : I_P(w_\delta) \to X$. (To see this, we recall (2.2.2) and use the fact that $\mathbb{D}I_P(y) = I_P(z_0.y) = I_{P^-}(y)$.) By the hypothesis and the fact that cap $I_P(w_\delta) = L(\delta)$ we conclude that $\mathbb{D}T$ is onto. This proves (i) and (ii) immediately follows from (2.4.7).

To complete the proof, it suffices to verify (2.5.3). We will use the Jacquet functor. Denote by $(\dots)^\sim$ the contragredient functor (see [**31**,(4.3.2)]). Define the category $\mathcal{O}'_\mathfrak{p}$ to be the full subcategory of all finitely generated $\mathcal{U}(\mathfrak{g})$-modules which are $\mathfrak{p}$-locally finite and have the same generalized infinitesimal character as χ. The Jacquet functor j from $\mathcal{HC}_\infty$ to $\mathcal{O}'_\mathfrak{p}$ is defined on objects by $j(X) = (X^\sim)^*_{\mathfrak{b}-finite}$, where $*$ is the full algebraic dual. The functor j is an exact covariant functor and by a result of Casselman [**31**,(4.1.5)] it is faithful. A key property of the Jacquet functor [**14**] is that

$$\text{(2.5.4)} \qquad Hom_{MA}\big(H_0(\mathfrak{n}^-, \mathbb{D}X), E(w_\delta)\big) = Hom_{\mathfrak{m}\oplus\mathfrak{a}}\big(H^0(\mathfrak{n}, j(X)), E(w_\delta)\big).$$

By the hypothesis and what is called the *support axiom* (see [**5**,§2]), the simple module in $\mathcal{O}'_\mathfrak{p}$ with highest weight $w_\mathfrak{p} w_\delta w_0 \lambda_\chi - \rho$ is a composition factor of $j(X)$ and $w_\mathfrak{p} w_\delta w_0 \lambda_\chi - \rho$ is a maximal weight space for $j(X)$. But this means that the right-hand Hom space in (2.5.4) is non-zero, as was to be proved. ♦

PROPOSITION 2.5.5. *Let γ and δ be distinct elements of $\mathcal{D}$ of equal length. Let X be a homomorphic image of $M(\gamma)$. Then $Ext^1_{\mathfrak{g},K}\big(M(\delta), X\big) = 0$.*

PROOF. Since $M(\gamma)$ contains a unique composition factor $L(\gamma) \neq L(\delta)$ of maximal length, we have

$$\begin{aligned} \dim_\mathbb{C} Hom_{\mathfrak{g},K}\big(M(\delta), X\big) &= \dim_\mathbb{C} Hom_{\mathfrak{g},K}(\mathbb{D}X, I_{P^-}(w_\delta)) \\ &= \dim_\mathbb{C} Hom_{MA}\big(H_0(\mathfrak{n}^-, \mathbb{D}X), E(w_\delta)\big) = 0 \end{aligned}$$

Additionally, reasoning as in the proof of the previous proposition actually shows

$$\dim_\mathbb{C} Hom_{MA}\big(H_1(\mathfrak{n}^-, \mathbb{D}X), E(w_\delta)\big) = \dim_\mathbb{C} Hom_{\mathfrak{m}\oplus\mathfrak{a}}\big(H^1(\mathfrak{n}, j(X)), E(w_\delta)\big) = 0.$$

The spectral sequence [**29**,(6.3.5)] now yields the result. ♦

REMARK 2.5.6. It is shown that $\dim_\mathbb{C} Ext^1_{\mathfrak{g},K}\big((M(\delta), M(\delta)\big) = \dim_\mathbb{C} \wedge^i \mathfrak{a}$ in [**12**], in contrast to (2.5.5).

2.6 Standard modules of dominant type.

We conclude this section with a discussion of some special properties of the standard module $M(\delta_{\max})$. Recall, we may associate to any finitely generated $\mathcal{U}(\mathfrak{g})$-module X its GK dimension $d(X)$ and multiplicity (or Bernstein number)

$e(X)$. (See [**22**] for further details.) Set $s = \dim_{\mathbb{R}} N$ for the subgroup N of G. Then it follows from [**27**, (6.5)] that

$$d\big(I_P(w)\big) = s, \text{ for all } w \in W^{\mathfrak{p}}. \tag{2.6.1}$$

Since $I_P(w_{\delta_{\min}}) = M(\delta_{\min}) = L(\delta_{\min})$, we have in particular that $s = d\big(L(\delta_{\min})\big)$. Combining (2.6.1) with Harish-Chandra's subquotient theorem [**31**, 3.5] and the fact that the GK dimension of a finite length module is the maximum of the GK dimensions of its composition factors yields:

$$\begin{aligned} &\text{(i) } d(X) \le s, \text{ for all } X \text{ in } \mathcal{HC}_\infty, \text{ and} \\ &\text{(ii) } s = \max\{d\big(L(\delta)\big) : \delta \in \mathcal{D}\}. \end{aligned} \tag{2.6.2}$$

We may now state the main result of this section.

THEOREM 2.6.3. *The socle of $M(\delta_{\max})$ is simple of GK dimension s and $M(\delta_{\max})/socM(\delta_{\max})$ has GK dimension strictly less than s.*

Our proof of Theorem 2.6.3 involves two ingredients. We begin by showing that the socle of $M(\delta_{\max})$ has GK dimension s. We then establish the uniqueness of a simple composition factor with GK dimension s. We need an equivalence relation on $\mathcal{D}$, which decomposes it into equivalence classes known as right cells. Given γ and δ in $\mathcal{D}$, we define $\gamma \le_R \delta$ if there is a sequence of elements $\gamma = \delta_1, \delta_2, \dots, \delta_r = \delta$ in $\mathcal{D}$ such that for each i there is a simple reflection $s(i)$ for which $L(\delta_{i+1})$ occurs as a composition factor in $\theta_{s(i)}L(\delta_i)$. For instance, $\delta_{\min} \le_R \delta$ for all δ. We define $\gamma \sim_R \delta$ if both $\gamma \le_R \delta$ and $\delta \le_R \gamma$. Then $\sim_R$ is the desired equivalence relation.

Consider the following three subsets of $\mathcal{D}$:

(2.6.4)

$$\begin{aligned} \mathcal{X} &= \{\delta \in \mathcal{D} : L(\delta) \text{ is socular}\}, \\ \mathcal{Y} &= \{\delta \in \mathcal{D} : \delta \sim_R \delta_{\min}\}, \\ \mathcal{Z} &= \{\delta \in \mathcal{D} : d\big(L(\delta)\big) = s\}. \end{aligned}$$

LEMMA 2.6.5. *$\mathcal{X} \subseteq \mathcal{Y} \subseteq \mathcal{Z}$.*

PROOF. To prove the first inclusion, we prove the following statement for all $\gamma \in \mathcal{D}$ by induction on the Bruhat order:

(*) If $L(\delta)$ is the summand of the socle of $M(\gamma)$, then $\delta \sim_R \delta_{\min}$.

If $\gamma = \delta_{\min}$, (*) is obvious. Assume (*) holds for all $\gamma' < \gamma$, and choose $s \in S$ such that $s \times \gamma < \gamma$. Suppose $L(\delta)$ is a summand of the socle of $M(\gamma)$. By (2.4.9), there is an exact sequence $0 \to M(\gamma) \to \theta_s M(\gamma) \to M(s \times \gamma) \to 0$. Since $L(\delta)$ is in the socle of $M(\gamma)$, it is in the socle of $\theta_s M(\gamma)$, which implies that $\theta_s L(\delta) \ne 0$ and that $\theta_s L(\delta)$ is a submodule of $\theta_s M(\gamma)$. Since the module $\theta_s L(\delta)$ does not have infinitesimal character (see the proof of [**28**,(3.7)]), and $M(\gamma)$ does, $\theta_s L(\delta)$ cannot lie wholly within $M(\gamma)$ and must contain a composition factor $L(\varepsilon)$ of

$socM(s \times \gamma)$. Thus $\delta \leq_R \varepsilon$, and by induction $\varepsilon \sim_R \delta_{\min}$. We may conclude that $\delta \leq_R \delta_{\min}$. But $\delta_{\min} \leq_R \delta$ always holds, so $\delta \sim_R \delta_{\min}$. This verifies $(*)$.

To prove the second inclusion, recall for any Harish-Chandra module X that $\theta_s X$ is a summand of a module obtained by tensoring X by a finite-dimensional module. Thus $d(\theta_s X) \leq d(X)$ and $d(L(\delta)) \leq d(X)$ for any composition factor $L(\delta)$ of $\theta_s X$. In particular, if $\gamma \leq_R \delta$, then $d\big(L(\delta)\big) \leq d\big(L(\gamma)\big)$. Thus, if $\delta \sim_R \delta_{\min}$, then $d\big(L(\delta)\big) = d\big(L(\delta_{min})\big)$. ♦

Notice that the inclusion $\mathcal{X} \subseteq \mathcal{Z}$ of Lemma 2.6.5 forces every composition factor of $socM(\delta_{\max})$ to have GK dimension s. When combined with our next result the proof of (2.6.3) is complete.

LEMMA 2.6.6. *The standard module $M(\delta_{\max})$ has a unique composition factor of GK dimension s. In other words, there are submodules X and Y of $M(\delta_{\max})$ with $X \subseteq Y$ such that X and $M(\delta_{\max})/Y$ have GK dimension strictly less than s and Y/X is simple of GK dimension s.*

Our proof will make use of the Jacquet functor j (see §2.4) to take us from $\mathcal{HC}_\infty$ to $\mathcal{O}'_\mathfrak{p}$, where we compute Bernstein multiplicities. By (2.6.2), s equals the maximum of the GK dimensions of modules in $\mathcal{HC}_\infty$; for similar reasons, s is the maximum of the GK dimensions of modules in $\mathcal{O}'_\mathfrak{p}$. Let us recall the point. For each $w \in W^\mathfrak{p}$, the generalized Verma module $M_\mathfrak{p}(w)$ is defined as $\mathcal{U}(\mathfrak{g}) \otimes_{\mathcal{U}(\mathfrak{p})} E(w)$ for $E(w)$ finite dimensional. We recall that $\mathfrak{n}^- = \theta(\mathfrak{n})$. We note $\dim_\mathbb{C} \mathfrak{n}^- = \dim_\mathbb{R} N = s$, and $M_\mathfrak{p}(w) = \mathcal{U}(\mathfrak{n}^-) \oplus_\mathbb{C} E(w)$ as a vector space. Thus we obtain

$$(2.6.7) \qquad d\big(M_\mathfrak{p}(w)\big) = s \text{ and } e\big(M_\mathfrak{p}(w)\big) = \dim_\mathbb{C} E(w).$$

Since every irreducible module in $\mathcal{O}'_\mathfrak{p}$ is a composition factor of a generalized Verma module, s bounds the GK dimensions of all modules in $\mathcal{O}'_\mathfrak{p}$.

Define $e_s(V)$ on $\mathcal{O}'_\mathfrak{p}$ to be $e(V)$ if $d(V) = s$ and 0 otherwise. From the definition of multiplicity, one sees that if $0 \to V' \to V \to V'' \to 0$ is a short exact sequence of modules in $\mathcal{O}'_\mathfrak{p}$, then $e_s(V) = e_s(V') + e_s(V'')$. Thus e_s induces a homomorphism from the Grothendieck group of $\mathcal{O}'_\mathfrak{p}$ to the integers. Combining e_s and j, we define e' on $\mathcal{HC}_\infty$ by $e'(X) = e_s\big(j(X)\big)$. Since j is exact, e' is an additive function on short exact sequences in $\mathcal{HC}_\infty$. The function e' behaves in a manner similar to e_s:

(2.6.8)

Let X be a Harish-Chandra module. Then

$$\begin{aligned} &e'(X) = 0 && \text{if } d(X) < s, \\ &e'(X) \text{ is a positive integer} && \text{if } d(X) = s. \end{aligned}$$

This follows because the GK dimensions of X and $j(X)$ are the same [**27**,(3.4), (4.5)].

The proof of 2.6.6 depends on the following calculation.

LEMMA 2.6.9. *Let $w \in W^{\mathfrak{p}}$ and let $E(w)$ be the inducing data for the principal series representation $I_P(w)$. Then* $e'(I_P(w)) = (\dim E(w))|W(A)|$.

PROOF. Recall from §2.2 the action of $W(A)$ on $W^{\mathfrak{p}}$. We begin with the Grothendieck group equality

$$[j(I_P(w))] = \sum_{z \in W(A)} [M_{\mathfrak{p}}(z.w)]. \tag{2.6.10}$$

To prove this, one uses the Osborne conjecture (a theorem in [**13**]) and work of Hecht-Schmid [**14**] to identify the distribution character of $I_P(w)$ with the character of $j(I_P(w))$ (on a suitable set); also, the character formula for an induced representation [**24**,§10.3] yields a precise formula for the distribution character of $I_P(w)$, which is formally the sum of the characters of the generalized Verma modules on the right side of (2.6.10). Thus (2.6.10) follows and $e'(I_P(w)) = \sum_{z \in W(A)} e_s(M_{\mathfrak{p}}(z.w))$. But $e_s(M_{\mathfrak{p}}(z.w)) = \dim_{\mathbb{C}} E(z.w)$ for all $z \in W(A)$ by (2.6.7). According to the definition of the action of $W(A)$ on W^P in §2.2, we see that $\dim_{\mathbb{C}} E(w) = \dim_{\mathbb{C}} E(z.w)$ for all $z \in W(A)$. This proves the Lemma. ♦

COROLLARY 2.6.11. *Let $\varepsilon \in \mathcal{D}$ and assume that the standard module $M(\varepsilon)$ is induced from a one dimensional representation . Then $M(\varepsilon)$ contains a unique composition factor of GK dimension s.*

PROOF. First observe for any $\delta \in \mathcal{D}$, by the additivity of e', that

$$e'(M(\delta)) = \sum_{\substack{\gamma \in \mathcal{D} \\ \gamma \leq \delta}} (M(\delta) : L(\gamma)) e'(L(\gamma)). \tag{2.6.12}$$

It follows by induction on the ordering of $\mathcal{D}$ and Lemma 2.6.9 that $e'(L(\gamma))$ is divisible by $|W(A)|$ for all $\gamma \in \mathcal{D}$. (The induction begins with the fact that $M(\delta_{\min}) = L(\delta_{\min})$.) In particular, by (2.6.8), if $d(L(\gamma)) = s$ then $e'(L(\gamma))$ is a positive integer multiple of $|W(A)|$. By Lemma 2.6.9 and the assumption on ε we have $e'(M(\varepsilon)) = |W(A)|$. Using (2.6.12) again, the Corollary follows.

PROOF OF LEMMA 2.6.6. Recall the fixed finite-dimensional representation F. If F is the one dimensional representation, then $M(\delta_{\max}) = I_P(w^{\mathfrak{p}})$, where $E(w^{\mathfrak{p}})$ is one dimensional, and the Lemma follows from Corollary 2.6.11. Otherwise, we may apply translation functors to pass from our given $\mathcal{HC}_\infty$ to the Harish-Chandra module category corresponding to the one dimensional F and pull back the desired result. [**29**,§7.2]. ♦

3. Character theory, filtrations, and the Hecke module

3.1 The Hecke module.

Every module in $\mathcal{HC}_\infty$ is of finite length, so the (integral) Grothendieck group Gr_∞ of $\mathcal{HC}_\infty$ is the free abelian group on the isomorphism classes of simple modules. Given a Harish-Chandra module X, let $[X]$ denote the associated element in Gr_∞; we call $[X]$ the *character* of X.

We will be considering certain polynomials in the ring $\mathbb{Z}[q^{1/2}, q^{-1/2}]$, where $q^{1/2}$ is taken to be an indeterminate with q as its square. Let $\wedge$ denote the involution on $\mathbb{Z}[q^{1/2}, q^{-1/2}]$ which switches $q^{1/2}$ and $q^{-1/2}$ and fixes $\mathbb{Z}$. The image of an element R under this involution will be denoted $R^\wedge$. Following Lusztig-Vogan [**25**, §4] and Vogan [**30**,§6], we have the following polynomials associated to $\mathcal{D}$.

DEFINITION 3.1.1. For γ and δ in $\mathcal{D}$, the *Lusztig-Vogan polynomial* $\mathcal{P}_{\gamma,\delta}$ in $\mathbb{Z}[q^{1/2}]$ is defined recursively as follows:

(i) $\mathcal{P}_{\gamma,\delta} = 0$ unless $\gamma \leq \delta$.
(ii) $\mathcal{P}_{\gamma,\delta} = 1$ if $\gamma = \delta$.
(iii) Assume $\gamma < \delta$. Then $\mathcal{P}_{\gamma,\delta}$ is defined by induction on $\ell(\delta)$ and then on $\ell(\delta) - \ell(\gamma)$. It is a polynomial of degree at most $(\ell(\delta) - \ell(\gamma) - 1)/2$, with the coefficient of $q^{(\ell(\delta)-\ell(\gamma)-1)/2}$ denoted by $\mu(\gamma, \delta)$. Assume $\delta \in \mathcal{D}$ and $s \in S$ satisfy $s \times \delta < \delta$ and let $\delta' = s \times \delta$. Then $\mathcal{P}_{\gamma,\delta}(q)$ is defined by the following recursion relation:

$$\mathcal{P}_{\gamma,\delta}(q) = R - \sum_{\substack{\varepsilon \in \mathcal{D} \\ \gamma \leq \varepsilon \leq \delta' \\ s\times\varepsilon < \varepsilon \text{ or } s\times\varepsilon = \varepsilon}} \mu(\varepsilon, \delta') \cdot q^{(\ell(\delta)-\ell(\varepsilon))/2} \cdot \mathcal{P}_{\gamma,\varepsilon}(q),$$

where $R =$

$$\begin{array}{ll} (1+q)\mathcal{P}_{\gamma,\delta'}(q) & \text{if } s \times \gamma = \gamma \\ \mathcal{P}_{\gamma,\delta'}(q) + q\mathcal{P}_{s\times\gamma,\delta'}(q) & \text{if } s \times \gamma > \gamma \\ q\mathcal{P}_{\gamma,\delta'}(q) + \mathcal{P}_{s\times\gamma,\delta'}(q) & \text{if } s \times \gamma < \gamma. \end{array}$$

REMARK 3.1.2. As discussed in §2.4, the structure on our poset $\mathcal{D}$ is analogous to that on a Bruhat poset. As a result, the above definition of Lusztig-Vogan polynomials is formally identical to the definition of *relative Kazhdan-Lusztig polynomials* given in [**4**] and [**8**] for the Bruhat poset associated to a complex, semisimple Lie algebra and its associated parabolic subalgebra. For a general G, one can still define Lusztig-Vogan polynomials, but several additional recursion formulas must be given; see [**30**,(6.14)].

We may also introduce what we call the *inverse Lusztig-Vogan polynomials.* Recall the notion of length for elements of $\mathcal{D}$ from (2.4.3). Call a total ordering $\delta_1, \ldots, \delta_r$ of $\mathcal{D}$ *good* if for any pair δ_a and δ_b with $\ell(\delta_a) > \ell(\delta_b)$ we have $a < b$. In particular we will have $\delta_1 = \delta_{\max}$ and $\delta_r = \delta_{\min}$. Let us fix a good total ordering $(\delta_1, \ldots, \delta_r)$. Then with respect to this ordering we may introduce the matrix $\mathcal{E}(q)$ with (γ, δ)-entry $(-1)^{\ell(\delta)-\ell(\gamma)} q^{(\ell(\delta)-\ell(\gamma))/2} \mathcal{P}^\wedge_{\gamma,\delta}(q)$. By definition, this matrix is unipotent, upper triangular, hence invertible. Define $\mathcal{Q}_{\gamma,\delta}(q)$ to be the polynomials such that $\mathcal{E}(q)^{-1}$ has (γ,δ)-entry $q^{(\ell(\delta)-\ell(\gamma))/2}\mathcal{Q}^{\wedge}_{\gamma,\delta}(q)$. One can also define these polynomials by recursion formulas.

We now introduce a free $\mathbb{Z}[q^{1/2},q^{-1/2}]$-module $\mathcal{M}$ with basis $\mathcal{D}$ and an action of the Hecke algebra $\mathcal{H}$ on $\mathcal{M}$ such that when q is set equal to 1 we recover Gr_∞. This was done by Lusztig and Vogan in [**25**] for any real semisimple matrix group and underlies their approach to the Kazhdan-Lusztig conjecture. Our $\mathcal{M}$ will differ from theirs and is adapted to the applications we have in mind. It is modelled after the Hecke module introduced for the categories $\mathcal{O}_{\mathfrak{p}}$ in [**19**].

Let $\mathcal{M}$ be the free $\mathbb{Z}[q^{1/2},q^{-1/2}]$-module on the basis $\{l(\delta):\delta\in\mathcal{D}\}$. We will refer to this set as the *simple basis* of $\mathcal{M}$. There is a surjection of abelian groups from $\mathcal{M}$ onto Gr_∞ with $q^il(\delta)$ sent to $[L(\delta)]$ for all i and δ. We wish to describe a basis $\{m(\delta):\delta\in\mathcal{D}\}$ of $\mathcal{M}$ as a free $\mathbb{Z}[q^{1/2},q^{-1/2}]$-module whose image in Gr_∞ is the standard basis. We use the inverse Lusztig-Vogan polynomials $\mathcal{Q}_{\gamma,\delta}(q)$ for this purpose.

DEFINITION 3.1.3. For $\delta\in\mathcal{D}$, let $m=\Sigma_{\gamma\in\mathcal{D}}q^{(\ell(\delta)-\ell(\gamma))/2}\mathcal{Q}^{\wedge}_{\gamma,\delta}(q)l(\gamma)$.

REMARK 3.1.4. It follows from the definition of the inverse polynomials that the set $\{m(\delta):\delta\in\mathcal{D}\}$ is a basis of $\mathcal{M}$, which we call the *standard basis*, and that

$$l(\delta)=\sum_{\gamma\in\mathcal{D}}(-1)^{(\ell(\delta)-\ell(\gamma))}q^{(\ell(\delta)-\ell(\gamma))/2}\mathcal{P}^{\wedge}_{\gamma,\delta}(q)m(\gamma).$$

We also wish to introduce an action of the operators θ_s on $\mathcal{M}$ as $\mathbb{Z}[q^{1/2},q^{-1/2}]$-module endomorphisms which is compatible with their action on Gr_∞. (Note: The notation 'θ_s' will ultimately have three roles; the desired interpretation will always be clear from context.)

DEFINITION 3.1.5. For each $s\in S$ let θ_s be the $\mathbb{Z}[q^{1/2},q^{-1/2}]$-module endomorphism of $\mathcal{M}$ defined for each $\delta\in\mathcal{D}$ by

$$\begin{aligned}\theta_s m(\delta)&=m(s\times\delta)+q^{-1/2}m(\delta) &&\text{if } s\times\delta>\delta\\ &=m(s\times\delta)+q^{1/2}m(\delta) &&\text{if } s\times\delta<\delta\\ &=0 &&\text{if } s\times\delta=\delta.\end{aligned}$$

The action of θ_s on the simple basis can be precisely described. The proof is a calculation using the defining relations of the Lusztig-Vogan polynomials. One may compare the proof of the analogous statement in [**19**, (2.2.1)].

THEOREM 3.1.6. *Given $s\in S$ and $\delta\in\mathcal{D}$, we have:*

$$\theta_s l(\delta)=(q^{1/2}+q^{-1/2})l(\delta)+l(s\times\delta)+\sum_{\substack{\gamma\in\mathcal{D}\\ s\times\gamma<\gamma \text{ or } s\times\gamma=\gamma}}\mu(\gamma,\delta)l(\gamma)$$

if $s\times\delta>\delta$ and 0 otherwise.

REMARK 3.1.7. Recall that there is associated to W the Hecke algebra $\mathcal{H}$, defined as the $\mathbb{Z}[q^{1/2}, q^{-1/2}]$-algebra with generators $\{T_s : s \in S\}$ and relations

(1) $T_s T_{s'} \ldots = T_{s'} T_s \ldots$ (n terms each) if $(ss')^n = e$ in W.
(2) $(T_s + 1)(T_s - q) = 0$ for all $s \in S$.

The action of the θ_s's on $\mathcal{M}$ provides $\mathcal{M}$ with the structure of an $\mathcal{H}$-module, where T_s acts as $q^{1/2}\theta_s - 1$. This may be proved as in [**19**,§2.5] (which follows the proof of Deodhar in [**8**]) and is the reason for the terminology 'Hecke module'.

Let us introduce still another basis of $\mathcal{M}$.

DEFINITION 3.1.8. For $\gamma \in \mathcal{D}$, let $p(\gamma) = \Sigma_{\delta \in \mathcal{D}} q^{(\ell(\delta) - \ell(\gamma))/2} \mathcal{Q}^{\wedge}_{\gamma,\delta}(q) m(\delta)$. Call $\{p(\gamma) : \gamma \in \mathcal{D}\}$ the projective basis of $\mathcal{M}$.

REMARK 3.1.9. Notice that the set $\{p(\gamma) : \gamma \in \mathcal{D}\}$ is a basis of $\mathcal{M}$, since in the definition of each $p(\gamma)$ the sum is taken only over those δ's with $\gamma \leq \delta$. In fact, by definition, the change of basis matrix from the standard basis to the projective basis is ${}^t\mathcal{E}^{-1}(q)$, the transpose of the change of basis matrix from the simple basis to the standard basis.

THEOREM 3.1.10. *Given $\gamma \in \mathcal{D}$ and $s \in S$, we have*

$$\begin{aligned} \theta_s p(\gamma) &= (q^{1/2} + q^{-1/2}) p(\gamma) && \textit{if } s \times \gamma > \gamma \\ &= p(s \times \gamma) + \sum_{\substack{\delta \in \mathcal{D} \\ \delta \geq \gamma, s \times \delta > \delta}} \mu(\gamma, \delta) p(\delta) && \textit{if } s \times \gamma < \gamma \\ &= \sum_{\substack{\delta \in \mathcal{D} \\ \delta \geq \gamma, s \times \delta > \delta}} \mu(\gamma, \delta) p(\delta) && \textit{if } s \times \gamma = \gamma \end{aligned}$$

PROOF. The proof is simply a matter of rewriting the $p(\gamma)$'s in the desired formula in terms of the standard basis. Comparing coefficients for the second and third formulas shows that the formula is equivalent to the polynomials $\mathcal{Q}_{\gamma,\delta}$ satisfying their defining relations. The details of the argument are given in [**19**,§3.1] for the case of the category $\mathcal{O}_{\mathfrak{p}}$. ♦

3.2. Filtered modules and the filtered Grothendieck group.

We wish to introduce a filtered version of $\mathcal{HC}_\infty$. First we define a *filtered Harish-Chandra module* X^* to be a Harish-Chandra module X along with a filtration $X = X^r \supseteq X^{r+1} \supseteq \ldots \supseteq X^S = (0)$ such that the successive factors X^i/X^{i+1} are semisimple. A *filtered homomorphism* $f^* : X^* \to Y^*$ is a homomorphism $f : X \to Y$ such that $f(X^i) \subseteq Y^i$. The *filtered category* $\mathcal{HC}^*_\infty$ is the category whose objects are filtered Harish-Chandra modules and whose morphisms are filtered homomorphisms. There is a *shift functor* σ defined on $\mathcal{HC}^*_\infty$ by shifting the filtration one step: $(\sigma X)^i = X^{i-1}$. Similarly one can define σ^t for any integer t. We need one more notion for filtered categories. Given a short exact sequence

$$0 \to X \underset{f}{\to} Y \underset{g}{\to} Z \to 0$$

in $\mathcal{HC}_\infty$ and a filtration Y^* on Y, the *induced filtrations* on X and Z from Y^* are given by $X^i = X \cap f^{-1}(Y^i)$ and $Z^i = g(Y^i)$. Let us say that a sequence of maps

$$0 \to X^* \underset{f}{\to} Y^* \underset{g}{\to} Z^* \to 0$$

in $\mathcal{HC}^*_\infty$ is *exact* if the unfiltered version is exact and if X^* and Z^* are the induced filtrations on X and Z. Also, we will refer to X^* as a submodule of Y^* if X is a submodule of Y and the filtration X^* is induced from Y^*.

Define the Grothendieck group Gr^*_∞ of $\mathcal{HC}^*_\infty$, or the *filtered Grothendieck group*, to be the free $\mathbb{Z}$-module with basis $\{[Y^*] : Y^* \text{ a filtered module}\}$ modulo the submodule generated by elements of the form $[Y^*] - [Z^*]$ if $Y^* \cong Z^*$ and of the form $[Y^*] - [X^*] - [Z^*]$ for exact sequences as above. For $\delta \in \mathcal{D}$, let $L(\delta)^*$ be the filtered module whose underlying module is $L(\delta)$, with $L(\delta)^0 = L(\delta)$ and $L(\delta)^1 = (0)$. The set $\{\sigma^i L(\delta)^* : i \in \mathbb{Z}, \delta \in \mathcal{D}\}$ consists of all the simple modules in $\mathcal{HC}^*_\infty$. Then Gr^*_∞ may be identified with the free $\mathbb{Z}[q^{1/2}, q^{-1/2}]$-module $\mathcal{M}$, where we associate $q^{i/2}1(\delta)$ to $\sigma^i L(\delta)^*$. Given a filtered Harish-Chandra module X^*, we call its *filtered character*, denoted $[X^*]$, the associated element in Gr^*_∞:

$$[X^*] = \sum_{i \in \mathbb{Z}} \sum_{\delta \in \mathcal{D}} (X^i / X^{i+1} : L(\delta)) q^{i/2} l(\delta).$$

We wish to associate to each standard module a canonical filtered module. Recall that the *radical* of a module X is the smallest submodule $radX$ such that $X/radX$ is semisimple, and that the *radical filtration* $\{rad^i X : i \geq 0\}$ is defined inductively by $rad^{i+1}X = rad(rad^i X)$.

DEFINITION 3.2.1. For $\delta \in \mathcal{D}$, let $M(\delta)^*$ be the filtered module with underlying module $M(\delta)$ and $M(\delta)^i = rad^i M(\delta)$.

THEOREM 3.2.2. *The filtered character of* $M(\delta)^*$ *is* $m(\delta)$.

PROOF. The Theorem is a re-statement of known results. Casian has proved in [**3**] that the radical filtration of a standard module coincides, up to shifts in indexing, with a weight filtration. But it follows from the work of Lusztig-Vogan in [**25**] that the weight filtration is described by the inverse Kazhdan-Lusztig polynomials, as discussed in [**6**]. ♦

Having identified the modules in $\mathcal{HC}^*_\infty$ which play the role of standard modules, we can also define the analogue of a standard flag.

DEFINITION 3.2.3. A *filtered standard flag* for a filtered module X^* in $\mathcal{HC}^*_\infty$ is a sequence of filtered submodules $0 \subseteq X^*_1 \subseteq \ldots \subseteq X^*_t = X^*$ such that for each i there is an integer $g(i)$ and a $\delta_i \in \mathcal{D}$ such that $X^*_i / X^*_{i-1} \cong \sigma^{g(i)} M(\delta_i)^*$.

Recall the duality functor $\mathbb{D}$ on $\mathcal{HC}_\infty$. One may also introduce a natural involution $\mathbb{D}$ on $\mathcal{M}$ which sends $q^{i/2} l(\delta)$ to $q^{-1/2} l(\delta)$ for any $\delta \in \mathcal{D}$ and $i \in \mathbb{Z}$.

PROPOSITION 3.2.4. *There is a functor $\mathbb{D}$ on $\mathcal{HC}^*_\infty$ such that, given a filtered Harish-Chandra module X^*, the underlying module of $\mathbb{D}X^*$ is $\mathbb{D}X$ and the filtered character of $\mathbb{D}X^*$ is $\mathbb{D}[X^*]$.*

A proof of this may be given as in [**19**,(6.1.6)]. We call the functor $\mathbb{D}$ the duality on $\mathcal{HC}^*_\infty$ and call a filtered module X^* *self-dual* if $\mathbb{D}X^* \cong X^*$. Notice in particular that $L(\delta)^*$ is self-dual for any $\delta \in \mathcal{D}$.

For each $s \in S$ we have a functor θ_s defined on $\mathcal{HC}_\infty$ and an endomorphism θ_s defined on $\mathcal{M}$ which are compatible in the sense that they induce the same action on the Grothendieck group Gr_∞. In fact one may define a functor θ_s on $\mathcal{HC}^*_\infty$ which is compatible with the other θ_s's.

THEOREM 3.2.5. *For each $s \in S$ there is a functor θ_s on $\mathcal{HC}^*_\infty$ such that given a filtered Harish-Chandra module X^*, the following holds:*

(i) *The underlying module of $\theta_s X^*$ is $\theta_s X$.*
(ii) $[\theta_s X^*] = \theta_s[X^*]$.

PROOF. The starting point in the proof is to show, given a module X, that the Loewy length of $\theta_s X$ is at most two more than the Loewy length of X. The proof of this is identical to the proof given in [**18**, appendix A], beginning with the case of Loewy length 1, which is (2.4.10)(iv). Given a filtered module X^*, we define the filtration $(\theta_s X)^*$ as follows: Let π_i be the composition of the canonical surjections of $\theta_s(X^i)$ to $\theta_s(X^i/X^{i+1})$ and of $\theta_s(X^i/X^{i+1})$ to its largest semisimple quotient. Let κ_{i-1} be the composition of the natural map of X^{i-1} to $\theta_s(X^{i-1})$ (arising via the adjoint functors of translation to and from the s-wall) and the map of $\theta_s(X^{i-1})$ into $\theta_s(X^i)$. Set $(\theta_s X)^i = Ker\pi_i + Im\kappa_{i-1}$. (Compare the diagram on page 77 of [**8**].) It is straightforward to check that $(\theta_s X)^i/(\theta_s X)^{i+1}$ is semisimple, using the fact that θ_s increases Loewy length by at most two, and that θ_s is functorial. ♦

In particular, the Theorem provides a canonical filtration on $\theta_s M(\delta)$ for each δ. Let us analyze this case more closely.

THEOREM 3.2.6. *Let $\delta \in \mathcal{D}$ and $s \in S$ satisfy $s \times \delta > \delta$. Then in $\mathcal{HC}^*_\infty$ there is a short exact sequence*

$$0 \to M(s \times \delta)^* \to \theta_s M(\delta)^* \to \sigma^{-1} M(\delta)^* \to 0.$$

The proof of this will require that we have a way to compare filtered characters.

DEFINITION 3.2.7. (1) Let $F(q)$ and $G(q)$ be two elements of $\mathbb{Z}[q^{1/2}, q^{-1/2}]$ with $F(q) = a_i q^{i/2} + a_{i+1} q^{(i+1)/2} + \dots$ and $G(q) = b_i q^{i/2} + b_{i+1} q^{(i+1)/2} + \dots$. Then we say that $F(q)$ dominates $G(q)$, written $F(q) \geq G(q)$, if the following three conditions hold:

(i) The coefficients of F and G are all non-negative.
(ii) $F(1) = G(1)$.
(iii) For each $j \geq i$, the inequality $a_i + a_{i+1} + \dots + a_j \leq b_i + b_{i+1} + \dots + b_j$ holds.

(2) Let m and m' be two elements of $\mathcal{M}$ with $m = \Sigma_{\delta\in\mathcal{D}}F(\delta,q)l(\delta)$ and $m' = \Sigma_{\delta\in\mathcal{D}}G(\delta,q)l(\delta)$. Then m dominates m', written $m \geq m'$, if $F(\delta,q) \geq G(\delta,q)$ for all $\delta \in \mathcal{D}$.

REMARK 3.2.8. (1) If $m_1 > m_2$ and $m_3 \geq m_4$ then $m_1 + m_3 > m_2 + m_4$.

(2) Let X^* be a filtered module such that $X^{i+t} = rad^i X$ for some integer t and all i and let Y^* be another filtration of X with $Y^t = X^t = X$. Then $[X^*] \geq [Y^*]$, and equality holds only if $X^* = Y^*$.

PROOF OF 3.2.6. By (2.4.9), we have a short exact sequence $0 \to M(s\times\delta) \to \theta_s M(\delta) \to M(\delta) \to 0$ in $\mathcal{HC}_\infty$. With respect to these homomorphisms and with $\theta_s M(\delta)^*$ as the middle term, let X^* be $M(s\times\delta)$ with the induced filtration and let Y^* be $M(\delta)$ with the induced filtration. Then

$$m(s\times\delta) + q^{-1/2}m(\delta) = \theta_s m(\delta) = [X^*] + [Y^*].$$

The coefficients of $l(s\times\delta)$ and $l(\delta)$ in $\theta_s m(\delta)$ are the same as their coefficients in $\theta_s l(\delta)$, namely, 1 and $(q^{1/2}+q^{-1/2})$, respectively. Thus we must have $(X^0/X^1 : L(s\times\delta)) = 1$ and $(Y^{-1}/Y^0 : L(\delta)) = 1$. Remark (6.3.4)(2) above then implies that $m(s\times\delta) \geq [X^*]$ and $q^{-1/2}m(\delta) \geq [Y^*]$. If either of these inequalities is strict, then we would have the strict inequality $m^*(s\times\delta) + q^{-1/2}m(\delta) > [X^*] + [Y^*]$, a contradiction. Thus the inequalities are equalities, proving the Theorem. ♦

COROLLARY 3.2.9. *Let $s \in S$ and let γ and δ be elements of $\mathcal{D}$ such that $s\times\delta > \delta$ and $s \in \tau(\gamma)$. Then $(\theta_s L(\delta) : L(\gamma)) = \mu(\gamma,\delta)$.*

PROOF. Using Theorem 3.2.6 and Corollary 3.2.9, we obtain the sequence of equalities:

$$\begin{aligned}\mu(\gamma,\delta) &= (M(\delta)^1/M(\delta)^2 : L(\gamma)) \\ &= (\theta_s(M(\delta)^0)/\theta_s(M(\delta)^1) : L(\gamma)) = (\theta_s L(\delta) : L(\gamma)).\end{aligned}$$

♦

In preparation for section 4, we need some additional results on the filtered category. We begin with a trivial observation.

LEMMA 3.2.10. *Let $f^* : X^* \to Y^*$ be a filtered homomorphism and let f be the underlying module homomorpism.*

(i) *Assume f is $1-1$ and let $X^\#$ be the filtration on X induced by f. Then $[X^\#] \leq [X^*]$.*
(ii) *Assume f is surjective and let $Y^\#$ be the filtration on Y induced by f. Then $[Y^*] \leq [Y^\#]$.* ♦

We wish to apply this to direct sums. Given two filtered modules X^* and Y^*, we define their direct sum in $\mathcal{HC}^*_\infty$ in the obvious way, with $(X\oplus Y)^i = X^i \oplus Y^i$. It is a direct sum in the categorical sense. Moreover, the Krull-Schmidt Theorem holds for $\mathcal{HC}^*_\infty$. (See [**20**].) One subtlety however is that it is not apparent that an indecomposable filtered module has as underlying module an indecomposable Harish-Chandra module.

Let $X = X_1 \oplus \ldots \oplus X_r$ be a Harish-Chandra module with a filtration X^*. Let X_i^* be the induced filtration on X_i for each i. We may also form the direct sum $X^{\#}$ of the X_i^*'s. Lemma 3.2.10 has the following consequence, which we will need in 4.2.

COROLLARY 3.2.11. $[X^{\#}] \geq [X^*]$.

PROOF. The direct sum of the embeddings $f_i : X_i^* \to X^*$ yields a filtered homomorphism f of $X^{\#}$ to X^*. Applying the Lemma yields the result. ♦

3.3 A Hecke Module map.

As described in [19], we can carry out the constructions of §3.2 to obtain a Hecke module $\mathcal{M}_{\mathcal{O}}$ associated to the category $\mathcal{O}_{\mathfrak{p}}$. In particular, we have the relative Kazhdan-Lusztig polynomials $\mathcal{P}^{\mathcal{O}}_{y,w}$ for y and $w \in W^{\mathfrak{p}}$, the $\mu_{\mathcal{O}}$-function, and the bases $\{m_{\mathcal{O}}(w) : w \in W^{\mathfrak{p}}\}$, $\{l_{\mathcal{O}}(w) : w \in W^{\mathfrak{p}}\}$ and $\{p_{\mathcal{O}}(w) : w \in W^{\mathfrak{p}}\}$ for $\mathcal{M}_{\mathcal{O}}$. With the Hecke modules $\mathcal{M}_{\mathcal{O}}$ and $\mathcal{M}$ in hand, it is possible to construct an interesting Hecke module map relating the two. We let Γ denote the left exact covariant functor on the category of $\mathfrak{g}$-modules defined by

(3.3.1) $$\Gamma(X) = \{v \in X \mid \dim_{\mathbb{C}} \mathfrak{U}(\mathfrak{k}).v < \infty \text{ and } \mathfrak{U}(\mathfrak{k}).v \text{ is a } K\text{-module}\}.$$

We call this the K-finite functor. (Note: Since M and K are both connected, this definition coincides with [29, 6.2.9].) Define $\mathcal{L} : \mathcal{O}_{\mathfrak{p}} \to \mathcal{HC}_\infty$ by

(3.3.2) $$\mathcal{L}(X) = (\Gamma(X^*_{\text{M-finite}}))^*_{\text{K-finite}}.$$

By [29, 6.3.5], we calculate

(3.3.3) $$\begin{aligned}\mathcal{L}(M_{\mathcal{O}}(w)) &= (\Gamma(\mathfrak{U}(\mathfrak{g}) \otimes_{\mathcal{U}(\mathfrak{p})} E(w))^*_{\text{M-finite}})^*_{\text{K-finite}} \\ &= (\Gamma(Hom_{\mathfrak{U}(\mathfrak{p})}(\mathfrak{U}(\mathfrak{g}), E(w)^*)))^*_{\text{K-finite}} \\ &= (I_P(E(w)^*))^*_{\text{K-finite}} \\ &= I_P(w).\end{aligned}$$

Given $w \in W^{\mathfrak{p}}$, recall that a projective module $P_{\mathcal{O}}(w)$ in $\mathcal{O}_{\mathfrak{p}}$ admits a filtration of the form $0 \subseteq Y_1 \subseteq \ldots \subseteq Y_n = P_{\mathcal{O}}(w)$, with $Y_i/Y_{i-1} = M_{\mathcal{O}}(w_i)$; we call this a (generalized) Verma flag for $P_{\mathcal{O}}(w)$. Using the $\mathcal{L}$-functor, we can build modules in $\mathcal{HC}_\infty$ having *principal series flags*.

LEMMA 3.3.4. *Fix $w \in W^P$ and let $0 \subseteq Y_1 \subseteq \ldots \subseteq Y_n = P_{\mathcal{O}}(w)$, with $Y_i/Y_{i-1} = M_{\mathcal{O}}(w_i)$, be a generalized Verma flag for the module $P_{\mathcal{O}}(w)$. Then the module $\mathcal{L}(P_{\mathcal{O}}(w))$ admits a filtration $0 \subseteq X_1 \subseteq \ldots \subseteq X_n = \mathcal{L}(P_{\mathcal{O}}(w))$ having the property that $X_i/X_{i-1} = I_P(w_i)$.*

PROOF. By [29, 6.3.5], $\mathcal{L}$ is exact on modules with a flag by modules of the form $M_{\mathcal{O}}(y)$. Now apply (3.5.3). ♦

We conclude this section by presenting the Hecke module analog of 3.3.4. Since the functor $\mathcal{L}$ is not exact, we may consider the Euler characteristic of its higher derived functors and get a $\mathbb{Z}$-linear map of integral Grothendieck groups

$$\chi(\mathcal{L}) : K(\mathcal{O}_P) \to K(\mathcal{HC}_\infty). \tag{3.3.5}$$

As described in [**6**, §5], there is a geometric construction of $\mathcal{L}$. Thus we can use standard geometric dictionaries and reduction mod p to replace our two $\mathfrak{g}$-module categories by categories of perverse sheaves carrying a Frobenius action. The version of $\mathcal{L}$ defined in characteristic p leads to an analog of (3.3.5), in which the Grothendieck groups are identified with our Hecke modules $\mathcal{M}_\mathcal{O}$ and $\mathcal{M}$. This gives us a $\mathbb{Z}$-linear map

$$\mathbb{L} : \mathcal{M}_\mathcal{O} \to \mathcal{M} \tag{3.3.6}$$

and [**6**, (5.38)] shows that this is a Hecke module map. There is a sleight of hand here, in that the Hecke modules of [**6**] and the ones we consider look quite different upon first glance. However, one can show that these Hecke modules are anti-isomorphic to the ones we have defined in this paper.

LEMMA 3.3.7. *Let $w \in W^\mathfrak{p}$ be a good representative for δ. Then $\mathbb{L}(m_\mathcal{O}(w)) = m(\delta)$ and $\mathbb{L}(p_\mathcal{O}(w)) = p(\delta) + \Sigma_{\varepsilon\in\mathcal{D},\,\varepsilon>\delta} k_\varepsilon p(\varepsilon)$, for non-negative integers k_ε.*

PROOF. The first statement follows from the main theorem in [**6**]. The second statement follows from the fact that $\mathbb{L}$ is a Hecke module homomorphism; in particular, $\mathbb{L}$ commutes with the actions of θ_s on $\mathcal{M}$ and $\mathcal{M}_\mathcal{O}$. Combining this with Theorem 3.1.10 and its analogue for $\mathcal{M}_\mathcal{O}$ yields the result. ♦

3.4 The $\mathcal{O}$-type hypothesis.

In this section, we formulate a condition which will allow the construction of pseudo-projective modules in the category $\mathcal{HC}_\infty$ in section 4. This condition will relate the parameter sets $W^\mathfrak{p}$ and $\mathcal{D}$ (associated to the categories $\mathcal{O}_\mathfrak{p}$ and $\mathcal{HC}_\infty$) in a manner which leads to a relationship between the $p_\mathcal{O}(w)$-basis of $\mathcal{M}_\mathfrak{p}$ and the $p(\delta)$-basis of $\mathcal{M}$. This relationship is realized via the Hecke module map $\mathcal{L}$.

Keeping the case of $\mathcal{O}_\mathfrak{p}$ in mind, it would be desirable if the principal series flag on $\mathcal{L}(P_\mathcal{O}(w))$ in (3.3.4) were actually a standard flag; i.e. each $I_P(w_i) = M(w_i)$. Similarly, it would be desirable if $\mathbb{L}(p_\mathcal{O}(w)) = p(\delta)$ for some good representative w of δ. This scenario is not always the case, but we will formulate a criterion in the next section which ultimately leads to it.

Recall that $\mathcal{D}$ consists of the $W(A)$-orbits in $W^\mathfrak{p}$. The sets $\mathcal{D}$ and $W^\mathfrak{p}$ can be related on an elementary level as follows: On the one hand, given $\delta = \{w_1, \dots, w_k\} \in \mathcal{D}$, recall from §2.2 that a good representative w_δ of δ is a maximal length element of δ. Given a $\delta \in \mathcal{D}$, the choice of good representative w_δ is not necessarily unique; however, given two good representatives w_δ and $w_{\delta'}$ of δ, we have $I_P(w_\delta) = I_P(w_{\delta'}) = M(\delta)$. On the other hand, given any $y \in W^\mathfrak{p}$, we define δ_y to be the unique $W(A)$-orbit in $W^\mathfrak{p}$ containing y; in particular, if w_δ

is a good representative for δ, then $\delta_{w_\delta} = \delta$. The main idea in what follows is to refine the notion of good representative so as to make the Lusztig-Vogan combinatorics of $\delta \in \mathcal{D}$ look like the analogous Kazhdan-Lusztig combinatorics of w_δ. Making this precise will involve a local condition relating the Kazhdan-Lusztig polynomials attached to δ and w_δ, and a global condition, which pastes these local conditions together in a manner which is compatible with the application of translation functors in the categories $\mathcal{O}_\mathfrak{p}$ and $\mathcal{HC}_\infty$.

We first describe a local condition relating $W^\mathfrak{p}$ and $\mathcal{D}$ parameters. Given $\delta \in \mathcal{D}$, we define

$$\mathcal{K}(\delta) = \{\varepsilon \in \mathcal{D} | \mu(\delta, \varepsilon) \neq 0\}, \tag{3.4.1}$$

which we refer to as the *δ-prism.* In an analogous way, using the function $\mu_\mathcal{O}$ associated to the category $\mathcal{O}_\mathfrak{p}$, we can define the *w-prism* $\mathcal{K}(w) \subseteq W^\mathfrak{p}$ associated to $w \in W^\mathfrak{p}$. We say that the prism $K(w)$ *s-covers* the prism $\mathcal{K}(\delta)$, denoted $\mathcal{K}(w) \xrightarrow{s} \mathcal{K}(\delta)$, if the following conditions hold:

(3.4.2)
(i) For every $z \in \mathcal{K}(w), z$ is a good representative for δ_z;
(ii) The map $z \to \delta_z$ is a bijection of $\mathcal{K}(w)$ onto $\mathcal{K}(\delta)$;
(iii) For every $z \in \mathcal{K}(w)$, $s \in \tau(z)$ if and only if $s \in \tau(\delta_z)$;
(iv) If $z \in \mathcal{K}(w)$ and $s \notin \tau(z)$, then $\mu_\mathcal{O}(w, z) \geq \mu(\delta_w, \delta_z)$.

Next, given $w \in W^\mathfrak{p}$, we define a *w-chain* to be a tuple $\mathbf{c}(w) = (w_1, w_2, \dots, w_u)$, with $w_1 = w$ and $w_u = w_{\max}$, satisfying the conditions

(3.4.3)
(i) $w_j \in W^\mathfrak{p}$ for $1 \leq j \leq u$;
(ii) $w_j < w_{j+1}$ for $1 \leq j \leq u-1$;
(ii) $w_{j+1} = s_j.w_j$for some $s_j \in S$.

Given $\delta \in \mathcal{D}$, we define a *δ-chain* to be a tuple $\mathbf{c}(\delta) = (\delta_1, \delta_2, \dots, \delta_v)$, with $\delta_1 = \delta$ and $\delta_v = \delta_{max}$, satisfying the conditions

(3.4.4)
(i) $\delta_j \in \mathcal{D}$ for $1 \leq j \leq v$;
(i) $\delta_j < \delta_{j+1}$ for $1 \leq j \leq v-1$;
(ii)$\delta_{j+1} = s'_j \times \delta_j$ for some $s'_j \in S$.

A chain $\mathbf{c}(w) = (w_1, w_2, \dots, w_u)$ is said to *cover* the chain $\mathbf{c}(\delta) = (\delta_1, \delta_2, \dots, \delta_v)$, denoted $\mathbf{c}(w) \to \mathbf{c}(\delta)$, if

(3.4.5)
(i) $u = v$ and $s_j = s'_j$, for all $1 \leq j \leq u-1$;
(ii) w is a good representative for δ_w.

Henceforth, denote this w by w_δ. Whenever $\mathbf{c}(w) \to \mathbf{c}(\delta)$, it easily follows that each w_i in the chain $\mathbf{c}(w)$ is a good representative for the element δ_{w_i} in $\mathbf{c}(\delta)$.

We now paste together the prism and chain condition to arrive at the notion of a weak $\mathcal{O}$-type.

DEFINITION 3.4.6. A parameter $\delta \in \mathcal{D}$ is of *weak $\mathcal{O}$-type* if there exists a four-tuple $(\delta, w_\delta, \mathbf{c}(\delta), \mathbf{c}(w_\delta))$, where w_δ is a good representative for δ and the chains $\mathbf{c}(\delta)$ and $\mathbf{c}(w_\delta)$ satisfy

(i) $\mathbf{c}(w_\delta) \to \mathbf{c}(\delta)$;

(ii) For every w_j in the chain $\mathbf{c}(w_\delta), \mathcal{K}(w_j) \xrightarrow{s_{j-1}} \mathcal{K}(\delta_{w_j})$.

If δ is of weak $\mathcal{O}$-type, we call $(\delta, w_\delta, \mathbf{c}(\delta), \mathbf{c}(w_\delta))$ an $\mathcal{O}$-type of δ. We remark that an $\mathcal{O}$-type for δ is not necessarily unique.

Recall from the introduction the notion of a standard flag of $\mathcal{O}$-type: Given a module X with a standard flag, one can use Proposition 2.5.5 to re-arrange the flag so that the new flag has submodules $Y \subseteq Z$ with all standard subquotients in Z/Y isomorphic to $M(\varepsilon)$ and no standard subquotients of Y or X/Z isomorphic to $M(\varepsilon)$. The standard flag on X is of $\mathcal{O}$-type if for every ε the resulting module X/Z is a direct sum of $M(\varepsilon)$'s.

DEFINITION 3.4.7. A parameter $\delta \in \mathcal{D}$ is of $\mathcal{O}$-type if it has an $\mathcal{O}$-type $(\delta, w_\delta, \mathbf{c}(\delta), \mathbf{c}(w_\delta))$ such that the standard flag on $\theta M(\delta_{\max})$ is of $\mathcal{O}$-type, where $\theta = \theta_{s_1} \circ \ldots \circ \theta_{s_{u-1}}$ and $s_1, \ldots, s_u$ are as in 3.4.3.

HYPOTHESIS 3.4.8. *Every element of $\mathcal{D}$ is of $\mathcal{O}$-type.*

Observe that if δ is of weak $\mathcal{O}$-type $(\delta, w_\delta, \mathbf{c}(\delta), \mathbf{c}(w_\delta))$, then every element γ in the chain $\mathbf{c}(\delta)$, is also of weak $\mathcal{O}$-type, with $\mathcal{O}$-type the appropriate truncation of $(\delta, w_\delta, \mathbf{c}(\delta), \mathbf{c}(w_\delta))$. If $\delta_{\min}$ is of $\mathcal{O}$-type, then one might initially hope that for every $\delta \in \mathcal{D}$ there exists an $\mathcal{O}$-type $(\delta_{\min}, w_\delta, \mathbf{c}(\delta), \mathbf{c}(w_\delta))$ such that δ lies in the chain $\mathbf{c}(\delta_{\min})$; i.e., there are enough $\mathcal{O}$-types associated to $\delta_{\min}$ to cover $\mathcal{D}$. If this were the case, then the hypothesis would hold. Unfortunately, this need not be the case. As an example, consider $SL(3, \mathbb{H})$, where $\mathcal{D}$ is parameterized in Figure 3. Then one can verify that (3.6.7) holds, but for every $\mathcal{O}$-type $(\delta_{15}, w_{\delta_{15}}, \mathbf{c}(\delta_{15}), \mathbf{c}(w_{\delta_{15}}))$, we find that $\delta_{14} \notin c(\delta_{15})$. The hypothesis is easy to verify in the case of a complex group; in this case, $W^{\mathfrak{p}} = W \times W$ and we may take our prisms and chains within the subgroup $1 \times W$ or the subgroup $W \times 1$. The hypothesis has also been verified for $Spin(2n+1, 1)$, $SL(3, \mathbb{H})$, and $E_{6,-26}$.

With the notion of $\mathcal{O}$-types available, we can return to the question raised at the end of 3.3.

LEMMA 3.4.9. *Assume δ has weak $\mathcal{O}$-type. Then $\mathcal{L}(P_{\mathcal{O}}(w_\delta))$ admits a standard module flag.*

PROOF. The Verma flag factors of $P_{\mathcal{O}}(w_\delta)$ are determined by elements in the prism of w_δ, all of which are good representatives by our hypothesis. ♦

The modules $\mathcal{L}(P_{\mathcal{O}}(w_\delta))$ are the pseudo-projective modules we desire. We will study them in more detail in §4, obtaining applications in §5.

4. The pseudo-projective modules

4.1 Construction of the pseudo-projectives..

In this section we will construct the desired pseudo-projective modules in $\mathcal{HC}_\infty$, under the hypothesis of 3.4.7. First we introduce some notation and definitions. Given a chain $\mathbf{c}(w)$ as in 3.4.4, denote by $|\mathbf{c}(w)|$ the product $s_1 s_2 \ldots s_{u-1}$. This is an element of W, with possibly many reduced factorizations, but we will always use the factorization associated to the given chain when we work with a specific reduced expression for $|\mathbf{c}(w)|$. We define $|\mathbf{c}(w)|$ analogously. Notice for an $\mathcal{O}$-type $(\delta, w_\delta, \mathbf{c}(\delta), \mathbf{c}(w_\delta))$ that $|\mathbf{c}(\delta)| = |\mathbf{c}(w_\delta)|$.

DEFINITION 4.1.1. (i) A module P in $\mathcal{HC}_\infty$ is a *pseudo-projective cover* of $L(\gamma)$ if

(i) P has simple cap $L(\gamma)$.
(ii) P has a standard flag of $\mathcal{O}$-type.
(iii) P satisfies BGG reciprocity: $(P : M(\delta)) = (M(\delta) : L(\gamma))$ for all $\delta \in \mathcal{D}$.

More generally, a module P is *pseudo-projective* if it is a direct sum of pseudo-projective covers of simple modules. We call P a *weak pseudo-projective cover* if (i) and (iii) hold and P has a standard flag (not necessarily of $\mathcal{O}$-type.) Let us note a projective property of pseudo-projectives.

LEMMA 4.1.2. *Let $P(\gamma)$ be a pseudo-projective cover of $L(\gamma)$. Let $\delta \in \mathcal{D}$. Then*

(i) $\dim Hom_{\mathcal{HC}}(P(\gamma), \mathbb{D}M(\delta)) = (\mathbb{D}M(\delta) : L(\gamma))$.
(ii) *Suppose $\gamma \neq \delta$. If $s \in S$ and $s \times \delta > \delta$, then* $\dim Hom_{\mathcal{HC}}(P(\gamma), \theta_s L(\delta)) = (\theta_s L(\delta) : L(\gamma))$.

PROOF. By BGG reciprocity and the self-duality of the simple modules, $(\mathbb{D}M(\delta) : L(\gamma)) = (P(\gamma) : M(\delta))$. But an argument similar to that in [**18**] or [**21**], using the fact that the standard flag on $P(\gamma)$ is of $\mathcal{O}$-type, shows that $(P(\gamma) : M(\delta)) = \dim Hom_{\mathcal{HC}}(P, \mathbb{D}M(\delta))$. For (ii), there are two cases. If $\gamma = s \times \delta$, then both sides are clearly 1. Otherwise, we may replace $\theta_s L(\delta)$ by the unique submodule X modulo which one obtains a length two module with $L(s \times \delta)$ as socle. Then X embeds in $\mathbb{D}M(\delta)$ and we are reduced to (i). ♦

REMARK 4.1.3. Note that the modules X_s and X_t for $SL(2, \mathbb{C})$ discussed in the introduction are both pseudo-projective covers of $L(\delta)$. In particular, pseudo-projective covers need not be unique. Also, the discussion shows that (ii) may fail if $\gamma = \delta$.

Let us also recall the indecomposable projective translation functors of Bernstein and Gelfand. Fix an infinitesimal character χ of $\mathfrak{g}$ and let F_χ be the finite-dimensional simple $\mathfrak{g}$-module with χ as infinitesimal character. Let $V(\chi)$ be the Verma module with F_χ as simple cap and let $Q(\chi)$ be its projective cover in $\mathcal{O}$. It follows from work of Bernstein and Gelfand on projective functors [**1**] that for each $w \in W$ there exists a functor θ_w on the category of finitely-generated

$\mathfrak{g}$-modules with generalized infinitesimal character χ with the following properties: (a) Let $t_1 \dots t_u$ be a reduced expression for w. Then θ_w is a summand of the composition of translation functors $\theta_{t_1} \circ \dots \circ \theta_{t_u}$. (b) If $s \in S$ and $ws > w$, then $\theta_s \circ \theta_w = \theta_{ws} \oplus \bigoplus_{z, zs<z, z<w} \mu_0(z,w)\theta_z$. (c) $\theta_w V(\chi) = Q(w.\chi)$. Here the μ_0-function is the Kazhdan-Lusztig μ-function for W.

Given $w \in W^{\mathfrak{p}}$ and a chain $\mathbf{c}(w)$, we have

$$\theta_{|\mathbf{c}(w)|} M_{\mathcal{O}}(w_{\max}) = P_{\mathcal{O}}(w). \tag{4.1.4}$$

This motivates the following definition:

DEFINITION 4.1.5. Let $\gamma \in \mathcal{D}$ and let $\mathbf{c}(\gamma)$ be a chain. Then $P(\gamma, |\mathbf{c}(\gamma)|)$ is the module $\theta_{|\mathbf{c}(\gamma)|} M(\delta_{\max})$.

REMARK 4.1.6. Because translation functors commute with the functor $\mathbb{L}$, we may also describe $P(\gamma, |\mathbf{c}(\gamma)|)$ as $\mathbb{L}(P_{\mathcal{O}}(|\mathbf{c}(\gamma)|)$. It is a natural candidate to be a pseudo-projective cover of $L(\gamma)$, for by 3.4.9 it has a standard flag and it clearly has $L(\gamma)$ in its cap. The key questions are whether the cap of $P(\gamma, |\mathbf{c}(\gamma)|)$ is simple, whether it satisfies BGG reciprocity, and whether the standard flag is of $\mathcal{O}$-type. Notice that by definition $p(\gamma)|_{q=1}$ is the character of a pseudo-projective cover of $L(\gamma)$, if one exists. Thus the problem of BGG reciprocity is related to the question raised at the beginning of §3.4 of whether $\mathbb{L}(p_{\mathcal{O}}(|\mathbf{c}(w_\gamma)|)) = p(\gamma)$. As noted there, this need not be the case. Nor need it be the case that $P(\gamma, |\mathbf{c}(\gamma)|)$ has simple cap or has a standard flag of $\mathcal{O}$-type. For example, if $G = SL(3, \mathbb{H})$ and $\delta = \delta_{10}$ as in the introduction, then with respect to the chain $\mathbf{c}(\delta_{10})$ discussed in the introduction, the module $P(\gamma, |\mathbf{c}(\gamma)|)$ has none of these properties.

THEOREM 4.1.7. *Let $\gamma \in \mathcal{D}$ be of (weak) $\mathcal{O}$-type, with $(\gamma, w_\gamma, \mathbf{c}(\gamma), \mathbf{c}(w_\gamma))$ the $\mathcal{O}$-type. Then*

(i) *$P(\gamma, |\mathbf{c}(\gamma)|)$ is a (weak) pseudo-projective cover of $L(\gamma)$.*
(ii) *$\mathbb{L}(p_{\mathcal{O}}(w_\gamma))) = p(\gamma)$.*

PROOF. Assume γ is of weak $\mathcal{O}$-type. We proceed by induction along the chain $\mathbf{c}(\gamma)$. Explicitly, let $\mathbf{c}(\gamma) = (\gamma_1, \dots, \gamma_k)$ with $\gamma_k = \delta_{\max}$. Each γ_i has an $\mathcal{O}$-type obtained by truncating the $\mathcal{O}$-type for γ, with chain $\mathbf{c}(\gamma_i)$. We will prove by downward induction on i that the Theorem holds for $P(\gamma_i, \mathbf{c}(\gamma_i))$. The induction begins with $\gamma = \delta_{\max}$, in which case $P(\gamma, \mathbf{c}(\delta_{\max})) = M(\delta_{\max})$ and the theorem holds trivially.

We may assume inductively that the Theorem holds for $\delta = \gamma_2$ and we wish to prove it for $\gamma = \gamma_1$. Let $s = s_1$ and $w = |\mathbf{c}(w_\delta)|$; then $\gamma = s \times \delta$ and $|\mathbf{c}(w_\gamma)| = ws$. The decomposition formula for $\theta_s \circ \theta_w$ yields

$$\theta_s P(\delta, \mathbf{c}(\delta)) = P(\gamma, \mathbf{c}(\gamma)) \oplus \bigoplus_{\substack{z \\ zs<z, z<w}} \mu_0(z,w)\theta_z M(\delta_{\max}) \tag{4.1.8}$$

We already know that $P(\gamma, \mathbf{c}(\gamma))$, as well as the other summands on the right side, has a standard flag. Each z in the sum lies in $W^{\mathfrak{p}}$, since $|\mathbf{c}(w_\delta)|$ does. Let

$z' = w_{\max} z w_0$ and let ε_z be the element of $\mathcal{D}$ containing z'. Then $\mu_0(z, w) = \mu_0(ww_0, zw_0) = \mu_{\mathcal{O}}(w', z')$. But by assumption the set $\{z' \in W^{\mathfrak{p}} : z's > z'$ and $\mu_0(w', z') > 0\}$ corresponds bijectively to the set $\{\varepsilon : s \times \varepsilon > \varepsilon$ and $\mu(\delta, \varepsilon) > 0\}$, with ε_z corresponding to z', and $\mu_{\mathcal{O}}(w', z') \geq \mu(\delta, \varepsilon_z)$. In particular, the number of summands on the right side of (4.1.8) is at least $1 + \sum_{z, zs<z, z<w} \mu(\delta, \varepsilon_z)$. On the other hand, using the self-adjointness of θ_s, Lemma 4.1.2, and Corollary 3.2.9, we obtain for $\varepsilon \in \mathcal{D}$ with $s \times \varepsilon > \varepsilon$ that: $(cap\theta_s P(\delta, \mathbf{c}(\delta)) : L(\varepsilon)) = (\theta_s L(\varepsilon) : L(\delta)) = \mu(\delta, \varepsilon)$. In other words, the number of summands of $\theta_s P(\delta)$ in (4.1.8) is at most the number of composition factors in $cap\theta_s P(\delta)$, which is $1 + \Sigma_{z, zs<z, z<w} \mu(\delta, \varepsilon_z)$. This forces each summand of (4.1.8) to be indecomposable with a simple cap. In particular, $P(\gamma, \mathbf{c}(\gamma))$ has a simple cap.

It remains to show that $P(\gamma, \mathbf{c}(\gamma))$ satisfies BGG reciprocity. For notational simplicity, given an element m of $\mathcal{M}$, let us write $m(1)$ for its image in Gr_∞. By Remark 4.1.6, we must show that the character $[P(\gamma, \mathbf{c}(\gamma))]$ equals $p(\gamma)(1)$. We have the formulas

$$\theta_s p(\delta) = p(\gamma) + \sum_{\substack{\varepsilon \\ s \times \varepsilon > \varepsilon}} \mu(\delta, \varepsilon) p(\varepsilon)$$

and

$$\theta_s p_{\mathcal{O}}(w_\delta) = p(w_\gamma) + \sum_{\substack{z' \\ z's > z'}} \mu_{\mathcal{O}}(w_\delta, z') p_{\mathcal{O}}(z').$$

By induction, $\mathbb{L}(p(w_\delta)) = p(\delta)$. Since $\mathbb{L}$ commutes with θ_s, we obtain

$$\theta_s p(\delta) = \mathbb{L}(p_{\mathcal{O}}(w_\gamma)) + \sum_{\substack{z' \\ z's > z'}} \mu_{\mathcal{O}}(w_\delta, z') \mathbb{L}(p_{\mathcal{O}}(z')).$$

Using the inequalities $\mu_{\mathcal{O}}(w', z') \geq \mu(\delta, \varepsilon_z)$ and Lemma 3.3.7, we conclude that $\mathbb{L}(p_{\mathcal{O}}(w_\gamma)) = p(\gamma)$ and that $\mathbb{L}(p_{\mathcal{O}}(z')) = p(\varepsilon_z)$ for each z in the sum. Using Lemmas 3.3.4 and 3.3.7, we see that $[P(\gamma, \mathbf{c}(\gamma))] = p(\gamma)(1)$ and also $[\theta_z M(\delta_{\max})] = p(\varepsilon_z)(1)$ for each z.

We have proved that $P(\gamma, \mathbf{c}(\gamma))$ is a weak pseudo-projective cover of $L(\gamma)$. Along the way, we have simultaneously shown that $\theta_z M(\delta_{\max})$ is a weak pseudo-projective cover of $L(\varepsilon_z)$. If γ is of $\mathcal{O}$-type, then clearly all these summands are pseudo-projective. This completes the proof. ♦

4.2 Filtrations on the pseudo-projective modules.

In this section we obtain a formula for the filtered character of a pseudo-projective module with respect to its radical filtration. For any indecomposable pseudo-projective module P, let P^* be the filtered module in $\mathcal{HC}^*_\infty$ with underlying module P and filtration $P^i = rad^i P$. We wish to prove the following result:

THEOREM 4.2.1. *Let $\gamma \in \mathcal{D}$ be of $\mathcal{O}$-type, with $\mathcal{O}$-type $(\gamma, w_\gamma, \mathbf{c}(\gamma), \mathbf{c}(w_\gamma))$. Then $[P(\gamma, \mathbf{c}(\gamma))^*] = p(\gamma)$. Equivalently, for $\delta \in \mathcal{D}$, we have $[P(\gamma, \mathbf{c}(\gamma))^* : \sigma^t M(\delta)^*] = (M(\delta)^* : \sigma^t L(\gamma)^*)$.*

We may regard the Theorem as a filtered form of BGG reciprocity. For its proof, we will need the following estimate:

LEMMA 4.2.2. *Let $\gamma \in \mathcal{D}$ and let $P(\gamma)$ be a pseudo-projective cover of $L(\gamma)$. Then $[P(\gamma)^*] \leq p(\gamma)$.*

PROOF. Let $r = (M(\delta) : L(\gamma))$. Then there is a standard flag for $P(\gamma)$ with a term K such that $P(\gamma)/K = P(\gamma, \delta)$ has a direct sum of r copies of $M(\delta)$ as a submodule, and such that all other standard flag factors $M(\varepsilon)$ satisfy $\varepsilon < \delta$. In particular, every composition factor $L(\delta)$ of $P(\gamma, \delta)$ is the cap of a standard flag factor. By Lemma 4.1.2(i) and its proof, $r = \dim Hom_{\mathcal{HC}}(P(\gamma, \delta), \mathbb{D}M(\delta)) = \mathbb{D}M(\delta) : L(\gamma))$. This yields for each i the equality $(P(\gamma, \delta)/rad^{i+1}P(\gamma, \delta) : L(\delta)) = (soc^i\mathbb{D}M(\delta) : L(\gamma)) = (M(\delta)/rad^{i+1}M(\delta) : L(\gamma))$. Equivalently, $(P(\gamma, \delta)^* : \sigma^i L(\delta)^*) = (M(\delta)^* : \sigma^i L(\gamma)^*)$ for all i, where $P(\gamma, \delta)^*$ is the radical filtration. By Theorem 3.2.2, and the definitions of $m(\delta)$ and $p(\gamma)$, we obtain that the caps of the $M(\delta)$'s occurring in a standard flag for $P(\gamma)$ lie in the layers of the radical filtration predicted by $p(\gamma)$, from which the lemma follows. ♦

PROOF OF 4.2.1. We proceed as in the proof of Theorem 4.1.7 by an induction along the chain $\mathbf{c}(\gamma)$. The induction begins, as there, with $\delta_{\max}$, for which the theorem is trivially true. Assume it is true for $\delta = \gamma_2$, and recall the notation of the proof of 4.1.7, which we carry over to this situation. In particular, (4.1.8) yielded a decomposition of $\theta_s P(\delta, \mathbf{c}(\delta))$ into a direct sum of pseudo-projective modules, which we may rewrite in the following form:

$$\theta_s P(\delta, \mathbf{c}(\delta)) = P(\gamma, \mathbf{c}(\gamma)) \oplus \bigoplus_i P(\varepsilon_i). \tag{4.2.3}$$

We will prove the theorem by obtaining the following filtered version of (4.2.3):

$$\theta_s P(\delta, \mathbf{c}(\delta))^* = P(\gamma, \mathbf{c}(\gamma))^* \oplus \bigoplus_i P(\varepsilon_i)^*. \tag{4.2.4}$$

Let f_i be the embedding of $P(\varepsilon_i)$ into $\theta_s P(\delta, \mathbf{c}(\delta))$ and let f be the embedding of $P(\gamma, \mathbf{c}(\gamma))$. With respect to these embeddings, the filtration on $\theta_s P(\delta, \mathbf{c}(\delta)))^*$ induces filtrations on each summand, making each f_i and f into filtered maps. Let us denote the resulting filtrations by $P(\gamma, \mathbf{c}(\gamma))^\#$ and $P(\varepsilon_i)^\#$ and let $P^\#$ in $\mathcal{HC}^*_\infty$ be the direct sum $P(\gamma, \mathbf{c}(\gamma))^\# \oplus \bigoplus_i P(\varepsilon_i)^\#$. Then $f \oplus \oplus_i f_i$ is a filtered map of $P^\#$ into $\theta_s P(\delta, \mathbf{c}(\delta))^*$. This yields the inequality $[\theta_s P(\delta, \mathbf{c}(\delta))^*] \leq [P^\#]$, as in Corollary 3.2.11.

Let us make an estimate for $[P^\#]$. Since $\theta_s L(\delta) = 0$, it is readily seen that the composition factors of $\theta_s P(\delta, \mathbf{c}(\delta))^*$ are of the form $\sigma^t L(\eta)$ with $t \geq 0$. Therefore each $P(\varepsilon_i)^\#$ satisfies $P(\varepsilon_i)^0 = P(\varepsilon_i)$ and similarly for $P(\gamma, \mathbf{c}(\gamma))^\#$. This allows us to apply Remark 3.2.8, concluding that $[P(\varepsilon_i)^\#] \leq [P(\varepsilon_i)^*]$ and $[P(\gamma, \mathbf{c}(\gamma))^\#] \leq [P(\gamma, \mathbf{c}(\gamma))^*]$. Combining this with Theorem 3.1.10, Theorem

3.2.5, and Lemma 4.2.2, we obtain:

$$p(\gamma)+\Sigma_i p(\varepsilon_i)=\theta_s p(\delta)=[\theta_s P(\delta,\mathbf{c}(\delta))^*]\le[P^\#]=[P(\gamma,\mathbf{c}(\gamma))^\#]+\Sigma_i[P(\varepsilon_i)^\#]$$
$$\le[P(\gamma,\mathbf{c}(\gamma))^*]+\sum_i[P(\varepsilon_i)^*]\le p(\gamma)+\Sigma_i p(\varepsilon_i).$$

We conclude that equality holds throughout, so that $[\theta_s P(\delta,\mathbf{c}(\delta))^*]=[P^\#]$. This implies that the filtered map $f\oplus\oplus_i f_i$ is an isomorphism, proving (4.2.4). At the same time, we obtain that $[P(\gamma,\mathbf{c}(\gamma))^*]=p(\gamma)$, proving the theorem, and that $[P(\varepsilon_i)^*]=p(\varepsilon_i)$ for each i. ♦

5. Applications of the pseudo-projective modules

5.1 Self-duality of some pseudo-projectives and characterization of socular simples.

Recall from §2.6 the subsets $\mathcal{X}$, $\mathcal{Y}$, $\mathcal{Z}$ of $\mathcal{D}$. Recall also the integer s, which is the GK dimension of $M(\delta_{\max})$. Let $\mathcal{U}=\{\delta\in\mathcal{D}:P(\delta)\cong\mathbb{D}P(\delta)$ for some pseudo-projective cover $P(\delta)$ of $L(\delta)\}$. It is clear that $\mathcal{U}\subseteq\mathcal{X}$: if $\delta\in\mathcal{U}$, then $L(\delta)=socP(\delta)$. But $P(\delta)$ has a standard flag, so $L(\delta)=socM(\varepsilon)$ for some ε. Thus, recalling Lemma 2.6.5, we have the inclusions $\mathcal{U}\subseteq\mathcal{X}\subseteq\mathcal{Y}\subseteq\mathcal{Z}$.

THEOREM 5.1.1. *Let $\gamma\in\mathcal{Z}$. If γ is of weak $\mathcal{O}$-type, then $\gamma\in\mathcal{U}$. More precisely, let $(\gamma,w_\gamma,\mathbf{c}(\gamma),\mathbf{c}(w_\gamma))$ be an $\mathcal{O}$-type for γ. Then $P(\gamma,\mathbf{c}(\gamma))$ is self-dual.*

PROOF. Our assumption on γ is that $d(L(\gamma))=s$. Let $x=|\mathbf{c}(\gamma)|$. By definition, $P(\gamma,\mathbf{c}(\gamma))=\theta_x M(\delta_{\max})$, which is a summand of $\theta_{s_1}\circ\ldots\circ\theta_{s_k}M(\delta_{\max})$ for a reduced expression $s_1\ldots s_k$ of x. By Theorem 2.6.3, the socle L of $M(\delta_{\max})$ has GK dimension s and $d(M(\delta_{\max})/L)<s$. Since translation cannot increase GK dimension, we find that $P(\gamma,\mathbf{c}(\gamma))$ lies inside the submodule $\theta_{s_1}\circ\ldots\circ\theta_{s_k}L$ of $\theta_{s_1}\circ\ldots\circ\theta_{s_k}M(\delta_{\max})$. In particular, $P(\gamma,\mathbf{c}(\gamma))$ is an indecomposable summand of $\theta_{s_1}\circ\ldots\circ\theta_{s_k}L$. The definition of θ_x implies that $\theta_x L$ is also a summand of $\theta_{s_1}\circ\ldots\circ\theta_{s_k}L$, implying that $P(\gamma,\mathbf{c}(\gamma))=\theta_x L$. Since each translation functor θ_s on $\mathcal{HC}_\infty$ commutes with the duality $\mathbb{D}$, an induction argument on $\ell(x)$ shows that θ_x commutes with $\mathbb{D}$ as well. Since L is self-dual, we conclude that $\theta_x L$ is self-dual, proving the theorem. ♦

COROLLARY 5.1.2. *(i) Let $\gamma\in\mathcal{D}$ have $\mathcal{O}$-type $(\gamma,w_\gamma,\mathbf{c}(\gamma),\mathbf{c}(w_\gamma))$. Then $L(\gamma)$ is socular if and only if $P(\gamma,\mathbf{c}(\gamma))$ is self-dual.*

(ii) If every $\gamma\in\mathcal{D}$ is of weak $\mathcal{O}$-type, then $\mathcal{U}=\mathcal{X}=\mathcal{Y}=\mathcal{Z}$. ♦

REMARK 5.1.3. We would like to be able to repeat the argument of Theorem 5.1.1 in the filtered category $\mathcal{HC}^*_\infty$, in order to obtain a filtered version. The desired result would be the following: Let t_G be as in §1.1. If $\gamma\in\mathcal{D}$ is socular and of $\mathcal{O}$-type, with $\mathcal{O}$-type $(\gamma,w_\gamma,\mathbf{c}(\gamma),\mathbf{c}(w_\gamma))$, then

(i) The filtered module $\sigma^{-t_G}P(\gamma,\mathbf{c}(\gamma))^*$ is self-dual in $\mathcal{HC}^*_\infty$.

(ii) In particular, $P(\gamma,\mathbf{c}(\gamma))$ has Loewy length $2t_G+1$.

(iii) $P(\gamma, \mathbf{c}(\gamma))$ is rigid.

The proof goes through to the point that one finds that $\sigma^{-t_G} P(\gamma, \mathbf{c}(\gamma))$ is a summand of $\theta_{s_1} \circ \ldots \circ \theta_{s_k} L^*$. Thus the Krull-Schmidt Theorem yields that $\mathbb{D}\sigma^{-t} P(\gamma)^* \cong \sigma^{-t} P(\gamma)^*$, as desired, or $P(\gamma, \mathbf{c}(\gamma))$ appears twice in a direct sum decomosition of $\theta_{s_1} \circ \ldots \circ \theta_{s_k} L$. Under additional hypotheses, which hold in all the cases for which we have verified Theorem 5.1.1, we can eliminate the latter possibility.

This filtered version of Theorem 5.1.1 has a number of consequences for standard modules, all of which we can obtain under hypotheses weaker than those required for the above argument. Thus we will not pursue this further.

5.2 A special class of standard modules.

Recall that t_G is defined by $\ell\ell M(\delta_{\max}) = t_G + 1$; this is an upper bound for the Loewy length of any standard module $M(\delta)$, as follows from the analogue of the discussion in [**17**,§4]. The values for t_G are given in the introduction. Our main goal in this section is the following result; we defer for a moment the definition of strong $\mathcal{O}$-type and the proof of (iii):

THEOREM 5.2.1. *Let $\gamma \in \mathcal{D}$ be socular of $\mathcal{O}$-type.*

(i) *There is a unique maximal element $\gamma^\#$ in the set $\{\delta \in \mathcal{D} : (M(\delta) : L(\gamma)) > 0\}$.*
(ii) *$M(\gamma^\#)$ has simple socle $L(\gamma)$ and is rigid.*
(iii) *If in addition γ is of strong $\mathcal{O}$-type, then the standard module $M(\gamma^\#)$ has Loewy length $t_G + 1$ and $P(\gamma, \mathbf{c}(\gamma))$ has Loewy length $2t_G + 1$.*

Let $\mathcal{T} = \{\delta \in \mathcal{D} : \delta = \gamma^\#$ for some socular γ of $\mathcal{O}$-type$\}$. Note that $\mathcal{T}$ contains $\delta_{\max}$ (if all γ's are of $\mathcal{O}$-type) and may be thought of as lying "near the top" of $\mathcal{D}$. Theorem 5.2.1 says that all standard modules attached to elements of $\mathcal{T}$ have features in common with $M(\delta_{\max})$.

PROOF. Let $(\gamma, w_\gamma, \mathbf{c}(\gamma), \mathbf{c}(w_\gamma))$ be an $\mathcal{O}$-type for γ. If $(M(\delta) : L(\gamma)) > 0$, then $M(\delta)$ occurs as a factor in a standard flag for $P(\gamma, \mathbf{c}(\gamma))$. If δ has maximal length with this property, then $M(\delta)$ occurs as the lowest term in some standard flag. Thus, if δ and δ' are two distinct maximal elements, there is a standard flag for $P(\gamma, \mathbf{c}(\gamma))$ whose second term is an extension of $M(\delta)$ by $M(\delta')$. By (2.4.5), this extension splits. In particular, the socle of $P(\gamma, \mathbf{c}(\gamma))$ is not simple. But the socularity of γ implies that $P(\gamma, \mathbf{c}(\gamma))$ is self-dual, with simple socle. It follows that there is a unique δ of maximal length, and the same argument shows that $M(\delta)$ has simple socle $L(\gamma)$. The rigidity of $M(\gamma^\#)$ follows from a general result of Casian, that the socle and weight filtrations of a Harish-Chandra module coincide if it has simple socle [**3**]. ♦

To prove (iii), we need an additional assumption on γ. Let us say γ is of *strong $\mathcal{O}$-type* if γ is of $\mathcal{O}$-type with $\mathcal{O}$-type $(\gamma, w_\gamma, \mathbf{c}(\gamma), \mathbf{c}(w_\gamma))$ and if in addition the condition (3.4.6)(ii) holds for every $s \in S$. This is the case for all the groups G for which we have verified that every γ is of $\mathcal{O}$-type.

LEMMA 5.2.2. *Let $\delta \in \mathcal{T}$ with $\delta = \gamma^{\#}$ and assume that γ is of strong $\mathcal{O}$-type. Let $s \in S$ satisfy $s \times \delta > \delta$. Then $s \times \delta \in \mathcal{T}$.*

PROOF. By Theorem 5.2.1, we have $(M(\delta) : L(\gamma)) = 1$ and $(M(s \times \delta) : L(\gamma)) = 0$. Thus $\theta_s L(\gamma) = 0$ and $s \in \tau(\gamma)$. A repeat of the argument in the proof of Theorem 4.1.7, using the hypothesis of strong $\mathcal{O}$-type on γ, yields that $\theta_s P(\gamma, \mathbf{c}(\gamma))$ is a direct sum of pseudo-projective modules $P(\varepsilon)$. Moreover, the self-duality of $P(\gamma, \mathbf{c}(\gamma))$ implies that $\theta_s P(\gamma, \mathbf{c}(\gamma))$ is self-dual. Therefore the Krull-Schmidt Theorem implies that each $P(\varepsilon)$ is isomorphic to some $\mathbb{D}P(\varepsilon')$. Since the $P(\varepsilon)$'s represent $\mathbb{Z}$-independent elements in Gr_∞, we must have $\varepsilon = \varepsilon'$ and each $P(\varepsilon)$ is self-dual. Since $M(\gamma)$ is the lowest flag factor of $P(\gamma, \mathbf{c}(\gamma))$, and $\theta_s M(\gamma)$ is an extension of $M(s \times \gamma)$ by $M(\gamma)$, one of the summands $P(\varepsilon)$ must have $M(s \times \gamma)$ at the bottom of its standard flag. We may conclude that $s \times \gamma = \varepsilon^{\#}$, proving the lemma. ♦

PROOF OF 5.2.1 (*iii*). Let $\delta \in \mathcal{T}$ and $s \in S$ satisfy $s \times \delta > \delta$. By Lemma 5.2.2, $s \times \delta \in \mathcal{T}$. If we prove that $\ell\ell M(\delta) = \ell\ell M(s \times \delta)$, then by continuing to the top of $\mathcal{D}$ we will obtain $\ell\ell M(\delta) = \ell\ell M(\delta_{\max})$. Suppose $\delta = \gamma^{\#}$; thus $socM(\delta) = L(\gamma)$. As in the proof of 5.2.2, we see that $\theta_s L(\gamma) = 0$, so that $\theta_s M(\delta) = \theta_s(M(\delta)/L(\gamma))$. Since θ_s increases Loewy length by at most two, as discussed in the proof of Theorem 3.2.5, we obtain $\ell\ell\theta_s M(\delta) \leq \ell\ell M(\delta) + 1$. But $M(s \times \delta)$ lies in $rad\theta_s M(\delta)$, so $\ell\ell M(s \times \delta) \leq \ell\ell M(\delta)$. The opposite inequality is obtained by a similar argument. (See [**17**].)

Let γ be as in the statement of Theorem 5.2.1. Theorem 4.2.1 yields a description of the radical filtration of $P(\gamma, \mathbf{c}(\gamma))$. Specifically, filtered BGG reciprocity holds: $P(\gamma, \mathbf{c}(\gamma))^*$ has a filtered standard flag with $\sigma^k M(\varepsilon)^*$ occurring $(M(\varepsilon)^* : \sigma^k L(\gamma)^*)$-many times. The properties of $\gamma^{\#}$ imply that there is a unique lowest flag factor, namely $\sigma^t M(\gamma^{\#})^*$, for $t + 1 = \ell\ell M(\gamma^{\#})$. Thus $\ell\ell P(\gamma, \mathbf{c}(\gamma))^* = 2t_G + 1$. ♦

We conclude the paper by showing that the standard modules attached to $\delta \in \mathcal{T}$ are exactly the standard modules having the maximal possible Loewy length, $t_G + 1$.

THEOREM 5.2.3. *Assume that every element of $\mathcal{D}$ is of strong $\mathcal{O}$-type. A standard module $M(\delta)$ has Loewy length $t_G + 1$ if and only if $\delta \in \mathcal{T}$.*

PROOF. By Theorem 5.2.1 it is enough to show that $M(\delta)$ has Loewy length at most t_G if $\delta \notin \mathcal{T}$. Suppose $\delta \notin \mathcal{T}$ and $M(\delta)$ has Loewy length $t_G + 1$. Then $rad^{t_G} M(\delta)$ is a non-zero submodule of $socM(\delta)$. Let $L(\gamma)$ be a summand of $radM(\delta)$ and let Y be a submodule of $M(\delta)$ maximal with respect to meeting $L(\gamma)$ in (0). Let $X = M(\delta)/Y$. Then X has Loewy length $t_G + 1$, simple cap $L(\delta)$, and simple socle $L(\gamma)$.

Let $(\gamma, w_\gamma, \mathbf{c}(\gamma), \mathbf{c}(w_\gamma))$ be an $\mathcal{O}$-type for γ and let $P(\gamma) = P(\gamma, \mathbf{c}(\gamma))$. Recall by Lemma 4.1.2 that $\dim Hom_{\mathcal{HC}}(P(\gamma), \mathbb{D}M(\delta)) = (\mathbb{D}M(\delta) : L(\gamma))$. In particular, since $\mathbb{D}X$ is a submodule of $\mathbb{D}M(\delta)$ with simple cap $L(\gamma)$, there is a surjection of $P(\gamma)$ onto $\mathbb{D}X$. But, as noted in the proof of 4.1.2, this surjection corresponds to an occurrence of $M(\delta)$ as a factor in a standard flag for $P(\gamma)$. Specifically, since $\mathbb{D}X$ has Loewy length t_G+1, the factor in question occurs in $rad^{t_G}P(\gamma)$. This yields an occurrence of $L(\gamma)$, the socle of $M(\delta)$, in $rad^{2t_G}P(\gamma)$. However, γ is socular, so $P(\gamma)$ has Loewy length $2t_G+1$ and simple socle $L(\gamma)$. Moreover, the socle occurs in the flag factor of the form $M(\gamma^{\#})$. By assumption, $\delta \neq \gamma^{\#}$. Thus we obtain two copies of $L(\gamma)$ in $socP(\gamma)$, a contradiction. ♦

REMARK. To be explicit, all the hypotheses used in this section have been verified for our group G if it is complex, $Spin(2n+1,1)$, $E_{6(-26)}$, or $SL(3,\mathbb{H})$. In addition, all the results on standard modules in this section have been verified for $SL(4,\mathbb{H})$, although all the hypotheses have not been. We conjecture that the statements on standard modules are true for all choices of G. The conjectural pseudo-projective modules, in addition to providing a means for proving the results on Loewy length of standard modules, also provide a context in which to understand why such results would be true.

EXAMPLES. We have provided an explicit parametrization and display of the set $\mathcal{D}$ in case G is either $SL(3,\mathbb{C})$, $Spin(2n+1,1)$, $SL(3,\mathbb{H})$, or $E_{6(-26)}$; see Figures 1 through 4.

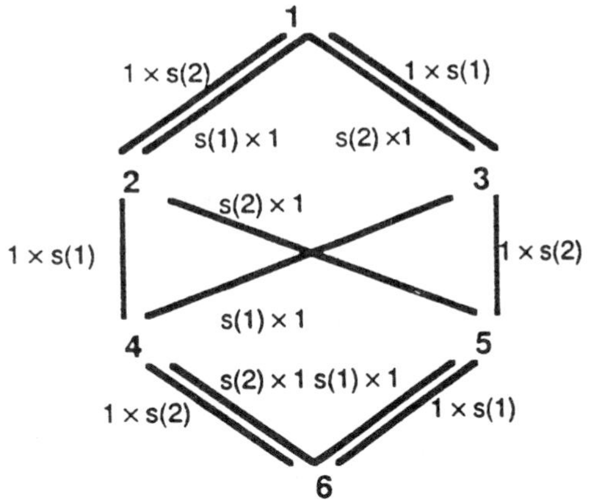

Figure 1: The $\mathcal{D}$ poset for SL(3,**C**)

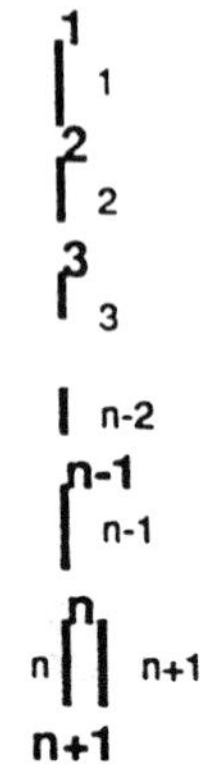

Figure 2: The $\mathcal{D}$ poset for Spin(2n+1,1)

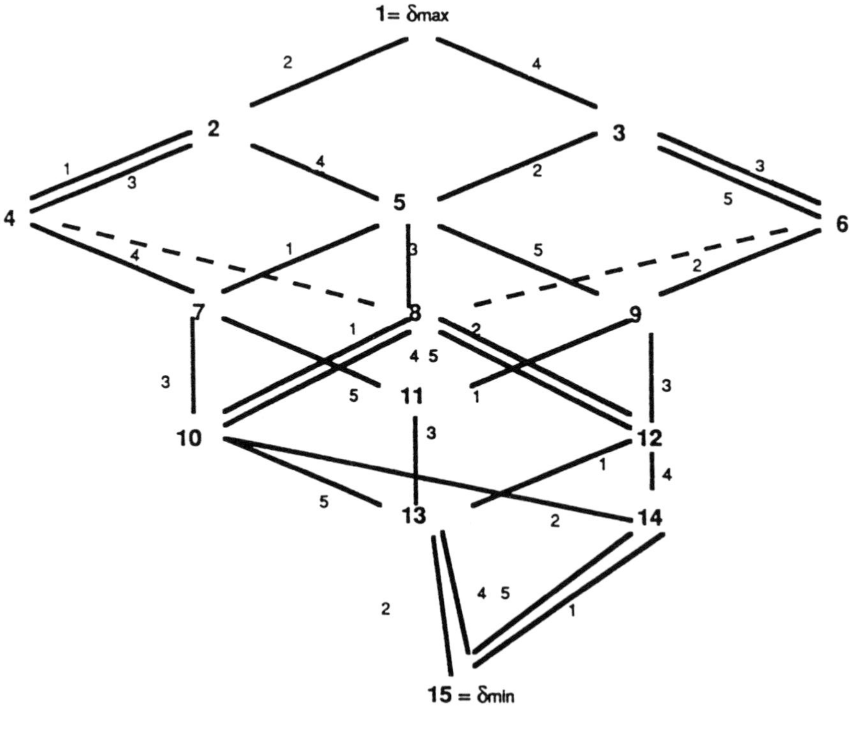

Figure 3: The $\mathcal{D}$ poset for SL(3,**H**)

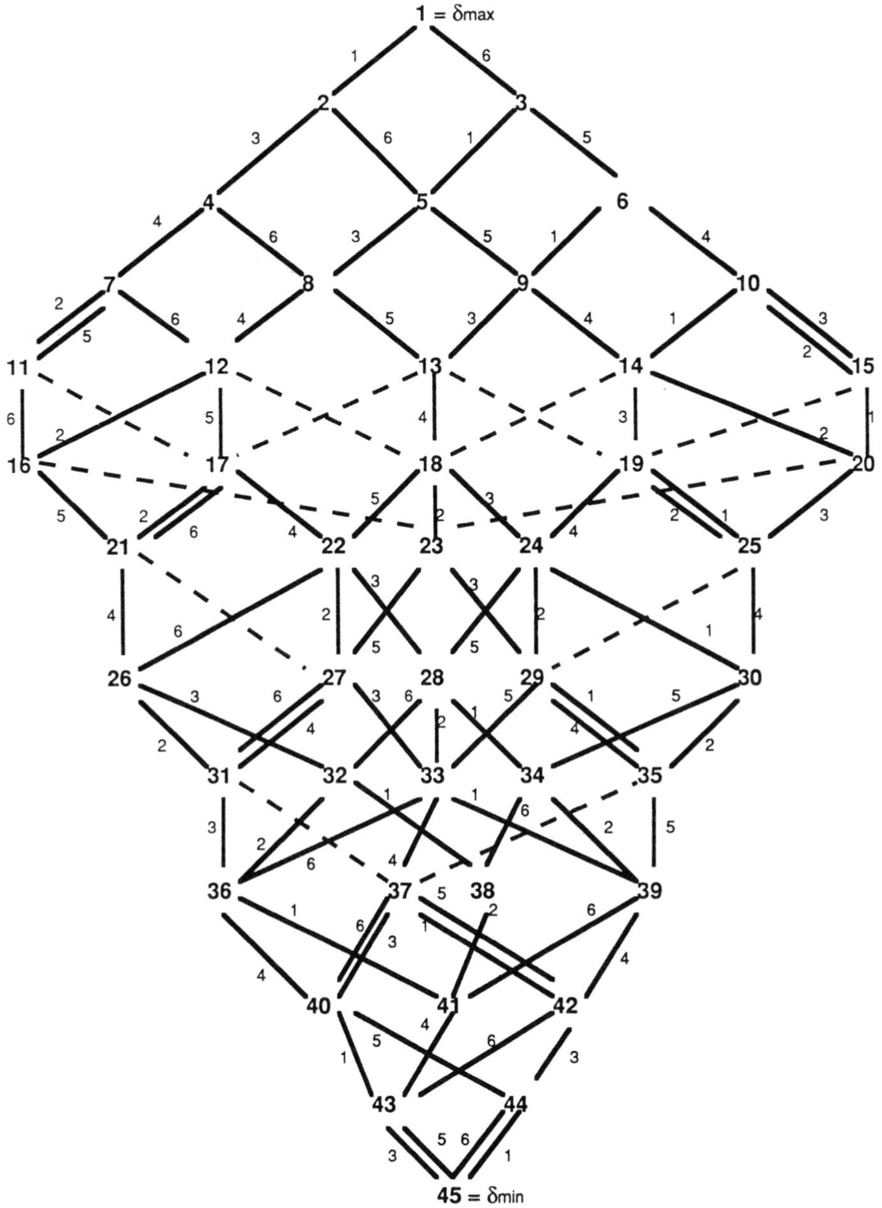

Figure 4: The $\mathcal{D}$ poset for $E_{6(-26)}$

REFERENCES

1. J. Bernstein and S. Gelfand, *Tensor products of finite and infinite dimensional representations of semisimple Lie algebras*, Comp. Math. 41 (1980), 245–285.
2. J. Bernstein, I. Gelfand, and S. Gelfand, *On a category of g-modules*, Funk. Anal. Appl. 10 (1976), 1–8.
3. L. Casian, *Graded characters of induced representations of real reductive Lie groups I*, J. Algebra 123 (1989), 289–326.
4. L. Casian and D. Collingwood, *The Kazhdan-Lusztig conjecture for generalized Verma modules*, Math. Z. 195 (1987), 581–600.
5. L. Casian and D. Collingwood, *Complex geometry and the asymptotics of real reductive Lie groups I*, Trans. Amer. Math. Soc. 300 (1987), 73–107.
6. L. Casian and D. Collingwood, *Weight filtrations for induced representations of real reductive Lie groups*, Adv. Math. 73 (1989), 79–146.
7. D. Collingwood, *Research Notes in Math. 137*, Pitman Publishing, London, 1985.
8. V. Deodhar, *On some geometric aspects of the Bruhat orderings II: The parabolic analogue of the Kazhdan-Lusztig polynomials*, J. Algebra 111 (1987), 483-506.
9. T. Enright, *Representations of complex semisimple Lie groups*, Tata Institute of Fundamental Research, Springer-Verlag, 1981.
10. O. Gabber and A. Joseph, *Towards the Kazhdan-Lusztig conjecture*, Ann. scient. Éc. Norm. Sup. 14 (1981), 261-302.
11. I. Gelfand, *The cohomology of infinite dimensional Lie algebras; some questions of integral geometry*, Proc. Int. Cong. Math., Nice (1970), 95-111.
12. A. Guichardet, *Cohomologie des groupes topologiques et des algebres de Lie*, Cedic/Fernand Nathan, Paris, 1980.
13. H. Hecht and W. Schmid, *Characters, asymptotics and n-homology of Harish-Chandra modules*, Acta Math. 151 (1983), 49–151.
14. H. Hecht and W. Schmid, *On the asymptotics of Harish-Chandra modules*, J. reine angew. Math. 343 (1983), 169–183.
15. J. Humphreys, *Introduction to Lie algebras and Representation Theory, Graduate Texts in Math. 9*, Springer-Verlag, New York, 1972.
16. R. Irving, *Projective modules in the category $\mathcal{O}_S$: self-duality*, Trans. Amer. Math. Soc. 291 (1985), 701–732.
17. R. Irving, *Projective modules in the category $\mathcal{O}_S$: Loewy series*, Trans. Amer. Math.Soc. 291 (1985), 733-754.
18. R. Irving, *The socle filtration of a Verma module*, Ann. Scient. Ê c. Norm. Sup. 21 (1988), 47–65.
19. R. Irving, *A filtered category $\mathcal{O}_S$ and applications*, Mem. Amer. Math. Soc. 41 (1990).
20. R. Irving, *Loewy filtrations of Weyl modules*, Trans. Amer. Math. Soc. - to appear.
21. R. Irving, *BGG algebras and the BGG reciprocity principle*, J. Algebra 135 (1990), 363–380.
22. J. Jantzen, *Einhü llende Algebren halbeinfacher Lie-Algebren*, Springer-Verlag, Berlin, Heidelberg, New York, Tokyo, 1983.
23. A. Joseph, *Dixmier's problem for Verma and principal series submodules*, Jour. Lond. Math. Soc. 20 (1979), 193–204.
24. A. Knapp, *Representation Theory of Semisimple Lie groups: An overview based on examples*, Princeton University Press, Princeton, 1986.
25. G. Lusztig and D. Vogan, *Singularities of closures of K orbits on flag manifolds*, Inv. Math. 71 (1983), 365-379.
26. N. Popescu, *Abelian categories with applications to rings and modules*, Academic Press, London, New York, 1973.
27. D. Vogan, *Gelfand-Kirillov dimension for Harish-Chandra modules*, Inv. Math. 48 (1978), 75–98.
28. D. Vogan, *Irreducible characters of semisimple Lie groups I*, Duke Math. J. 46, 61–108.
29. D. Vogan, *Representations of Real Reductive Lie Groups Progress in Math. 15*, Birkhäuser, Boston, 1981.

30. D. Vogan, *Irreducible characters of semisimple Lie groups III: Proof of the Kazhdan-Lusztig conjecture in the integral case*, Inv. Math. 71 (1983), 381–417.
31. N. Wallach, *Real Reductive Groups I, Pure and applied Math. 132*, Academic Press, Boston, 1988.

DEPARTMENT OF MATHEMATICS, UNIVERSITY OF WASHINGTON, SEATTLE, WASHINGTON 98195

E-mail address: colling@math.washington.edu, irving@math.washington.edu

Contemporary Mathematics
Volume **139**, 1992

A Simple Proof of a Necessary and Sufficient Condition for the Existence of Nontrivial Global Sections of a Line Bundle on a Schubert Variety

ROMUALD DABROWSKI

ABSTRACT. In this paper we give a simple proof of a criterion for the existence of global sections for line bundles on Schubert varieties. Unlike previous proofs, which use the nontrival fact that Schubert varieties are normal, our proof is based on an elementary geometric property of root systems and their Weyl groups.

Let G be a semi-simple algebraic group over an algebraically closed field k of arbitrary characteristic. As usual, let B denote a fixed Borel subgroup of G, T a maximal torus in B, Φ the root system associated with T, Φ_+ the set of positive roots relative to B, Δ the set of simple roots in Φ_+, W the Weyl group of Φ generated by the set of reflections $s_\alpha, \alpha \in \Phi_+$. For any $w \in W$ let $\ell(w)$ denote the length of w relative to Φ_+. We also let $\langle \mu, \alpha^\vee \rangle$ denote the usual pairing between the weights and coroots of Φ. For any w in W, let $X_w = \overline{BwB}/B$ be the corresponding Schubert variety. Finally, for any character μ of B we let L_μ denote the corresponding line bundle on X_w. Then

$$H^0(X_w, L_\mu) = \{f \in k[\overline{BwB}] \mid f(xb) = \mu^{-1}(b)f(x) \text{ for all } x \in \overline{BwB}, b \in B\}$$

(see e.g. [**J, Part 2, Chapter 14**]). The aim of this paper is to give an elementary proof of the following theorem.

1980 *Mathematics Subject Classification* (1985 *Revision*). Primary, 20G15; Secondary, 141M15, 17B20.

The author would like to thank M. Brion, A. Daszkiewicz, and P. Polo for useful discussions concerning the nonvanishing criterion. The author also wishes to thank the referee for many suggestions which considerably improved the original manuscript. It should also be mentioned that the idea of the above proof occurred to the author as he was trying to generalize the results of [S] on Kloosterman sums for the general linear group

This paper is in final form and no version of it will be published elsewhere.

THEOREM. *Let w be an element of the Weyl group and let μ be a character of B. Then the following conditions are equivalent:*

(i) $H^0(X_w,L_\mu) \neq \{0\}$,
(ii) $\langle \mu,\alpha^\vee \rangle \leq 0$ *for all* $\alpha \in \Phi_+$ *such that* $\ell(ws_\alpha) = \ell(w) - 1$,
(iii) $\langle \mu,\alpha^\vee \rangle \leq 0$ *for all* $\alpha \in \Phi_+$ *such that* $w(\alpha) \in -\Phi_+$.

Remark. The equivalence of (i) and (ii) was first proved in [**P, Corollaire 2.3**] by using the nontrivial fact that Schubert varieties are normal. We show that the theorem follows from an elementary proposition about root systems and their Weyl groups. We will prove the proposition first and next we will prove the theorem.

We let E denote the real vector space spanned by Φ and let $(\cdot,\cdot)$ be a fixed positive definite bilinear form on E such that

$$\langle \alpha,\beta^\vee \rangle = 2(\alpha,\beta)/(\beta,\beta)$$

for all roots α, β. We will need some basic definitions and facts about convex cones in E. We recall that the *dual cone* X° of a subset X of E is defined by the formula

$$X^\circ = \{x \in E \mid (x,y) \geq 0 \text{ for all } y \in X\}.$$

A *ray* spanned by a vector $x \in E$, $x \neq 0$ is the set $\{rx \mid r \geq 0\}$. A subset C of E is called a *convex cone* if for any x, $y \in C$ and r, s nonnegative real numbers, $rx + sy \in C$. A convex cone C is called *polyhedral* if it is the convex hull of a finite set of rays. A ray spanned by a vector $z \in C$, $z \neq 0$ is called an *extreme ray* of C if z is not a positive linear combination of two linearly independent vectors in C . Finally, a subset F of a polyhedral cone C is called a *wall* of C if F spans a linear subspace of E of codimension 1 in the linear subspace spanned by C and $F = \{x \in E \mid (x,u) = 0\}$ for some nonzero vector u in the dual cone C°. We will need the following facts about polyhedral cones. Suppose that $C \neq \{0\}$ is a polyhedral cone. Then:

(1) C° is a polyhedral cone and $C = (C^\circ)^\circ$,
(2) If C does not contain any nontrivial linear subspace then the set of extreme rays of C is finite and C is the convex hull of its extreme rays,
(3) Suppose that the linear span of C is E. Then the extreme rays of C° are precisely the rays in C° which are perpendicular to the walls of C,
(4) Suppose that C contains a basis of E and let $C = \bigcup\{C_i \mid i = 1,2,\ldots,n\}$, where C_i, $i = 1,2,\ldots,n$ are polyhedral cones. Then at least one of the cones C_i contains a basis of E.

Proofs of (1) and (2) can be found in [**F, Chapter I, Theorems 3, 14**]. Property (3) follows from (1) and (2). Indeed, if $u \in C^\circ$ is a vector perpendicular to a wall F of C, and u is a positive linear combination of vectors u' and u'' in C°, then it is easy to see that u' and u'' are both perpendicular to F. Since the linear space spanned by F is of codimension 1, u' and u'' are linearly dependent. Hence the ray spanned by u is extreme in C°. Suppose now that $\{ru \mid r \geq 0\}$, $u \neq 0$, is an extreme ray in C° and let $x_1,x_2,\ldots,x_q$, be nonzero vectors in C whose nonnegative linear combinations form C. We can assume that for some p,

$1 \leq p \leq q$, $(x_i,u) = 0$ for $1 \leq i \leq p$ and $(x_i,u) > 0$ for $i > p$. We claim that the linear span of $\{x_1,x_2,\dots,x_p\}$ is of codimension 1 in E. If not, we can find a vector v, perpendicular to $x_1,x_2,\dots,x_p$, such that u and v are linearly independent. Then for sufficiently small $\varepsilon > 0$ linearly independent vectors $u - \varepsilon v$, $u + \varepsilon v$ are in C°. But this violates the assumption that u spans an extreme ray in C°. It is now clear that $\{x \in C \mid (x,u) = 0\}$ is a wall of C to which u is perpendicular.

Clearly (4) holds if dimension of E is 1 or 2. Suppose that E is of arbitrary dimension. For any $i = 1, 2, \dots, n$, let E_i denote the linear subspace spanned by C_i. It is enough to show that the union of subspaces E_i is E. Let $x \in E$. Since the linear span of C is E, we must have $x = y - z$ for some y, z in C. Let E' be the linear space spanned by y and z. Then (4) applied to E', $C' = C \cap E'$ and to the collection of cones $C_i' = C_i \cap E'$, $i = 1,2,\dots,n$, implies that x belongs to the union of the linear subspaces spanned by C_i' which is contained in $\bigcup\{E_i \mid i = 1,2,\dots,n\}$.

We need the following notation. Let D denote the dual cone of the set of positive roots. Let w_0 denote the unique element of W such that $w_0(\Phi_+) = -\Phi_+$. For each $w \in W$ we introduce the following sets:

$$\begin{aligned}
P(w) &= \{\alpha \in \Phi_+ \mid w(\alpha) \in -\Phi_+\} \\
P_1(w) &= \{\alpha \in \Phi_+ \mid \ell(ws_\alpha) = \ell(w) - 1\} \\
C(w) &= \text{convex cone spanned by } P(w) \\
D(w) &= \bigcup\{\theta(D) \mid \theta \in W, P(w) \cap P(\theta^{-1}) = \emptyset\}
\end{aligned}$$

The following well-known facts can be easily proved by induction on $\ell(w)$.

(a) For any $w \in W$ and $\alpha \in \Delta$, $\ell(ws_\alpha) = \ell(w) - 1$ if and only if $\alpha \in P(w)$.

(b) For any $w \in W$, $P(w) = \{\alpha \in \Phi_+ \mid \ell(ws_\alpha) < \ell(w)\}$ and therefore $P_1(w) \subseteq P(w)$.

(c) For any $w,\theta \in W$, $P(w) \cap P(\theta^{-1}) = \emptyset$ if and only if $\ell(w\theta) = \ell(w) + \ell(\theta)$.

Our proof of the theorem is based on the following proposition. It appears that this result, interesting in its own right, has been overlooked in the literature.

PROPOSITION. *Let w be any element of W. Then:*

(i) *the extreme rays of $C(w)$ are precisely the rays spanned by the elements of $P_1(w)$ and therefore the convex cones spanned by $P_1(w)$ and $P(w)$ are the same,*

(ii) *$P(w)^\circ = P_1(w)^\circ = D(w)$ and*

$$P_1(w) = \bigcup\{P(w) \cap \theta(\Delta) \mid \theta \in W, P(w) \cap P(\theta^{-1}) = \emptyset\}.$$

PROOF. (i) We will first show that the elements of $P_1(w)$ span extreme rays in $C(w)$. Let $\alpha \in P(w)$. Set

$$A = \{\beta \in P(ws_\alpha) \mid s_\alpha(\beta) \in \Phi_+\}$$

and

$$B = \{\beta \in P(ws_\alpha) \mid s_\alpha(\beta) \in -\Phi_+\}.$$

It is easy to see that $s_\alpha(A) \cap B = \emptyset$ and $\alpha \notin s_\alpha(A) \cup B$. Moreover, we claim that $s_\alpha(A) \cup B \subseteq P(w)$. The inclusion $s_\alpha(A) \subseteq P(w)$ is obvious. Suppose that $\beta \in B$. Then $s_\alpha(\beta) = \beta - \langle\beta,\alpha^\vee\rangle\alpha \in -\Phi_+$ and consequently $\langle\beta,\alpha^\vee\rangle > 0$. Hence $w(\beta) = ws_\alpha(\beta) + \langle\beta,\alpha^\vee\rangle w(\alpha)$ belongs to $-\Phi_+$ and therefore $\beta \in P(w)$. We have proved that if $\alpha \in P(w)$ then the disjoint union $\{\alpha\} \cup s_\alpha(A) \cup B$ is contained in $P(w)$. Then it is easy to see that for any $\alpha \in P(w)$, $\alpha \in P_1(w)$ if and only if $\{\alpha\} \cup s_\alpha(A) \cup B = P(w)$. Suppose that $\alpha \in P_1(w)$ is a positive linear combination of some elements in $P(w) \setminus \{\alpha\}$. Therefore $\alpha = s_\alpha(\beta) + \gamma$, for some β in the convex cone spanned by A and some γ in the convex cone spanned by B. Then $-\alpha = s_\alpha(\alpha) = \beta + \gamma - \langle\gamma,\alpha^\vee\rangle\alpha$, or equivalently, $(\langle\gamma,\alpha^\vee\rangle - 1)\alpha = \beta + \gamma$, where $\langle\gamma,\alpha^\vee\rangle - 1 > 0$. But this is impossible since $ws_\alpha(\alpha)$ is a positive root and $ws_\alpha(\beta+\gamma)$ is a positive linear combination of negative roots. To complete the proof of (i) we have to show that extreme rays of $C(w)$ are spanned by the elements of $P_1(w)$. Clearly the extreme rays in $C(w)$ are spanned by some elements of $P(w)$. Suppose that $\alpha \in P(w)$ spans an extreme ray in $C(w)$ and $\alpha \notin P_1(w)$. Then there exists $\beta \in P(w)$ such that $\beta \notin \{\alpha\} \cup s_\alpha(A) \cup B$ (A,B as above). Therefore $-s_\alpha(\beta) \in P(w)$ and $\langle\beta,\alpha^\vee\rangle\alpha = -s_\alpha(\beta) + \beta$ where $\langle\beta,\alpha^\vee\rangle > 0$. This contradicts our assumption that the ray spanned by α is extreme in $C(w)$. Since $P(w)$ does not contain any nontrivial linear subspace of E, $C(w)$ is spanned by its extreme rays. Therefore $C(w)$ coincides with the convex cone spanned by $P_1(w)$.

(ii) It follows from (i) that $P(w)^\circ = P_1(w)^\circ$. The definition of $D(w)$ implies that $D(w) \subseteq P(w)^\circ$. Therefore it is enough to prove $P(w)^\circ \subseteq D(w)$. We first show that $D(w)$ is a polyhedral cone. Since $D(w)$ is a union of polyhedral cones it is enough to prove that $D(w)$ is a convex cone. Suppose that this is not the case. Then there exist $\lambda_1,\lambda_2 \in D$ and $\theta_1,\theta_2 \in W$, such that $P(w) \cap P(\theta_i^{-1}) = \emptyset$, $i = 1,2$, and $\mu = \theta_1(\lambda_1) + \theta_2(\lambda_2)$ is not in $D(w)$. Since $E = \bigcup\{\theta(D) \mid \theta \in W\}$ we have $\mu = \theta(\lambda)$ for some $\theta \in W$ and $\lambda \in D$. Moreover, by a continuity argument we can assume that λ is *regular*, that is $(\lambda,\alpha) > 0$ for all $\alpha \in \Phi_+$. Since we have assumed $\mu = \theta(\lambda) \notin D(w)$, $P(w) \cap P(\theta^{-1}) \neq \emptyset$. Let $\gamma \in P(w) \cap P(\theta^{-1})$. Hence

$$(\mu,\gamma) = (\theta(\lambda),\gamma) = (\lambda,\theta^{-1}(\gamma)) < 0.$$

But this is impossible since $\gamma \notin P(\theta_1^{-1}) \cup P(\theta_2^{-1})$ implies

$$(\mu,\gamma) = (\theta_1(\lambda),\gamma) + (\theta_2(\lambda),\gamma) = (\lambda,\theta_1^{-1}(\gamma)) + (\lambda,\theta_2^{-1}(\gamma)) \geq 0.$$

Since we now know that $D(w)$ is a polyhedral cone, the duality property (1) of polyhedral cones implies that in order to show $P(w)^\circ \subseteq D(w)$ it is enough to show $D(w)^\circ \subseteq C(w)$. Since $D(w)$ contains a basis of E, properties (2) and (3) of polyhedral cones imply that $D(w)^\circ$ is spanned by some vectors which are perpendicular to the walls of $D(w)$. Let F be a wall of $D(w)$ and let $\beta \in D(w)^\circ$ be perpendicular to F. Since $F = \bigcup\{F \cap \theta(D) \mid \theta \in W, P(w) \cap P(\theta^{-1}) = \emptyset\}$, property (4) of polyhedral cones implies that there exists $\theta \in W$, $P(w) \cap P(\theta^{-1}) = \emptyset$, such that $F \cap \theta(D)$ is a wall of the polyhedral cone $\theta(D)$. Therefore $\theta^{-1}(F) \cap D$ is a wall of D to which $\theta^{-1}(\beta)$ is perpendicular. Hence $\theta^{-1}(\beta)$ must be a multiple of a simple root α. Therefore we can assume $\theta^{-1}(\beta) = \pm\alpha$.

We claim that $\beta \in P(w)$ and therefore $\theta^{-1}(\beta) = \alpha$, since $P(w) \cap P(\theta^{-1}) = \emptyset$. The claim follows since $P(w) \cap P(w_0 w) = \emptyset$ and consequently $D(w)^\circ \cap \Phi \subseteq w^{-1}w_0(D)^\circ \cap D^\circ \cap \Phi = P(w)$. Therefore we have proved that each extreme ray in $D(w)^\circ$ is spanned by a root in $P(w)$ of the form $\theta(\alpha)$ for some $\alpha \in \Delta$, $\theta \in W$ such that $P(w) \cap P(\theta^{-1}) = \emptyset$. Therefore we have the desired inclusion $D(w)^\circ \subseteq C(w)$ and by (i)

$$P_1(w) \subseteq \bigcup\{P(w) \cap \theta(D) \mid \theta \in W, P(w) \cap P(\theta^{-1}) = \emptyset\}.$$

Suppose that $\beta = \theta(\alpha) \in P(w)$ for some $\alpha \in \Delta$ and some $\theta \in W$ such that $P(w) \cap P(\theta^{-1}) = \emptyset$ (or equivalently $\ell(w\theta) = \ell(w) + \ell(\theta)$). Then $w(\beta) = w\theta(\alpha) \in -\Phi_+$ implies $\ell(w\theta s_\alpha) = \ell(w\theta) - 1$ and therefore

$$\begin{aligned} \ell(w) > \ell(ws_\beta) &= \ell(w\theta s_\alpha \theta^{-1}) \geq \ell(w\theta s_\alpha) - \ell(\theta) \\ &= \ell(w\theta) - 1 - \ell(\theta) = \ell(w) - 1. \end{aligned}$$

Hence $\ell(ws_\beta) = \ell(w) - 1$ and $\beta \in P_1(w)$. Therefore

$$P_1(w) = \bigcup\{P(w) \cap \theta(D) \mid \theta \in W, P(w) \cap P(\theta^{-1}) = \emptyset\}$$

as required. □

Remarks.

1. The proof of (i) presented here is due to V. Deodhar. A different proof of (i) was communicated to the author by M. Dyer.
2. We say that a root β in a subset C of Φ is *indecomposable* in C if β is not a sum of two roots in C. Then it can be showed using essentially the same argument as in the proof of (i) that if Φ is simply laced (i.e. $|\langle\alpha,\beta^\vee\rangle| \leq 1$ for all roots α, $\beta \in \Phi$) then the indecomposable roots of $P(w)$ are precisely the elements of $P_1(w)$. This is not the case in general. For example if

$$\Phi = \{\pm\alpha_1, \pm\alpha_2, \pm(\alpha_1 + \alpha_2), \pm(2\alpha_1 + \alpha_2)\}$$

is the root system of type B_2 (e.g. **[Bo, Planche** X], then

$$P(w_0 s_{\alpha_1}) = \Phi_+ \setminus \{\alpha_1\} = \{\alpha_2, \alpha_1 + \alpha_2, 2\alpha_1 + \alpha_2\},$$

$\beta = \alpha_1 + \alpha_2$ is indecomposable in $P(w_0 s_{\alpha_1})$, and $\beta \notin P_1(w_0 s_{\alpha_1})$. It is an interesting problem to give an accessible description of the set of indecomposable roots in $P(w)$ which holds for all root systems.

Proof of the theorem. The equivalence of (ii) and (iii) follows directly from the proposition. Therefore it is enough to show that (i) and (iii) are equivalent. The easier part of the proof is to show that (i) implies (iii). This is proved in **[P, p. 289]** and it does not require the proposition. We recall the argument for completeness. Suppose that $H^0(X_w, L_\mu) \neq \{0\}$. Let $f \neq 0$ be a B-eigenvector for the action of B on $H^0(X_w, L_\mu)$ by left translations. Such an f exists by the

Lie-Kolchin theorem. Restricting f to BwB, one easily sees that f is completely determined by its value on a fixed element $n \in N_G(T)$, such that $\pi(n) = w$. Here π denotes the natural homomorphism from $N_G(T)$ onto W. Let α be a root and let $\Psi_\alpha: SL(2,k) \to G$ be the corresponding algebraic group homomorphism such that for any $x \in k$, $\varepsilon \in k^*$,

$$u_\alpha(x) = \Psi_\alpha \begin{pmatrix} 1 & x \\ 0 & 1 \end{pmatrix} \in U, \quad \text{the unipotent radical of } B,$$

$$u_{-\alpha}(x) = \Psi_\alpha \begin{pmatrix} 1 & 0 \\ x & 1 \end{pmatrix} \in U_-, \quad \text{the unipotent radical of the Borel subgroup opposite to } B,$$

$$t_\alpha(\varepsilon) = \Psi_\alpha \begin{pmatrix} \varepsilon & 0 \\ 0 & \varepsilon^{-1} \end{pmatrix} \in T,$$

and

$$\sigma_\alpha = \Psi_\alpha \begin{pmatrix} 0 & 1 \\ -1 & 0 \end{pmatrix}$$

satisfies $\pi(\sigma_\alpha) = s_\alpha$, where s_α is the reflection corresponding to α. Suppose now that $\alpha \in P(w)$. Then

$$\begin{aligned} n\sigma_\alpha u_{-\alpha}(\varepsilon) &= n\sigma_\alpha u_\alpha(\varepsilon^{-1})\sigma_\alpha^{-1} t_\alpha(\varepsilon) u_\alpha(\varepsilon^{-1}) \\ &= n u_{-\alpha}(-\varepsilon^{-1}) n^{-1} n t_\alpha(\varepsilon) u_\alpha(\varepsilon^{-1}) \in BwB \end{aligned}$$

and $f(n\sigma_\alpha u_{-\alpha}(\varepsilon)) = f(n)\varepsilon^{-\langle\mu,\alpha^\vee\rangle}$. Since the map $x \mapsto f(n\sigma_\alpha u_{-\alpha}(x))$ is a regular function on the 1-dimensional affine space, we must have $\langle\mu,\alpha^\vee\rangle \le 0$.

We will now show that (iii) implies (i). Suppose that μ is a character of B satisfying $\langle\mu,\alpha^\vee\rangle \le 0$ for all $\alpha \in P(w)$. We will use the proposition to construct explicitly a nonzero element in $H^0(X_w, L_\mu)$. By the proposition $-\mu = \theta(\lambda)$ for some $\lambda \in D$ and some $\theta \in W$, such that $P(w) \cap P(\theta^{-1}) = \emptyset$. Let V_λ be the irreducible representation of G with highest weight λ, and let v_λ be a highest weight vector in V_λ. For any $\tau \in W$, let $\underline{\tau}$ be a fixed representative of τ in $N_G(T)$ and set $v_{\tau\lambda} = \underline{\tau} v_\lambda$. Then it is known that $v_{\tau\lambda}$ spans the T-weight space corresponding to weight $\tau(\lambda)$. Fix a basis of V_λ consisting of eigenvectors of T and containing the vectors $v_{\tau\lambda}, \tau \in W$. For any $\tau \in W$, let $v_{\tau\lambda}^*$ denote the dual vector to $v_{\tau\lambda}$ in the contragredient representation V_λ^*. We define a regular function f on G by the formula

$$f(g) = v_{w\theta\lambda}^*(g v_{\theta\lambda}).$$

It is clear that $f(\underline{w}) \neq 0$. We claim that f satisfies

$$(*) \qquad f(xb) = \mu(b^{-1}) f(x) \text{ for all } x \in \overline{BwB},\ b \in B$$

and therefore it defines a nonzero element of $H^0(X_w, L_\mu)$.

FIRST PROOF OF $(*)$. We observe that to show $(*)$ it is enough to show

$$(**) \qquad f(xb) = \mu(b^{-1}) f(x) \text{ for all } x \in BwB,\ b \in B.$$

Indeed, if this is the case, then for a fixed $b \in B$ the regular functions $x \mapsto f(xb)$ and $x \mapsto \mu(b^{-1})f(x)$, $x \in G$, agree on BwB. Hence they must also agree on $\overline{BwB}$ and therefore (*) holds. Set $U_1 = U \cap w^{-1}(U)$ and $U_2 = U \cap w^{-1}(U_-)$. Then U_2 is the subgroup of U generated by the elements $u_\alpha(a)$, $\alpha \in P(w)$, $a \in k$. Let $x \in BwB$. Then $x = u_1 n u_2$, where n is a representative of w in $N_G(T)$ and u_1, $u_2 \in U$. We can write u_2 in the form $u_2 = u'u''$ where $u' \in U_1$ and $u'' \in U_2$. We observe that $P(w) \cap P(\theta^{-1}) = \emptyset$ implies $\underline{\theta}^{-1}u''\underline{\theta} \in \theta^{-1}(U_2) \subseteq U$. Therefore

$$u''v_{\theta\lambda} = u''\underline{\theta}v_\lambda = \underline{\theta}\,\underline{\theta}^{-1}u''\underline{\theta}v_\lambda = \underline{\theta}v_\lambda = v_{\theta\lambda}$$

and consequently,

$$f(x) = f(u_1nu') = f(u_1nu'n^{-1}n) = f(n)$$

since $nu'n^{-1} \in U$ and $nv_{\theta\lambda}$ is a nonzero multiple of $v_{w\theta\lambda}$. Let $b \in B$. Then $b = ut$ for some $t \in T$, $u \in U$ and

$$f(xb) = \theta\lambda(t)f(u_1nu_2u) = \theta\lambda(t)f(n) = \theta\lambda(b)f(x) = \mu(b^{-1})f(x).$$

SECOND PROOF OF (*) (proposed by the referee). For each $y \in W$ denote by $F_{y\lambda}$ the B-submodule of V_λ generated by $v_{y\lambda}$ and set $F^+_{y\lambda} = \{v \in F_{y\lambda} \mid v^*_{y\lambda}(v) = 0\}$. It is easy to show that $F^+_{y\lambda}$ is a B-submodule of $F_{y\lambda}$, k-linearly spanned by elements of the form $uv_{y\lambda} - v_{y\lambda}, u \in U$. Therefore (*) implies that for any $y \leq w$ ($\leq$ is the Chevalley order in W) $v_{w\theta\lambda}$ does not belong to the B-module generated by $\underline{y}F^+_{y\lambda}$. In fact a stronger result can be derived directly. This result can then be used to prove (*).

For each $y \in W$ denote by $\Omega(y)$ the set of weights of $F_{y\lambda}$. Then one has:

LEMMA. *Let x, $y \in W$.*

(i) *If $x \leq y$ then $F_{x\lambda} \subseteq F_{y\lambda}$. Moreover, if λ is regular then $x \leq y$ if and only if $F_{x\lambda} \subseteq F_{y\lambda}$.*
(ii) *$\Omega(y)$ is contained in the convex hull of $\{z\lambda \mid z \in W, z \leq y\}$.*

PROOF. For a proof of (i) see [**BGG, Theorem 2.9**]. Part (ii) is proved by induction on $\ell(y)$. The claim clearly holds if $y = id$. So we can assume $y = s_\alpha x$, $\ell(x) = \ell(y) - 1$ (α is a simple root) and result proved for x. Since $y^{-1}\alpha \in -\Phi_+$, $y\lambda - m\alpha = y(\lambda - my^{-1}\alpha)$ is not a weight of V_λ for any positive integer m. Hence $u_{-\alpha}(\varepsilon)v_{y\lambda} = v_{y\lambda}$ for all $\varepsilon \in k$. Therefore, if P_α is the minimal parabolic subgroup of G corresponding to α, then $F_{y\lambda}$ is the P_α-submodule of V_λ generated by $F_{x\lambda}$. Hence $F_{y\lambda}$ is the k-span of the vectors $u_{-\alpha}(\varepsilon)v$, where $v \in F_{x\lambda}$ and $\varepsilon \in k$. Therefore, if $\nu \in \Omega(y)$, then $\nu = \eta - m\alpha$, for some $\eta \in \Omega(x)$, and a nonnegative integer m. But also $s_\alpha\nu \in \Omega(y)$, hence $s_\alpha\nu = \eta' - m'\alpha$, so that $\nu = s_\alpha\eta' + m'\alpha = \eta - m\alpha$. Hence ν belongs to the segment $[s_\alpha\eta', \eta]$. By induction hypothesis η and $s_\alpha\eta'$ belong, respectively, to the convex hull of $\{z\lambda \mid z \leq x\}$ and $\{s_\alpha z\lambda \mid z \leq x\}$. It follows that ν belongs to the convex hull of $\{z\lambda \mid z \leq s_\alpha x = y\}$, as required.

Now let $\nu \in \Omega(\theta)$ and $y \leq w$. By part (ii) of the lemma, $y\nu$ belongs to the convex hull of $\{y\theta'\lambda \mid \theta' \leq \theta\}$. Since $\ell(w\theta) = \ell(w) + \ell(\theta)$, then $y\theta' \leq w\theta$ for all $y \leq w$ and $\theta' \leq \theta$. Hence by part (i) of the lemma the weights $y\theta'\lambda$ belong to $\Omega(w\theta)$. It follows that $y\nu - w\theta\lambda$ belongs to the convex cone spanned by the positive roots, and $y\nu = w\theta\lambda$ if and only if $\nu = \theta\lambda$. From this one obtains $f(xb) = \theta\lambda(b)f(x) = \mu(b^{-1})f(x)$ for all $x \in \overline{BwB}$ and $b \in B$. $\square$

References

[BGG] Bernstein–Gelfand–Gelfand, *Schubert cells and cohomology of the spaces G/P*, Russian Math. Surveys **28** (1973), 1–26.

[Bo] N. Bourbaki, *Groupes et algèbres de Lie*, Ch. 4–6, Hermann, Paris, 1968.

[J] J. C. Jantzen, *Representations of Algebraic groups*, Academic Press, New York, 1987.

[F] W. Fenchel, *Convex Cones, Sets and Functions*, Princeton University Press, Princeton, NJ, 1951.

[P] P. Polo, *Variétés de Schubert et excellentes filtrations*, Astérisque **173–174** (1989), 281-311.

[S] G. Stevens, *Poincaré series on GL(r) and Kloosterman sums*, Math. Ann. **277** (1986), 25–51.

DEPARTMENT OF MATHEMATICS, INDIANA UNIVERSITY, BLOOMINGTON, INDIANA 47401

Contemporary Mathematics
Volume **139**, 1992

Kazhdan-Lusztig Bases and Isomorphism Theorems for q-Schur Algebras

JIE DU

ABSTRACT. This paper gives a basis for q-Schur algebras like the 'C'_w' basis by Kazhdan-Lusztig for the Hecke algebras. We also obtain useful results, such as the isomorphism theorem for q-Schur algebras which is an analogue of Tits' theorem in the Hecke algebra case.

0. Introduction

In their influential paper [**KL1**] in 1979, Kazhdan and Lusztig introduced some remarkable bases for the Hecke algebra of an arbitrary Coxeter group. These bases play an important role in the representation theory of algebraic groups and quantum groups. As an application of Kazhdan-Lusztig theory, Lusztig constructed an explicit isomorphism between the Hecke algebra of a Weyl group over $\mathbb{Q}(q^{\frac{1}{2}})$ and the group algebra over the same field of the Weyl group ([**L1**]). The existance of such an isomorphism was conjectured by Iwahori, and proved by Tits over the algebraic closure $\bar{\mathbb{Q}}$ of $\mathbb{Q}$ and by Benson and Curtis over $\mathbb{Q}$ (except for type E_7 and E_8). Some supplements to Lusztig's isomorphism theorem are given by Curtis in [**C1**].

The main result of this paper is an explicit construction of an isomorphism of the q-Schur algebra over $\mathbb{Q}(q^{\frac{1}{2}})$, introduced by Dipper and James in [**DJ2**], with the corresponding Schur algebra over $\mathbb{Q}(q^{\frac{1}{2}})$. To do this, the author first introduces in §2 some new bases for the q-Schur algebras, which behave in a similar way as the Kazhdan-Lusztig bases for the Hecke algebras. This work is motivated by [**C1, Theorem (1.10)**]. Some interesting properties for such bases are discussed in §3. We shall prove the isomorphism theorem in §4 as an application of these new bases. In a future paper, the author will study some

1991 *Mathematics Subject Classification.* Primary 16S50, 16S80; Secondary 17B37.

The research in this paper was carried out during the half year program on representations of finite groups and related topics at MSRI in Berkeley in 1990.

This paper is in final form and no version of it will be submitted for publication elsewhere.

further applications of these bases, such as the "asymptotic algebra" of a q-Schur algebra.

By using intersection cohomology approach, Beilinson, Lusztig and MacPherson constructed briefly in [**BLM**] some similar bases for q-Schur algebras (see also [**GL**]). We shall prove in an appendix that the bases we introduced in this paper are exactly the same as the Beilinson-Lusztig-MacPherson's bases.

The author would like to thank Professor Leonard Scott for many helpful conversations. He also wishes to thank Professor Charles Curtis for the reprints of his lecture notes [**C2**].

1. Kazhdan-Lusztig Bases for Hecke Algebras

1.1 Let $t = q^{\frac{1}{2}}$ be an indeterminate and let $A = \mathbb{Z}[t, t^{-1}]$ be the ring of Laurent polynomials in t. We set $A^- = \mathbb{Z}[t^{-1}]$, $A^+ = \mathbb{Z}[t]$.

Let (W, S) be a Coxeter group, where S is the corresponding set of the simple reflections. The Hecke algebra $\mathcal{H}$ over A corresponding to W, following [**KL1**], is a free A-module with basis $\{T_w\}_{w\in W}$ satisfying

$$\begin{array}{rll} T_wT_{w'} = T_{ww'} & \text{if} & l(ww') = l(w) + l(w'), \\ (T_s + 1)(T_s - q) = 0 & \text{if} & s \in S; \end{array}$$

here $l(w)$ is the length of w.

The ring A admits an involution $a \to \bar{a}$ defined by $\bar{t} = t^{-1}$. This extends to an involutive automorphism $h \to \bar{h}$ of $\mathcal{H}$, defined by $\overline{\sum a_w T_w} = \sum \bar{a}_w T_{w^{-1}}^{-1}$. In [**KL1**], Kazhdan and Lusztig showed, for any $w \in W$, there is a unique element $C_w \in \mathcal{H}$ such that $\overline{C_w} = C_w$ and

$$C_w = \sum_{y \le w} (-1)^{l(w)-l(y)} q^{l(w)/2-l(y)} P_{y,w}(q^{-1}) T_y. \tag{1.a}$$

where $\le$ is the usual Bruhat order on W and $P_{y,w}$ is a polynomial in q of degree $\le \frac{1}{2}(l(w) - l(y) - 1)$ for $y < w$, and $P_{w,w} = 1$. Also, the set $\{C_w\}_{w\in W}$ forms a basis of $\mathcal{H}$.

1.2 Let j be the involution of the ring $\mathcal{H}$ given by

$$j\left(\sum a_w T_w\right) = \sum (-1)^{l(w)} \bar{a}_w q^{-l(w)} T_w.$$

Let $B_w = (-1)^{l(w)} j(C_w)$, for each $w \in W$. Note that B_w is denoted C'_w in [**KL1**]. We have $\overline{B_w} = B_w$ and

$$B_w = q^{-l(w)/2} \sum_{y \le w} P_{y,w}(q) T_y. \tag{1.b}$$

$\{C_w\}_{w\in W}$ and $\{B_w\}_{w\in W}$ are usually called *Kazhdan-Lusztig bases.*

We define for any $x, y, z \in W$ some $g_{x,y,z}$ in A so that

$$B_x B_y = \sum_z g_{x,y,z} B_z.$$

According to [**Lu2, (3.1.2)**], $g_{x,y,z}$ has ≥ 0 coefficients as a polynomial in t, t^{-1}, and $g_{x,y,z}(t) = g_{x,y,z}(t^{-1})$. In particular, we have for $s \in S$,

$$(1.c) \qquad B_s B_y = \begin{cases} B_{sy} + \sum_{x<y, sx<x} \mu(x,y) B_x \text{ if } sy > y, \\ (t+t^{-1}) B_y \text{ if } sy < y, \end{cases}$$

where $\mu(x,y)$ is the coefficient of $t^{l(y)-l(x)-1}$ in $P_{x,y}$ for $x < y$.

For $z \in W$, we define

$$\mathbf{a}(z) = \max_{x,y \in W} \deg(g_{x,y,z})$$

where the degree is taken with respect to t. In fact, $\mathbf{a}(z)$ is the least nonnegative integer satifying

$$t^{\mathbf{a}(z)} g_{x,y,z} \in A^+, \forall x, y \in W.$$

This function on W is usually called the **a**-*function.*

1.3 We define a preorder $\leq_L$ on W as follows: $x \leq_L y$ if there exists $h \in \mathcal{H}$ such that the coefficient of B_x in the linear expression of hB_y is nonzero. The perorder $\leq_R$ is defined by $x \leq_R y$ if $x^{-1} \leq_L y^{-1}$. We denote by $\leq_{LR}$ the preorder generated by $\leq_L, \leq_R$. Note that these definitions coincide with that in [**KL1**] by (1.c). The equivalence classes in W defined by the preorders $\leq_L, \leq_R$ and $\leq_{LR}$ are called *left cells, right cells* and *two-sided cells.*

One remarkable property of the **a**-function is:

$$z \leq_{LR} z' \quad \Rightarrow \quad \mathbf{a}(z) \geq \mathbf{a}(z').$$

Therefore, it takes a constant value on a two-sided cell.

1.4 Let I, J be subsets of S, and let $\mathcal{D}_{IJ}$ denote the set of distinguished representatives of minimal length in the double cosets $W_I \backslash W / W_J$, where W_K is the subgroup of W generated by $K \subseteq S$. On the other hand, we denote by $\mathcal{D}^+_{IJ}$ the set of representatives of maximal length in the double cosets $W_I \backslash W / W_J$. Thus, each double coset $D \in W_I \backslash W / W_J$ is characterized as the set

$$D = \{y \in W : x \leq y \leq x^+\}$$

where $x \in D \cap \mathcal{D}_{IJ}$ and $x^+ \in D \cap \mathcal{D}^+_{IJ}$. We denote the obvious bijection from $\mathcal{D}_{IJ}$ to $\mathcal{D}^+_{IJ}$ by $x \to x^+$, and let $\mathcal{D}_I = \mathcal{D}_{I\emptyset}$.

We define, following [**C1**], an A-submodule $\mathcal{H}_{IJ}$ of $\mathcal{H}$ which is spanned by the standard basis elements

$$T_D = \sum_{x \in D} T_x, \quad D \in W_I \backslash W / W_J.$$

We have the following lemma (see [**C1, (1.9),(1.10)**]).

LEMMA 1.5. *Maitain the above notation.*

(1) *The A-submodule* $\mathcal{H}_{IJ}$ *of* $\mathcal{H}$ *is characterized as the set*

$$\{h \in \mathcal{H}; T_s h = qh = hT_{s'}, \forall s \in I, s' \in J\}.$$

(2) *The elements* $\{B_w; w \in \mathcal{D}_{IJ}^+\}$ *form an A-basis of* $\mathcal{H}_{IJ}$.

(3) $\overline{\mathcal{H}_{IJ}} = \mathcal{H}_{IJ}$. □

1.6 Let $\Phi : \mathcal{H} \to \mathcal{H}$ be the ring homomorphism defined by $\Phi(t) = -t$, $\Phi(T_x) = (-q)^{l(x)}T_{x^{-1}}^{-1}$, $x \in W$. Then $\Phi^2 = 1$ and $C_x = \Phi(B_x)$. Applying Φ to $\mathcal{H}_{IJ}$, we have

LEMMA 1.7. *Let* $\tilde{\mathcal{H}}_{IJ} = \Phi(\mathcal{H}_{IJ})$. *Then*

(1) $\tilde{\mathcal{H}}_{IJ}$ *can be characterized as the set*

$$\{h \in \mathcal{H}; T_s h = -h = hT_{s'}, \forall s \in I, s' \in J\}.$$

(2) *The elements* $C_x, x \in \mathcal{D}_{IJ}^+$ *form a basis of* $\tilde{\mathcal{H}}_{IJ}$.

(3) *The elements*

$$\tilde{T}_D = \sum_{w \in D} (-q)^{-l(w)} T_w, D \in W_I \backslash W / W_J$$

also form a basis of $\tilde{\mathcal{H}}_{IJ}$.

PROOF. Since $T_s^{-1} = q^{-1}T_s + (q^{-1} - 1)$ for $s \in S$, it follows that

$$T_s\Phi(h) = \Phi(-qT_s^{-1}h) = -q\Phi(T_s^{-1}h) = -\Phi(h)$$

for any $h \in \mathcal{H}_{IJ}$ and $s \in I$. Similarly, for $s' \in J$, we have $\Phi(h)T_{s'} = -\Phi(h)$. So (1) is proved. The proof of (2) is obvious, and (3) follows from the fact $j(T_D) = \tilde{T}_D$ and (2). □

§2. Kazhdan-Lusztig Bases for q-Schur Algebras

Form now on, we let $W = \mathfrak{S}_r$ be the symmetric group on r letters, and let S be the set of basic transpositions

$$(1, 2), (2, 3), \cdots, (r - 1, r).$$

2.1 Let λ be a composition of r, namely,

$$\lambda = (\lambda_1, \cdots, \lambda_m)$$

with $\lambda_1 + \cdots + \lambda_m = r$ and $\lambda_i \geq 0, \forall i$. λ determines naturally a subset $I(\lambda)$ of S. We will use the standard notation $W_\lambda, \mathcal{D}_{\lambda\mu}, \cdots$ instead of $W_{I(\lambda)}, \mathcal{D}_{I(\lambda),I(\mu)}, \cdots$ for the symmetric groups.

For a positive integer n, let $\Lambda(n,r)$ be the set of all compositions of r into n parts. We put $x_\lambda = T_{W_\lambda}$ and $y_\lambda = j(x_\lambda)$ (see 1.2). By 1.4 the right ideal $x_\lambda\mathcal{H}$ of $\mathcal{H}$ has a basis $\{x_\lambda T_z; z \in \mathcal{D}_\lambda\}$. Let

$$\mathcal{S}_q(n,r) = \mathrm{End}_{\mathcal{H}}\left(\bigoplus_{\lambda\in\Lambda(n,r)} x_\lambda\mathcal{H}\right).$$

$\mathcal{S}_q(n,r)$ is called a *q-Schur algebra*, following Dipper and James [**DJ2**]. If we specialize q to 1, then $\mathcal{S}_1(n,r)$ is the Schur algebra over $\mathbb{Z}$ (see [**G**]).

For $\lambda,\mu \in \Lambda(n,r), w \in \mathcal{D}_{\lambda\mu}$, define $\phi^w_{\lambda,\mu} \in \mathcal{S}_q(n,r)$ by

$$\phi^w_{\lambda,\mu}(x_\nu h) = \delta_{\mu,\nu} \sum_{x\in W_\lambda w W_\mu} T_x h, \tag{2.a}$$

where $\delta_{\mu,\nu}$ is the Kronecker δ. Dipper and James proved [**DJ2**] that the set $\{\phi^w_{\lambda,\mu}; \lambda,\mu \in \Lambda(n,r), w \in \mathcal{D}_{\lambda\mu}\}$ forms a basis of $\mathcal{S}_q(n,r)$. This is called the *standard basis* of $\mathcal{S}_q(n,r)$.

2.2 Let

$$\mathcal{B} = \{(\lambda,w,\mu); \lambda,\mu \in \Lambda(n,r), w \in \mathcal{D}_{\lambda\mu}\}.$$

We now define a new basis for $\mathcal{S}_q(n,r)$. Recall from 1.4 and 1.5 that for $w \in \mathcal{D}_{\lambda\mu}$, $B_{w^+} \in \mathcal{H}_{\lambda\mu}$. Write

$$B_{w^+} = \sum_{z\in\mathcal{D}_{\lambda\mu}} \alpha_{z,w} T_{W_\lambda z W_\mu}. \tag{2.b}$$

The explicit expression of $\alpha_{z,w}$ will be given in Lemma 3.8. We define for $b = (\lambda,w,\mu) \in \mathcal{B}$,

$$\theta_b = t^{l(w_{0,\mu})} \sum_{z\in\mathcal{D}_{\lambda\mu}} \alpha_{z,w}\phi^z_{\lambda,\mu}, \tag{2.c}$$

where $w_{0,\mu}$ is the longest element of W_μ.

THEOREM 2.3. (1) *The elements $\{\theta_b\}_{b\in\mathcal{B}}$ form a basis of $\mathcal{S}_q(n,r)$.*

(2) *If $b = (\lambda,w,\mu)$ and viewing θ_b as a map from $x_\mu\mathcal{H}$ to $x_\lambda\mathcal{H}$, then we have $\theta_b(B_{w_{0,\mu}}) = B_{w^+}$.*

PROOF. Let $\phi_\lambda = \phi^1_{\lambda,\lambda}$. Then ϕ_λ is an idempotent of $\mathcal{S}_q(n,r)$ and

$$\mathcal{S}_q(n,r) = \bigoplus_{\lambda,\mu\in\Lambda(n,r)} \phi_\lambda\mathcal{S}_q(n,r)\phi_\mu.$$

Obviously, for $\lambda,\mu \in \Lambda(n,r)$, we have

$$\phi_\lambda\mathcal{S}_q(n,r)\phi_\mu \cong \mathrm{Hom}_{\mathcal{H}}(x_\mu\mathcal{H}, x_\lambda\mathcal{H}),$$

and $\phi^w_{\lambda,\mu}$, $w \in \mathcal{D}_{\lambda\mu}$ form a basis of $\phi_\lambda\mathcal{S}_q(n,r)\phi_\mu$. Now, the map

$$\mathrm{Hom}_{\mathcal{H}}(x_\mu\mathcal{H}, x_\lambda\mathcal{H}) \to \mathcal{H}_{\lambda\mu}$$

given by sending $\phi^w_{\lambda,\mu}$ to $\phi^w_{\lambda,\mu}(x_\mu)$ is an A-module isomorphism by 1.4 and (2.a). Therefore, we obtain that the elements

$$\sum_{z\in\mathcal{D}_{\lambda\mu}} \alpha_{z,w}\phi^z_{\lambda,\mu}, w \in \mathcal{D}_{\lambda\mu}$$

form a basis of $\mathrm{Hom}_{\mathcal{H}}(x_\mu\mathcal{H}, x_\lambda\mathcal{H})$ by (1.5) and (2.b). Therefore, $\{\theta_b\}_{b\in\mathcal{B}}$ is a basis of $\mathcal{S}_q(n,r)$. (1) is proved.

Note from (1.b) that $B_{w_{0,\mu}} = t^{-l(w_{0,\mu})}x_\mu$. Thus, we have

$$\begin{aligned}\theta_b(B_{w_{0,\mu}}) &= t^{-l(w_{0,\mu})}\theta_b(x_\mu)\\ &= \sum_{z\in\mathcal{D}_{\lambda\mu}} \alpha_{z,w}\phi^z_{\lambda,\mu}(x_\mu)\\ &= B_{w^+}.\end{aligned}$$

So, (2) is proved. □

2.4 Recall from 1.6 the map Φ. Since $\Phi(q) = q$ and $\Phi(T_w) = (-q)^{l(w)}T^{-1}_{w^{-1}}$, Φ is a $\mathbb{Z}[q,q^{-1}]$-algebra automorphism of $\mathcal{H}$, and is the same map as defined in [**DJ1**, §**2**] (denoted by #). Thus, it is known that Φ induces an algebra isomorphism from $\mathcal{S}_q(n,r)$ to the algebra

$$\tilde{\mathcal{S}}_q(n,r) = \mathrm{End}_{\mathcal{H}}\left(\bigoplus_{\lambda\in\Lambda(n,r)} y_\lambda\mathcal{H}\right).$$

According to [**DJ0**] or [**Du1**], $\tilde{\mathcal{S}}_q(n,r)$ has a standard basis elements $\psi^w_{\lambda,\mu}, \forall\lambda,\mu \in \Lambda(n,r), w \in \mathcal{D}_{\lambda\mu}$, where

$$\psi^w_{\lambda,\mu}(y_\nu h) = \delta_{\mu,\nu} \sum_{x\in W_\lambda wW_\mu} (-q)^{-l(x)}T_x h,$$

for all $h \in \mathcal{H}$, $\nu \in \Lambda(n,r)$.

2.5 As in 2.2, we define for each $b = (\lambda, w, \mu) \in \mathcal{B}$,

$$\tilde{\theta}_b = (-t)^{-l(w_{0,\mu})} \sum_{z\in\mathcal{D}_{\lambda\mu}} \tilde{\alpha}_{z,w}\psi^z_{\lambda,\mu}$$

where $\tilde{\alpha}_{z,w}$ is defined by

$$C_{w^+} = \sum_{z\in\mathcal{D}_{\lambda\mu}} \tilde{\alpha}_{z,w}\tilde{T}_{W_\lambda wW_\mu}.$$

This definition makes sense by 1.7(3).

THEOREM 2.6. *Maintain the above notation.*

(1) *The elements* $\{\tilde{\theta}_b\}_{b\in\mathcal{B}}$ *form a basis for* $\tilde{\mathcal{S}}_q(n,r)$;
(2) *For each* $b = (\lambda, w, \mu) \in \mathcal{B}$, $\tilde{\theta}_b(C_{w_{0,\mu}}) = C_{w^+}$;
(3) $\Phi(\theta_b) = \tilde{\theta}_b, \forall b \in \mathcal{B}$.

PROOF. The proof of (1) and (2) is similar to that of Theorem 2.3. We now prove (3).

Since $C_{w_{0,\mu}} = (-t)^{l(w_{0,\mu})} y_\mu$ by (1.a), we have

$$\Phi(x_\mu) = \Phi(t^{l(w_{0,\mu})} B_{w_{0,\mu}}) \\ = (-t)^{l(w_{0,\mu})} C_{w_{0,\mu}} \\ = q^{l(w_{0,\mu})} y_\mu. \tag{2.d}$$

On the other hand, if $\phi^w_{\lambda,\mu}(x_\mu) = h_w x_\mu$ for some $h_w \in \mathcal{H}$ then

$$\Phi(\phi^w_{\lambda,\mu})(y_\mu) = \Phi(h_w) y_\mu \tag{2.e}$$

by definition [**DJ1, §2**]. Thus, for $b = (\lambda, w, \mu) \in \mathcal{B}$, we have by (2.c)

$$\begin{aligned} \Phi(\theta_b)(C_{w_{0,\mu}}) &= (-t)^{l(w_{0,\mu})} \sum_{z \in \mathcal{D}_{\lambda\mu}} \Phi(\alpha_{z,w}) \Phi(\phi^z_{\lambda,\mu})(C_{w_{0,\mu}}) \\ &= (-t)^{l(w_{0,\mu})} \sum_{z \in \mathcal{D}_{\lambda\mu}} \Phi(\alpha_{z,w}) \Phi(h_z) C_{w_{0,\mu}} \quad \text{by (2.e)} \\ &= (-t)^{2l(w_{0,\mu})} \sum_{z \in \mathcal{D}_{\lambda\mu}} \Phi(\alpha_{z,w}) \Phi(h_z) y_\mu \\ &= \sum_{z \in \mathcal{D}_{\lambda\mu}} \Phi(\alpha_{z,w}) \Phi(h_z) \Phi(x_\mu) \quad \text{by (2.d)} \\ &= \Phi(B_{w^+}) \\ &= C_{w^+} \\ &= \tilde{\theta}_b(C_{w_{0,\mu}}). \end{aligned}$$

Therefore, $\Phi(\theta_b) = \tilde{\theta}_b$. □

We call $\{\theta_b\}_{b \in \mathcal{B}}$ and $\{\tilde{\theta}_b\}_{b \in \mathcal{B}}$ the *Kazhdan-Lusztig bases* of the q-Schur algebras.

REMARK 2.7. As mentioned in [**Du1**], the isomorphism

$$\Phi : \mathcal{S}_q(n, r) \to \tilde{\mathcal{S}}_q(n, r)$$

does not map the standard basis elements $\phi^w_{\lambda,\mu}$ to the standard basis elements $\psi^w_{\lambda,\mu}$. The previous theorem shows that it does map θ_b to $\tilde{\theta}_b$. In fact, the isomorphism

$$j : \mathcal{S}_q(n, r) \to \tilde{\mathcal{S}}_q(n, r)$$

induced by the automorphism j of $\mathcal{H}$ maps $\phi^w_{\lambda,\mu}$ to $\psi^w_{\lambda,\mu}$. The details are left to the reader (see 1.7).

§3. Basic Properties

For simplicity, we denote $\phi^w_{\lambda,\mu}$ by ϕ_b where $b = (\lambda, w, \mu)$.

3.1 We first extend the involution $^-$ on $\mathcal{H}$ to the q-Schur algebras. Recall Lemma 1.5 (3). For any $b = (\lambda, w, \mu) \in \mathcal{B}$, we have $\phi_b(B_{w_{0,\mu}}) \in \mathcal{H}_{\lambda\mu}$, and hence $\overline{\phi_b(B_{w_{0,\mu}})} \in \mathcal{H}_{\lambda\mu}$. So, we may write

$$\overline{\phi_b(B_{w_{0,\mu}})} = \sum_{z \in \mathcal{D}_{\lambda\mu}} \alpha_{z,b} T_{W_\lambda z W_\mu}.$$

We now define a map $^-$: $\mathcal{S}_q(n,r) \to \mathcal{S}_q(n,r)$ such that

$$\bar{t} = t^{-1},$$
$$\overline{\phi_b} = \sum_{z \in \mathcal{D}_{\lambda\mu}} \alpha_{z,b} \phi^z_{\lambda,\mu}.$$

PROPOSITION 3.2. (1) *The map* $^-$ *is an algebra automorphism of order 2.*

(2) *If* $n \geq r$, *then the restriction of* $^-$ *on* $\mathcal{H}$ *as a subalgebra of* $\mathcal{S}_q(n,r)$ *coincides with the involution* $^-$ *on* $\mathcal{H}$.

(3) $\overline{\theta_b} = \theta_b$ *for all* $b \in \mathcal{B}$.

PROOF. (1) It is suffices to prove that $^-$ is an algebra homomorphism. We need to prove that $\overline{\phi_a \phi_b} = \overline{\phi_a}\,\overline{\phi_b}$ for all $a, b \in \mathcal{B}$. Without loss of generality, we may assume $a = (\lambda, y, \mu)$, $b = (\mu, w, \nu)$. Now,

$$\begin{aligned}
\overline{\phi_a}\,\overline{\phi_b}(B_{w_{0,\nu}}) &= \overline{\phi_a}(\overline{\phi_b(B_{w_{0,\nu}})}) \\
&= \overline{\phi_a}\left(t^{l(w_{0,\nu})}\overline{\phi_b(x_\nu)}\right) \quad \text{by (1.b)} \\
&= \overline{\phi_a}\left(t^{l(w_{0,\nu})}\overline{x_\mu h}\right) \quad \text{where } \phi_b(x_\nu) = x_\mu h \\
&= \overline{\phi_a}\left(t^{l(w_{0,\nu}) - l(w_{0,\mu})} B_{w_{0,\mu}} \bar{h}\right) \\
&= t^{l(w_{0,\nu}) - l(w_{0,\mu})} \overline{\phi_a}(B_{w_{0,\mu}}) \bar{h} \\
&= t^{l(w_{0,\nu}) - l(w_{0,\mu})} \overline{\phi_b(B_{w_{0,\mu}})h} \\
&= t^{l(w_{0,\nu})} \overline{\phi_b(x_\mu)h} \\
&= \overline{\phi_a \phi_b(B_{w_{0,\mu}})}.
\end{aligned}$$

Therefore, (1) is proved. To prove (2), we note that

$$\mathcal{H} \cong \phi_\omega \mathcal{S}_q(n,r) \phi_\omega$$

given by $T_x \to \phi^x_{\omega\omega}$, where $\omega = (1^r) \in \Lambda(n,r)$. So the rest of the proof is obvious. Finally, the proof of (3) is straightforward by using Theorem 2.3(2). $\square$

3.3 For each $b = (\lambda, w, \mu) \in \mathcal{B}$, b determines uniquely the element $w^+ \in \mathcal{D}^+_{\lambda\mu}$. Let

$$\mathcal{B}' = \coprod_{\lambda,\mu \in \Lambda(n,r)} \mathcal{D}^+_{\lambda\mu},$$

be a disjoint union. Then we have a bijection between $\mathcal{B}$ and $\mathcal{B}'$. So, we may identify $\mathcal{B}$ with $\mathcal{B}'$. Thus, the notation $\mathbf{a}(b)$ for $b \in \mathcal{B}$ makes sense. Using this identification and noting that if

$$B_x B_y = \sum_z g_{x,y,z} B_z$$

for $x \in \mathcal{D}^+_{\lambda\mu}, y \in \mathcal{D}^+_{\mu\rho}$, then $g_{x,y,z} \neq 0$ only if $z \in \mathcal{D}^+_{\lambda\rho}$, we introduce an algebra $\mathcal{S}$ which is free A-module with basis $\{\beta_a\}_{a\in\mathcal{B}}$ satisfying

$$\beta_a\beta_b = \sum_{c\in\mathcal{B}} g_{a,b,c}\beta_c \tag{3.a}$$

for any $a = (\lambda, x, \mu)$, $b = (\nu, y, \rho) \in \mathcal{B}$, where $g_{a,b,c}$ is defined in 1.2 for $\mu = \nu$, and $g_{a,b,c} = 0$ if $\mu \neq \nu$. We see that the structure constants for $\mathcal{S}$ come from that for $\mathcal{H}$.

For any $a, b \in \mathcal{B}$, we define elements $f_{a,b,c}$ in A by

$$\theta_a\theta_b = \sum_{c\in\mathcal{B}} f_{a,b,c}\theta_c.$$

The following result shows that the structure constants for $\mathcal{S}_q(n, r)$ are also determined by that for $\mathcal{H}$.

PROPOSITION 3.4. *Maintain the notation in* 3.3. *For any* $a, b, c \in \mathcal{B}$, *there exists a Laurent polynomial* $h_{a,b} \in A$ *such that*

$$h_{a,b} f_{a,b,c} = g_{a,b,c}.$$

PROOF. Let $a = (\lambda, x, \mu)$, $b = (\nu, y, \rho)$. If $\mu \neq \nu$, then

$$f_{a,b,c} = g_{a,b,c} = 0 \qquad \text{for all } c \in \mathcal{B},$$

and we take $h_{a,b} = 0$. We now assume $\mu = \nu$. According to (1.c) and Lemma 1.5, we have

$$B_{w_{0,\mu}} B_{y^+} = h_\mu B_{y^+} \quad \text{where} \quad h_\mu = t^{-l(w_{0,\mu})} \sum_{w\in W_\mu} t^{2l(w)}.$$

Now, by noting the identification in 3.3, we have

$$\begin{aligned}
\sum_{c\in\mathcal{B}} f_{a,b,c} B_c &= \theta_a\theta_b(B_{w_{0,\rho}}) \\
&= \theta_a(B_b) \\
&= \theta_a(h_\mu^{-1} B_{w_{0,\mu}} B_b) \\
&= h_\mu^{-1}\theta_a(B_{w_{0,\mu}}) B_b \\
&= h_\mu^{-1} B_a B_b \\
&= \sum_{c\in\mathcal{B}} h_\mu^{-1} g_{a,b,c} B_c.
\end{aligned}$$

Therefore, let $h_{a,b} = h_\mu$, then $h_{a,b} f_{a,b,c} = g_{a,b,c}$ for all $c \in \mathcal{B}$, and the conclusion follows. □

COROLLARY 3.5. *Let K be a field containing A. Then the algebra $\mathcal{S}^K$ is isomorphic to the q-Schur algebra $\mathcal{S}_q(n,r)^K$ if K is large enough, where $(-)^K = - \otimes_A K$.*

PROOF. Suppose $\sqrt{h_\lambda} \in K$ for all $\lambda \in \Lambda(n,r)$. Put for each $b = (\lambda, w, \mu)$,

$$\beta'_b = \frac{\beta_b}{\sqrt{h_\lambda h_\mu}}.$$

Then, by Proposition 3.4, the map $\beta'_b \mapsto \theta_b$ from $\mathcal{S}^K$ to $\mathcal{S}_q(n,r)^K$ gives an algebra isomorphism. □

3.6 Let M be a free A-module and let $^- : M \to M$ be a $\mathbb{Z}$-linear map of order 2 such that $\overline{tm} = t^{-1}\overline{m}$ for all $m \in M$. Pick an A^--lattice $\mathcal{L} \subseteq M$ with basis $\{m_i\}_{i \in I}$. Then $\overline{\mathcal{L}}$ is an A^+-lattice of M. In general, the following diagram (compare [**L5**])

$$\text{(3.b)}\qquad \begin{array}{ccccc} \mathcal{L}/t^{-1}\mathcal{L} & \xleftarrow{\pi'} & \mathcal{L}\cap\bar{\mathcal{L}} & \xrightarrow{i} & \bar{\mathcal{L}} \\ \| & & \downarrow^{j} & & \| \\ \mathcal{L}/t^{-1}\mathcal{L} & \xleftarrow[\pi]{} & \mathcal{L} & \xrightarrow[-]{} & \bar{\mathcal{L}} \end{array}$$

is not necessarily commutative, where i, j is the inclusion map and π is the natural projection and $\pi' = \pi j$. In fact, such a commutativity relates to the existence of certain interesting basis of M. More precisely, we have (see [**L4**]),

DEFINITION 3.7. Maintain the notation above. If, for each $i \in I$, there is a unique element $c_i \in \mathcal{L}$ such that $\bar{c}_i = c_i$ and $\pi(c_i) = \pi(m_i)$, then the basis $\{c_i\}_{i\in I}$ is called the *IC basis* of M with respect to $(\mathcal{L}, ^-)$.

Obviously, the diagram (3.b) is commutative if the IC basis of M exists.

As Lusztig pointed out in [**L4**] [**L5**], the basis $\{B_w\}_{w\in W}$ of a Hecke algebra, or the canonical basis of a quantum eveloping algebras [**L3**] is an IC basis in the sense above by choosing an appropriate lattice $\mathcal{L}$ and the map $^-$. We now show this is also true for the basis $\{\theta_b\}_{b\in\mathcal{B}}$ of the q-Schur algebra.

LEMMA 3.8. *For each $w \in \mathcal{D}_{\lambda\mu}$, if*

$$B_{w^+} = \sum_{z\in\mathcal{D}_{\lambda\mu}} \alpha_{z,w} T_{W_\lambda z W_\mu},$$

Then $z^+ \leq w^+$ and $\alpha_{z,w} = t^{-l(w^+)}P_{z^+,w^+}$. Therefore, the degree in t of $\alpha_{z,w}$ is at most $-l(z^+) - 1$.

PROOF. This follows immediately from (1.b) and Lemma 1.5. □

For any $b = (\lambda, w, \mu) \in \mathcal{B}$, we put

$$\text{(3.c)}\qquad \hat{\phi}_b = t^{-l(w^+)+l(w_{0,\mu})}\phi_b.$$

By Lemma 3.8 and (2.c), we have

$$\begin{aligned}\theta_b &= t^{l(w_{0,\mu})} \sum_{z\in\mathcal{D}_{\lambda\mu}, z^+\le w^+} t^{-l(w^+)} P_{z^+,w^+} \phi_{(\lambda,z,\mu)} \\ &= \sum_{z\in\mathcal{D}_{\lambda\mu}, z^+\le w^+} t^{-l(w^+)+l(z^+)} P_{z^+,w^+} \hat{\phi}_{(\lambda,z,\mu)}.\end{aligned}$$

Therefore,

$$\theta_b \in \hat{\phi}_b + t^{-1} \sum_{z\in\mathcal{D}_{\lambda\mu}, z^+\le w^+} A^- \hat{\phi}_{(\lambda,z,\mu)}.$$

Thus, we have

THEOREM 3.9. *Keep the above notation. Let $\mathcal{L}$ be the A^--lattice of $\mathcal{S}_q(n,r)$ with basis $\hat{\phi}_b$, $b\in\mathcal{B}$ and let $^-$ be defined as in* 3.1. *Then $\{\theta_b\}_{b\in\mathcal{B}}$ is an IC basis.*

PROOF. By 3.2(3) and the previous argument, we have $\bar{\theta}_b = \theta_b$, and θ_b and $\hat{\phi}_b$ have the same image in $\mathcal{L}/t^{-1}\mathcal{L}$. The uniqueness follows from a similar argument as in [**KL1, 2.2**]. □

3.10 The Kazhdan-Lusztig bases for Hecke algebras and the canonical bases for quantum enveloping algebras have the positivity properties for structure constants (see 1.2 and [**L3**]). It is natural to conjecture that such a property holds also for the structure constants $f_{a,b,c}$ defined in 3.3. It is very likely that this would follow from Beilison-Lusztig-MacPherson's construction and an argument similar to that in [**Sp, §2**].

§4. Isomorphism Theorems

In this section we choose a second indeterminate t' and put $A' = \mathbb{Z}[t', t'^{-1}]$. Also, we put $A^{\mathbb{Q}} = \mathbb{Q}[t, t^{-1}]$; $A'^{\mathbb{Q}} = \mathbb{Q}[t', t'^{-1}]$. We shall use the same notation $\mathcal{S}_q(n,r)$ to denote the q-Schur algebras over $A^{\mathbb{Q}}$.

4.1 In [**L1**], G. Lusztig constructed explicitly an isomorphism between the Hecke algebra $\mathcal{H}^{\mathbb{Q}(t)}$ over $\mathbb{Q}(t)$ and the group algebra $\mathbb{Q}(t)\mathfrak{S}_r$. In this section, we are going to establish an isomorphism theorem between the q-Schur algebra $\mathcal{S}_q(n,r)^{\mathbb{Q}(t)}$ and the Schur algebra $\mathcal{S}_1(n,r)^{\mathbb{Q}(t)}$, by using the Kazhdan-Lusztig bases for q-Schur algebras. The proof of this result is essentially similar to that of [**C2, (8.2)**] which is actually based on the proof due to Lusztig ([**L1**]). We will see from 4.10 that when $n \ge r$, the restriction of this isomorphism to the Hecke algebra coincides with the Lusztig's isomorphism.

4.2 Let $\mathcal{S}_{q'}(n,r)$ be the q-Schur algebra over $A'^{\mathbb{Q}}$, where we have put $q' = t'^2$. Denote by θ'_b, $b\in\mathcal{B}$ the Kazhdan-Lusztig basis of $\mathcal{S}_{q'}(n,r)$ as (2.c), and denote by the elements $f'_{a,b,c}$ defined as in 3.3 with respect to t' , for all $a,b,c\in\mathcal{B}$, the structure constants of $\mathcal{S}_{q'}(n,r)$.

As in [**L1**], we consider a free A-module M with basis $\{m_b\}_{b\in\mathcal{B}}$. We let $M_{t,t'} = M\otimes A^{\mathbb{Q}}_{t,t'}$, where $A_{t,t'} = \mathbb{Z}[t,t^{-1},t',t'^{-1}]$. We define a left action of $\mathcal{S}_q(n,r)$ and a right action of $\mathcal{S}_{q'}(n,r)$ on $M_{t,t'}$ as follows:

$$\theta_a m_b = \sum_{c\in\mathcal{B}} f_{a,b,c} m_c, \quad m_b\theta'_c = \sum_{d\in\mathcal{B}} f'_{b,c,d} m_d \tag{4.a}$$

where $a, b, c, d \in \mathcal{B}$. These actions do not necessarily commute.

LEMMA 4.3. *For all* $a, b, c \in \mathcal{B}$, *we have*

$$(\theta_a m_b)\theta'_c - \theta_a(m_b\theta'_c) \in \sum_{d\in\mathcal{B},\mathbf{a}(d)>\mathbf{a}(b)} A^{\mathbb{Q}}_{t,t'} m_d$$

where $\mathbf{a}(-)$ *is the* $\mathbf{a}$*-function defined in* 1.3.

PROOF. By the definition, it suffices to prove the lemma for

$$a = (\lambda, x, \mu), b = (\mu, y, \nu), c = (\nu, z, \rho) \in \mathcal{B}.$$

Consider the Hecke algebra $\mathcal{H}_{t,t'} = \mathcal{H} \otimes A^{\mathbb{Q}}_{t,t'}$ over $A^{\mathbb{Q}}_{t,t'}$, and define a second multiplication $*$ on $\mathcal{H}_{t,t'}$ by setting for any $u, v \in W$

$$B_u * B_v = \sum_{w\in W} g'_{u,v,w} B_w$$

where $g'_{u,v,w}$ is defined as in 1.2 with respect to t'. Thus, by **[C2, (7.3)]**, we have

$$(B_u B_v) * B_w - B_u(B_v * B_w) \in \sum_{v'\in W,\mathbf{a}(v')>\mathbf{a}(v)} A^{\mathbb{Q}}_{t,t'} B_{v'}.$$

It follows that

$$\sum_{u'} g_{u,v,u'} g'_{u',w,v'} = \sum_{w'} g_{u,w',v'} g'_{v,w,w'}$$

whenever $\mathbf{a}(v') = \mathbf{a}(v)$.

Now, take $(u, v, w) = (x^+, y^+, z^+)$. According to 3.3, the above equation becomes the form

$$\sum_{a'\in\mathcal{D}^+_{\lambda\nu}} g_{a,b,a'} g'_{a',c,b'} = \sum_{c'\in\mathcal{D}^+_{\mu\rho}} g_{a,c',b'} g'_{b,c,c'}.$$

whenever $\mathbf{a}(b') = \mathbf{a}(b)$ where $b' \in \mathcal{D}^+_{\lambda\rho}$.

Applying Proposition 3.4 and noting $h_{a,b} = h_{a,c'}$, $h'_{a',c} = h'_{b,c}$, we obtain

$$\sum_{a'} f_{a,b,a'} f'_{a',c,b'} = \sum_{c'} f_{a,c',b'} f'_{b,c,c'}$$

whenever $\mathbf{a}(b') = \mathbf{a}(b)$. So, the lemma is proved. □

4.4 Recall from 1.3 the preorder $\leq_{LR}$. It induces a preorder $\leq$ on the set of two-sided cells, namely, if Ω, Ω' are two-sided cells we define $\Omega \leq \Omega'$ if there exist $w \in \Omega$, $w' \in \Omega'$ such that $w \leq_{LR} w'$.

For each two-sided cell Ω, let $M(\Omega)$ be the $A^{\mathbb{Q}}$-submodule of M generated by m_b, $b \in \Omega$. $M(\Omega)$ is not a (left) $\mathcal{S}_q(n, r)$-submodule. However, by **[L2, (4.3.3)]**, the $A^{\mathbb{Q}}$-submodule

$$M^{\leq}_{\Omega} = \sum_{\Omega'\leq\Omega} M(\Omega')$$

is a left $\mathcal{S}_q(n, r)$- and right $\mathcal{S}_{q'}(n, r)$-submodule of M.

Let $\mathcal{S}(n,r)$ be the Schur algebra over $A^{\mathbb{Q}}$, namely, $\mathcal{S}(n,r) \cong \mathcal{S}_1(n,r) \otimes A^{\mathbb{Q}}$. Let $\mathcal{S}(n,r)$ act on M from right by setting $t' = 1$ in the action of $\mathcal{S}_{q'}(n,r)$ defined in 4.1. Obviously, $M_{\Omega}^{\leq}$ is also a right $\mathcal{S}(n,r)$-submodule of M, and if we order the set of two-sided cells as $\Omega_1, \Omega_2, \cdots, \Omega_m$, where $\Omega_i \leq \Omega_j$ implies $i < j$, we obtain a filtration of M:

$$0 \subseteq M_1 \subseteq M_2 \subseteq \cdots \subseteq M_m = M, \tag{4.b}$$

where $M_i = \sum_{j \leq i} M_{\Omega_j}^{\leq}$ for all i. Let gr$M$ be the graded module associated with this filtration. It is a left $\mathcal{S}_q(n,r)$-module and a right $\mathcal{S}(n,r)$-module. By Lemma 4.3 we have

LEMMA 4.5. *The graded $A^{\mathbb{Q}}$-module grM is a $\mathcal{S}_q(n,r)$-$\mathcal{S}(n,r)$-bimodule.* □

We now can prove the following isomorphism theorem.

THEOREM 4.6. (1) *There is a unique homomorphism $\eta : \mathcal{S}_q(n,r) \to \mathcal{S}(n,r)$ such that for each $b \in \mathcal{B}$, and $k \in \mathcal{S}_q(n,r)$, we have*

$$km_b - \eta(k)m_b = \sum_{a \not\sim_{LR} b} r_a m_a$$

where $r_a \in A^{\mathbb{Q}}$.

(2) *The extended map $\eta \otimes 1 : \mathcal{S}_q(n,r)^{\mathbb{Q}(t)} \to \mathcal{S}(n,r)^{\mathbb{Q}(t)}$ is an isomorphism as $\mathbb{Q}(t)$-algebras.*

PROOF. As mentioned at the beginning of this section, the proof sketched below is essentially due to G. Lusztig.

Firstly, by (4.5), the $\mathcal{S}_q(n,r)$-$\mathcal{S}(n,r)$-bimodule on gr$M$ determines uniquely an $A^{\mathbb{Q}}$-algebra homomorphism

$$\alpha : \mathcal{S}_q(n,r) \to \mathrm{End}_{\mathcal{S}(n,r)}(\mathrm{gr}M).$$

Secondly, by specializing q to 1 as in 4.4, it is clear that M, hence gr$M$, becomes an $\mathcal{S}(n,r)$-bimodule which affords in fact the two-sided regular representation of $\mathcal{S}(n,r)$. So the $A^{\mathbb{Q}}$-algebra homomorphism

$$\beta : \mathcal{S}(n,r) \to \mathrm{End}_{\mathcal{S}(n,r)}(\mathrm{gr}M)$$

defined by the two-sided action of $\mathcal{S}(n,r)$ on gr$M$ is an isomorphism. Now, let $\eta = \beta^{-1} \circ \alpha$. The first statement follows.

To show $\eta \otimes 1$ is an isomorphism, we note that $\mathcal{S}_q(n,r)^{\mathbb{Q}(t)}$ is a semi-simple $\mathbb{Q}(t)$-algebra ([**DJ0**]), and

$$\dim_{\mathbb{Q}(t)} \mathcal{S}_q(n,r)^{\mathbb{Q}(t)} = \dim_{\mathbb{Q}(t)} \mathcal{S}(n,r)^{\mathbb{Q}(t)}.$$

Now, a standard argument completes the proof. □

4.7 Recall from the proof of Theorem 2.3 that the elements ϕ_λ, $\lambda \in \Lambda(n,r)$ are idempotents of $\mathcal{S}_q(n,r)$, and the proof of 3.5 shows that the map

$$J_\lambda : \mathcal{H}_{\lambda\lambda} \to \phi_\lambda \mathcal{S}_q(n,r) \phi_\lambda$$

sending B_{w^+} to $h_\lambda\theta_{(\lambda,w,\lambda)}$ is an $A^{\mathbb{Q}}$-algebra homomorphism. Similarly, let $\xi_\lambda = \phi_\lambda|_{t=1}$, the specialization of ϕ_λ when $t=1$. Then the elements ξ_λ, $\lambda \in \Lambda(n,r)$, are idempotents of $\mathcal{S}(n,r)$, and the map

$$j_\lambda = J_\lambda|_{t=1} \otimes 1 : H(W, W_\lambda) \to \xi_\lambda \mathcal{S}(n,r)\xi_\lambda$$

is an $A^{\mathbb{Q}}$-algebra isomorphism, where $H(W, W_\lambda)$ is the subalgebra of group ring $A^{\mathbb{Q}}W$ with $A^{\mathbb{Q}}$-basis

$$\{\tau_D\}, \quad \text{where} \quad \tau_D = \sum_{x\in D} x, \quad D \in W_\lambda\backslash W/W_\lambda$$

(see [**C1,** §**1**]), and $\xi_\lambda = \xi_\lambda \otimes 1$ by abuse of notation.

The $\phi_\lambda\mathcal{S}_q(n,r)\phi_\lambda$-module $\phi_\lambda M$ has a filtration obtained by applying ϕ_λ to (4.b). The associated graded $A^{\mathbb{Q}}$-module $\mathrm{gr}\phi_\lambda M$ is clearly isomorphic to $\phi_\lambda\mathrm{gr}M$ as $\phi_\lambda\mathcal{S}_q(n,r)\phi_\lambda$-$\mathcal{S}(n,r)$-bimodules. Thus, the induced $A^{\mathbb{Q}}$-algebra homomorphism

$$\alpha_\lambda : \phi_\lambda\mathcal{S}_q(n,r)\phi_\lambda \to \mathrm{End}_{\mathcal{S}(n,r)}(\mathrm{gr}\phi_\lambda M)$$

is acturally the restriction of α to $\phi_\lambda\mathcal{S}_q(n,r)\phi_\lambda$.

Similarly, the $\xi_\lambda\mathcal{S}(n,r)\xi_\lambda$-$\mathcal{S}(n,r)$-bimodule structure of $\mathrm{gr}\xi_\lambda M$ determines an $A^{\mathbb{Q}}$-algebra isomorphism

$$\beta_\lambda : \xi_\lambda\mathcal{S}(n,r)\xi_\lambda \to \mathrm{End}_{\mathcal{S}(n,r)}(\mathrm{gr}\xi_\lambda M)$$

which is acturally the restriction of β to $\xi_\lambda\mathcal{S}(n,r)\xi_\lambda$. Now, by the definition in (4.a), it is clear that $\phi_\lambda M = \xi_\lambda M$, and hence $\mathrm{gr}\phi_\lambda M = \mathrm{gr}\xi_\lambda M$. Therefore, we have

PROPOSITION 4.8. *Keep the notation in* 4.6.

(1) *The restriction of* η *to* $\phi_\lambda\mathcal{S}_q(n,r)\phi_\lambda$ *induces the* $A^{\mathbb{Q}}$*-algebra homomorphism*

$$\eta_\lambda = \beta_\lambda^{-1} \circ \alpha_\lambda : \phi_\lambda\mathcal{S}_q(n,r)\phi_\lambda \to \xi_\lambda\mathcal{S}(n,r)\xi_\lambda.$$

(2) *The extended map* $\eta_\lambda \otimes 1 : (\phi_\lambda\mathcal{S}_q(n,r)\phi_\lambda)^{\mathbb{Q}(t)} \to (\xi_\lambda\mathcal{S}(n,r)\xi_\lambda)^{\mathbb{Q}(t)}$ *is an* $\mathbb{Q}(t)$*-algebra isomorphism.* □

We are now going to compare the homomorphism η_λ with those homomorphisms defined in [**C1, (2.4)**].

Let E_λ be the $A^{\mathbb{Q}}$-submodule of $\mathcal{S}$ (see 3.3) generated by $\{\beta_b\}_{b\in\mathcal{D}^+_{\lambda\lambda}}$. $\mathcal{H}_{\lambda\lambda}$ acts on E_λ from both sides via (3.a) by viewing $\mathcal{H}_{\lambda\lambda}$ as a subalgebra of $\mathcal{S}$. This induces a left and a right action of $H(W, W_\lambda)$ by specializing t to 1. We denote the left $H(W, W_\lambda)$-action on E_λ by $*$. Clearly, E_λ is isomorphic to $E'_{I(\lambda)}$, the free $A^{\mathbb{Q}}$-module defined in [**C1,** §**1**], as a right $H(W, W_\lambda)$- or a left $\mathcal{H}_{\lambda\lambda}$- module. By the map J_λ above, we define an $A^{\mathbb{Q}}$-module isomorphism

$$J_\lambda : E_\lambda \to \xi_\lambda M\xi_\lambda$$

by sending β_a to $h_\lambda m_a$, for each $a \in \mathcal{D}^+_{\lambda\lambda}$. Thus, we have

$$J_\lambda(k\beta_a) = J_\lambda(k)J_\lambda(\beta_a), \quad J_\lambda(g * \beta_a) = j_\lambda(g)J_\lambda(\beta_a)$$

for all $k \in \mathcal{H}_{\lambda\lambda}$, $g \in H(W, W_\lambda)$ and $a \in \mathcal{D}^+_{\lambda\lambda}$.

THEOREM 4.9. *Let $\varphi_\lambda : \mathcal{H}_{\lambda\lambda} \to H(W, W_\lambda)$ be the map defined as in* [**C1, (2.4)**]. *Then, the diagram*

$$\begin{array}{ccc} \mathcal{H}_{\lambda\lambda} & \xrightarrow{\varphi_\lambda} & H(W, W_\lambda) \\ {\scriptstyle J_\lambda}\downarrow & & \downarrow{\scriptstyle j_\lambda} \\ \phi_\lambda \mathcal{S}_q(n,r)\phi_\lambda & \xrightarrow{\eta_\lambda} & \xi_\lambda \mathcal{S}(n,r)\xi_\lambda \end{array}$$

is commutative, that is, we have $\varphi_\lambda = j_\lambda^{-1} \circ \eta_\lambda \circ J_\lambda$.

In particular, when $n \geq r$ and $\lambda = \omega = (1^r)$, J_λ is an $A^{\mathbb{Q}}$-algebra isomorphism. So we have $\eta_\omega = \varphi_\omega$.

PROOF. The homomorphism φ_λ is characterized by the condition that for each $k \in \mathcal{H}_{\lambda\lambda}$ and $a \in \mathcal{D}^+_{\lambda\lambda}$,

$$k\beta_a - \varphi_\lambda(k) * \beta_a = \sum_{b \in \mathcal{D}^+_{\lambda\lambda}, b \not\sim_{LR} a} r_b \beta_b$$

for some $r_b \in A$. Applying J_λ to this equation, we have

$$J_\lambda(k)h_\lambda m_a - j_\lambda(\varphi_\lambda(k))h_\lambda m_a = \sum_b r_b h_\lambda m_b.$$

Therefore, by Theorem 4.6 we obtain

$$\eta_\lambda(J_\lambda(k)) = j_\lambda(\varphi_\lambda(k))$$

as desired. □

REMARK 4.10. (1) As pointed out by Curtis in [**C1**], the isomorphism φ_λ is conjugate, in a certain sense, to the isomorphism defined in [**L1**]. However, if we replace $\mathcal{S}_q(n,r)$, θ_b, etc. by $\tilde{\mathcal{S}}_q(n,r)$, $\tilde{\theta}_b$, etc., a parallel argument shows that all results in §3 and §4 can be translated into the corresponding results for $\tilde{\mathcal{S}}_q(n,r)$. In particular, the isomorphism theorem holds. If $\tilde{\eta}$ denotes the corresponding $A^{\mathbb{Q}}$-algebra homomorphism from $\tilde{\mathcal{S}}_q(n,r)$ to $\mathcal{S}(n,r)$, and $\tilde{\eta}_\lambda$ is the restriction as in 4.8, then $\tilde{\eta}_\omega$ (for $n \geq r$) is exactly the same isomorphism defined in [**L1**]. We leave the details to the reader.

(2) Let K be a field, and let $A^{\mathbb{Q}} \to K$ be a specialization such that $\mathcal{S}_q(n,r)^K$ is semisimple, that is, the image of t is not a root of unity. Then, it is clear that

$$\eta \otimes 1_K : \mathcal{S}_q(n,r)^K \to \mathcal{S}(n,r)^K$$

is a K-algebra isomorphism. This isomorphism induces an isomorphism between the categories of $\mathcal{S}_q(n,r)^K$-modules and $\mathcal{S}(n,r)^K$-modules, hence an isomorphism between the categories of rational modules for the quantum linear group $GL_q(n,K)$ (see [**PW**]) and general linear group $GL(n,K)$.

When the image of t is a root of unity, say, an ℓ-th primitive root of 1 with ℓ odd, we still have some beautiful relationship between the q-Schur and Schur algebras. In fact, we can see from [**Du2**] that a Schur algebra is a homomorphic image of a certain q-Schur algebra.

APPENDIX

In this appendix, we will show that the new basis for the q-Schur algebra defined in §2 is exactly the same as the basis defined by Berlinson, Lusztig and MacPherson in [**BLM**]. We shall follow the notation in [**BLM**], and denote the ring $\mathbb{Z}[t, t^{-1}]$ by $\mathcal{A}$ in this appendix.

The author is indebted to Professor E. Cline and Dr. M. Reeder for a number of discussions regarding some geometric facts in this appendix.

Geometric interpretation of q-Schur algebras. Let V be a vector space of dimension r over a field F, and let $\mathcal{F}$ be the set of all n-step flags

$$V_1 \subseteq V_2 \subseteq \cdots \subseteq V_n = V.$$

The group $G = GL(V)$ acts naturally on $\mathcal{F}$, and the G-orbits on $\mathcal{F}$ are the fibres of the map from $\mathcal{F}$ to $\Lambda(n, r)$ given by

$$(V_1 \subseteq V_2 \subseteq \cdots \subseteq V_n) \mapsto (\mathrm{dim}V_1, \mathrm{dim}V_2/V_1, \cdots, \mathrm{dim}V_n/V_{n-1}).$$

If $\mathcal{F}_\lambda$ denotes the inverse image of $\lambda \in \Lambda(n, r)$, then $\mathcal{F} = \cup_{\lambda \in \Lambda(n,r)} \mathcal{F}_\lambda$ is a disjoint union. Let $M_{n,r}$ be the set of $n \times n$ matrices with nonnegative integer entries so that the sum of the entries is r. Then there is a (surjective) map from $\mathcal{F} \times \mathcal{F}$ to $M_{n,r}$, defined in [**BLM, 1.1**], such that the G-orbits on $\mathcal{F} \times \mathcal{F}$ (diagonal action) are fibres of this map. We denote by $\mathcal{O}_A$ the fibre at $A \in M_{n,r}$.

For each $(\lambda, w, \mu) \in \mathcal{B} = \{(\lambda, w, \mu) \,|\, \lambda, \mu \in \Lambda(n, r), w \in \mathcal{D}_{\lambda\mu}\}$ (2.2), we define a matrix $A = (v_{ij}) \in M_{n,r}$ such that $W_\lambda^w \cap W_\mu = W_\nu$ with

$$\nu = (v_{11}, v_{21}, \cdots, v_{n1}, v_{12}, \cdots, v_{nn})$$

where $\nu^{(i)} = (v_{1i}, \cdots, v_{ni}) \in \Lambda(n, \mu_i)$ and $\mu = (\mu_1, \cdots, \mu_n)$ (see [**Du2, 1.3**]). This defines a bijective map τ from $\mathcal{B}$ to $M_{n,r}$. If G_λ denotes the standard parabolic subgroup of G corresponding to $\lambda \in \Lambda(n, r)$, then we may identify $\mathcal{F}_\lambda$ with G/G_λ. Thus, the G-orbits on $\mathcal{F} \times \mathcal{F}$ has the representatives (G_λ, wG_μ) with $(\lambda, w, \mu) \in \mathcal{B}$, and if $\tau(\lambda, w, \mu) = A$ then $\mathcal{O}_A$ ($A \in M_{n,r}$) contains the element (G_λ, wG_μ).

For any $A, A', A'' \in M_{n,r}$ and any q', a power of prime number, we take $F = \mathbb{F}_{q'}$, a finite field with q' elements, and let

$$g_{A,A',A'',q'} = \#\{f \in \mathcal{F} \,|\, (f_1, f) \in \mathcal{O}_A, (f, f_2) \in \mathcal{O}_{A'}\} = c_0 + c_1 q' + \cdots + c_m {q'}^m$$

where $(f_1, f_2) \in \mathcal{O}_{A''}$, and put

$$g_{A,A',A''} = c_0 + c_1 t^2 + \cdots + c_m t^{2m}.$$

Following [**BLM**] we define $\mathbf{K}_{n,r}$ to be the free $\mathcal{A}$-module ($\mathcal{A} = \mathbb{Z}[t, t^{-1}]$!) with basis e_A ($A \in M_{n,r}$) satisfying

$$e_A e_{A'} = \sum_{A'' \in M_{n,r}} g_{A,A',A''} e_{A''}.$$

When specializing t^2 to q', one sees easily from the definition that

$$\mathbf{K}_{n,r}|_{t^2=q'} = \mathrm{End}_{FG(q')}\left(\bigoplus_{\lambda \in \Lambda(n,r)} (1_{G_\lambda(q')})^{G(q')} \right),$$

where $(1_{G_\lambda(q')})^{G(q')}$ is the permutation representation of $G(q')$ over the right cosets of $G_\lambda(q')$. So, $\mathbf{K}_{n,r}$ is isomorphic to the q-Schur algebra (see [**DJ1, 2.24**]), and if we denote $\phi^w_{\lambda\mu}$ by ϕ_A for $\tau(\lambda, w, \mu) = A$, we obtain by a straightforward argument the explicit isomorphism as follows.

LEMMA A.1. *The map $e_A \to \phi_A$ is an algebra isomorphism from $\mathbf{K}_{n,r}$ to $\mathcal{S}_q(n,r)$.* $\square$

Dimension of the orbit $\mathcal{O}_A$. We shall, from now on, assume that F is an algebraically closed field. Fix $\tau(\lambda, w, \mu) = A$. Then $\mathcal{O}_A$ is represented by (G_λ, wG_μ). Since the stablizer of this element is $G^w_\lambda \cap G_\mu$, we have $\mathcal{O}_A \cong G/G^w_\lambda \cap G_\mu$ as a variety, and hence

$$d(A) := \dim \mathcal{O}_A = \dim G/G^w_\lambda \cap G_\mu.$$

The dimension $d(A)$ of $\mathcal{O}_A$ can be computed in terms of the entries of the matrix A (see [**BLM, §2**]). However, the following result gives another description of $d(A)$. Let w_A be the longest word in $W_\lambda w W_\mu$.

LEMMA A.2. *Keep the above notation. We have*

$$d(A) = \ell(w_0) + \ell(w_A) - \ell(w_{0,\lambda}) - \ell(w_{0,\mu}),$$

where $w_{0,\nu}$ is the longest word in W_ν and $w_0 = w_{0,(r)}$.

PROOF. We first note that $G_\lambda w G_\mu / G_\lambda$ is the G_μ-orbit of $G_\lambda w / G_\lambda$ on G/G_λ. Since $\mathrm{Stab}_{G_\mu}(G_\lambda w/G_\lambda) = G^w_\lambda \cap G_\mu$, we have as a variety

$$G_\lambda w G_\mu / G_\lambda \cong G_\mu / G^w_\lambda \cap G_\mu.$$

Secondly, we note that

$$G_\lambda w G_\mu = \bigcup_{y \in W_\lambda w W_\mu} ByB,$$

where $B = G_{(1^r)}$ is the Borel subgroup (see e.g. [**Ca, (2.8.1)**]). Therefore, dim $G_\lambda w G_\mu / B = \ell(w_A)$. Now, we have

$$\begin{aligned} d(A) &= \dim G/G_\mu + \dim G_\mu / G_\lambda^w \cap G_\mu \\ &= \dim G/G_\mu + \dim G_\lambda w G_\mu / G_\lambda \\ &= \dim G/G_\mu + \dim G_\lambda w G_\mu / B - \dim G_\lambda / B \\ &= \ell(w_0) - \ell(w_{0,\mu}) + \ell(w_A) - \ell(w_{0,\lambda}) \end{aligned}$$

as desired. □

Let $pr_1 : \mathcal{O}_A \to \mathcal{F}$ be the projection on the first factor. It is easy to see that $pr_1(\mathcal{O}_A) = G/G_\lambda$. Therefore,

(A.3) $$d_1(A) := \dim pr_1(\mathcal{O}_A) = \ell(w_0) - \ell(w_{0,\lambda}).$$

Bases of Beilinson, Lusztig and MacPherson. We define $A' \leq A$ for $A', A \in M_{n,r}$ if $\mathcal{O}_{A'}$ is contained in the Zariski closure $\bar{\mathcal{O}}_A$ of $\mathcal{O}_A$. In fact, one sees easily that $A' \leq A$ iff $w_{A'} \leq w_A$ (here $\leq$ is the Bruhat ordering on W), and

$$\bar{\mathcal{O}}_A = \bigcup_{A' \leq A} \mathcal{O}_{A'},$$

which gives rise to a stratification of $\bar{\mathcal{O}}_A$.

Let $IC(\bar{\mathcal{O}}_A)$ for $A \in M_{n,r}$ be the intersection cohomology complex of Deligne-Goresky-MacPherson of $\bar{\mathcal{O}}_A$ with coefficients in F ([**KL2**]). Note that the restriction of $IC(\bar{\mathcal{O}}_A)$ to $\mathcal{O}_A$ is the constant sheaf complex F (no shift!). Let $IC_A = j_{A,*} IC(\bar{\mathcal{O}}_A)$ be the extension by 0 to $\mathcal{F} \times \mathcal{F}$ where $j_A : \bar{\mathcal{O}}_A \to \mathcal{F} \times \mathcal{F}$. The i-th cohomology $\mathcal{H}^i IC_A$ of IC_A restricted to an orbit $\mathcal{O}_{A'}$ with $A' \leq A$ is a trivial local system on $\mathcal{O}_{A'}$, since the isotropy group of a points in $\mathcal{O}_{A'}$ are connected. Let $(\mathcal{H}^i IC_A)_{A'}$ denote the stalk at any point of $\mathcal{O}_{A'}$, and define for $A' \leq A$

$$\tilde{P}_{A',A} = \sum_{i \geq 0} \dim (\mathcal{H}^i IC_A)_{A'} t^i.$$

Note that $\tilde{P}_{A',A} t^{-d(A)+d(A')} = P_{A',A}$, the polynomial defined in [**BLM, 1.4**]. Following [**BLM**], we define a new basis for $\mathbf{K}_{n,r}$ by

$$\{A\} = \sum_{A' \leq A} P_{A',A}[A'],$$

where $[A'] = t^{-d(A')+d_1(A')} e_A$.

Recall from 2.2 the Kazhdan-Lusztig basis $\{\theta_b\}_{b \in \mathcal{B}}$ of the q-Schur algebra $\mathcal{S}_q(n,r)$. Now, we have the following.

THEOREM A.4. *Under the identification of* (A.1), *we have* $\theta_b = \{A\}$ *where* $\tau(b) = A$ *for all* $b \in \mathcal{B}$.

PROOF. By (A.2) and (A.3), we have $-d(A') + d_1(A') = -\ell(w_{A'}) + \ell(w_{0,\mu})$ if $A' = (\lambda, w', \mu)$. So $[A'] = \hat{\phi}_{A'}$ by (3.c). Since $-d(A) + d(A') = -\ell(w_A) + \ell(w_{A'})$

by (A.2) again, it suffices from (3.9) to prove

$$\tilde{P}_{A',A} = P_{w_{A'},w_A}.$$

To prove this, we consider the canonical projection

$$\pi : G/B \times G/B \to \mathcal{F}_\lambda \times \mathcal{F}_\mu.$$

Let $\mathcal{O}_w$ be the G-orbit in $G/B \times G/B$ corresponding to $w \in W$, and let $IC(\bar{\mathcal{O}}_w)$ be the intersection cohomology complex of $\bar{\mathcal{O}}_w$ whose restriction to $\mathcal{O}_w$ is the complex F. Then π induces a map $\pi : \bar{\mathcal{O}}_{w_A} \to \bar{\mathcal{O}}_A$. By **[GM, 5.4.2]** and noting the difference of our definition for intersection cohomology complexes from that in [**GM**], we have $\pi^*(IC(\bar{\mathcal{O}}_A)) \cong IC(\bar{\mathcal{O}}_{w_A})$. Therefore, $\pi^*(IC_A) \cong IC_{w_A}$, where IC_{w_A} is the extension by 0 of $IC(\bar{\mathcal{O}}_{w_A})$ to $G/B \times G/B$. Since π^* is exact, it follows that

$$(\mathcal{H}^i IC_{w_A})_y \cong (\pi^* \mathcal{H}^i IC_A)_y \cong (\mathcal{H}^i IC_A)_{\pi(y)},$$

where $(\mathcal{H}^i IC_{w_A})_y$ denotes the stalk at any point in $\mathcal{O}_y$. Now,

$$P_{w_{A'},w_A} = \sum_{i \geq 0} \dim \mathcal{H}^i(IC_{w_A})_{w_{A'}} t^i$$

(see **[KL2, 4.3]**, **[Sp, 2.4]**). Consequently, $P_{w_{A'},w_A} = \tilde{P}_{A',A}$ as desired. □

We remark that it is very likely that, by imitating the proof given in **[KL2, 4.3]** and noting the work in [**De**], one might obtain a second proof of the result (A.4).

REFERENCES

[BLM] A. A. Beilinson, G. Lusztig and R. MacPherson, *A geometric setting for the quantum deformation of GL_n*, Duke Math. J. **61** (1990), 655–677.

[Ca] R. Carter, *Finite groups of Lie type*, John Wiley, Now York, 1985.

[C1] C. W. Curtis, *On the Lusztig's isomorphism theorem for Hecke algebras*, J. Algebra **92** (1985), 348–365.

[C2] C. W. Curtis, *Representations of Hecke algebras*, Astérisque 168 (1988), 13–60.

[De] V. Deodhar, *On some geometric aspects of the Bruhat orderings II. The parabolic analogue of Kazhdan-Lusztig polynomials*, J. Algebra **111** (1987), 483-506.

[DJ0] R. Dipper and G. James, *Blocks and idempotents of Hecke algebras of general linear groups*, Proc. London Math. Soc. **53** (1987), 57–82.

[DJ1] R. Dipper and G. James, *The q-Schur algebra*, Proc. London Math. Soc. **59** (1989), 23–50.

[DJ2] R. Dipper and G. James, *q-Tensor spaces and q-Weyl modules*, Trans. Amer. Math. Soc. **727** (1991), 251–282.

[Du1] Jie Du, *The modular representation theory of q-Schur algebras*, Trans. Amer. Math. Soc. **329** (1992), 253–271.

[Du2] Jie Du, *The modular representation theory of q-Schur algebras II*, Math. Z. **208** (1991), 503–536.

[G] J. A. Green, *Polynomial Representations of GL_n*, Lecture Notes in Math. 830, Springer Verlag, Berlin Heidelberg New York, 1980.

[GL] I. Grojnowski and G. Lusztig, *On bases of irreducible representations of quantum GL_n*, preprint.

[GM] M. Goresky and R. MacPherson, *Intersection Homology II*, Invent. Math. **71** (1983), 77–129.

[KL1] D. Kazhdan and G. Lusztig, *Representations of Coxeter groups and Hecke algebras*, Invent. Math. **53** (1979), 165–184.

[KL2] D. Kazhdan and G. Lusztig, *Schubert varieties and Poincaré duality*, Proc. Sympos. Pure Math. **36** 185–203, Amer. Math. Soc. 1980.

[L1] G. Lusztig, *On a theorem of Benson and Curtis*, J. Algebra **71** (1981), 490–498.

[L2] G. Lusztig, *Cells in the affine Weyl groups*, Algebraic Groups and Related Topics, Adv. Stud. Pure Math. **6**, (1985) 255–287.

[L3] G. Lusztig, *Canonical bases arising from quantized enveloping algebras*, J. Amer. Math. Soc. **3** (1990), 447–498.

[L4] G. Lusztig, *Intersection cohomology methods in representation theory*, Proc. of ICM, Kyoto (1991), 155–174.

[L5] G. Lusztig, *Canonical bases for enveloping algebras*, Lecture in MSRI, Berkeley (1990).

[PW] B. Parshall and J.-p. Wang, *Quantum linear groups*, Mem. Amer. Math. Soc. **439** (1991).

[Sp] T. A. Springer, *Quelques applications des la cohomology d'intersection*, Sém. Bourbaki **589**, Astérisque **92**–**95** (1981–82) 249–273.

MATHEMATICAL SCIENCES RESEARCH INSTITUTE, BERKELEY, CALIFORNIA

Current address: School of Mathematics and Statistics, University of Sydney, Sydney, N. S. W. 2006, Australia

E-mail address: du_j@maths.su.oz.au

Contemporary Mathematics
Volume **139**, 1992

Hecke Algebras and Shellings of Bruhat Intervals II; Twisted Bruhat Orders

M.J.DYER

ABSTRACT. We study a class of partial orders on a Coxeter group, and some related modules for the Hecke algebra. Some properties of Bruhat order are extended to these more general orders, notably the existence of Kazhdan-Lusztig polynomials, and relations between the structure of intervals in the order (for instance, shellability) and structure constants for the associated Hecke algebra module. The generalized Kazhdan-Lusztig polynomials are used to formulate a Kazhdan-Lusztig type conjecture for Kac-Moody Lie algebras.

Introduction

Let (W, S) be a Coxeter system and $T = \cup_{w \in W} wSw^{-1}$ denote its set of reflections. Let $A = W_J \cap T$ where W_J is the parabolic subgroup of W generated by some $J \subseteq S$. For $w \in W$, let $wA = N(w) + wAw^{-1} \subseteq T$ where $N(w) = \{ t \in T \mid l(tw) < l(w) \}$ (l is the length function), and $+$ denotes symmetric difference.

Define $v \leq_A w$ ($v, w \in W$) if there exist $t_1, \dots, t_n \in T$ with $v = t_n \dots t_1 w$ and $t_i \in t_{i-1} \dots t_1 wA$ for $i = 1, \dots, n$. If $\mathcal{A} = \emptyset$, $\leq_A$ is Bruhat order; if $A = T$, $\leq_A$ is reverse Bruhat order. For any $J \subseteq S$, $\leq_A$ is a locally finite partial order on W; these orders are the objects of study in this paper. We show they share many properties with Bruhat order. For example, the simplicial complex of an open interval is shown to be a shellable sphere (3.10) and we define Kazhdan-Lusztig polynomials ([**14**]) for these intervals in (4.2)(3) by using constructions from [**9**]. For $\mathcal{A} = \emptyset$, one obtains the Kazhdan-Lusztig polynomials $P_{x,y}$; for $A = T$, one gets the inverse Kazhdan-Lusztig polynomials $Q_{x,y}$. The polynomials are related to structure constants for the generic Hecke algebra (§4). It is to be expected that for crystallographic Coxeter systems these polynomials will play a role in

1991 *Mathematics Subject Classification.* Primary 20F55; Secondary 06A08, 16A99, 17B67.
Partially supported by the N.S.F.
This paper is in final form and no version of it will be submitted for publication elsewhere.

the study of the geometry of Kac-Moody groups and the representation theory of Kac-Moody Lie algebras (the construction produces nothing new for finite W). For example, we show in §5 that the polynomial associated to an interval $[\tau, \sigma]$ in $\leq_A$ (where $\tau, \sigma \in W^J$, $A = W_J \cap T$, $W^J = \{ w \in W \mid N(w^{-1}) \cap W_J = \emptyset \}$) coincides with the "parabolic" Kazhdan-Lusztig $P^J_{\tau,\sigma}$ defined by Deodhar in [**4**] (see also [**3**]) and thus has an interpretation in terms of the intersection cohomology of Schubert varieties in G/P (where G is a Kac-Moody group with Weyl group W and P is the standard parabolic subgroup of G corresponding to W_J).

The order $\leq_A$ defined above is of interest more generally, in fact when A is any initial section of a reflection order of T [**9**]. In this generality, intervals in $\leq_A$ need not be finite. However, our results and constructions apply to any finite interval in $\leq_A$ in which every closed length two subinterval has cardinality four. In §6, we indicate a relation between such intervals and certain posets which arise in the study of Verma modules for a Kac-Moody Lie algebra $\mathfrak{g}$ with symmetrizable Cartan matrix.

More specifically, let $\mathfrak{h}$ be the Cartan subalgebra of $\mathfrak{g}$. For $\lambda \in \mathfrak{h}^*$, let $M(\lambda)$ be the Verma module with highest weight λ and let $L(\lambda)$ denote the unique irreducible quotient of $M(\lambda)$. There is a (locally finite) partial order $\leq$ on $\mathfrak{h}^*$ such that for $\lambda, \mu \in \mathfrak{h}^*$, one has $[M(\lambda): L(\mu)] \neq 0$ iff $\lambda \leq \mu$ [**13**]. In §6, it is shown that certain intervals $[\lambda, \mu]$ (for which there do not exist $\nu_1, \nu_2 \in \mathfrak{h}^*$ such that $\lambda \leq \nu_1 < \nu_2 \leq \mu$ and $\nu_2 - \nu_1$ is an integral multiple of an imaginary root) are isomorphic to ("quotients" of) intervals in $\leq_A$, for appropriate $\leq_A$. It is natural to conjecture that the value at $q = 1$ of the corresponding Kazhdan-Lusztig polynomial is $[M(\lambda): L(\mu)]$ (see (6.14)); this extends the conjecture of [**6**, (5.16)]. I am grateful to George Lusztig for suggesting the possible relevance of these polynomials in this context.

1 Twisted orders on a Coxeter group

Let (W, S) be a Coxeter system and $l: W \to \mathbb{N}$ denote the corresponding length function. Let $T = \cup_{w \in W} wSw^{-1}$ and regard the power set $\mathcal{P}(T)$ as an abelian group under symmetric difference. Define $N: W \to \mathcal{P}(T)$ by $N(w) = \{ t \in T \mid l(tw) < l(w) \}$; by [**7**, §2], N is characterized by $N(s) = \{s\}$ $(s \in S)$ and $N(xy) = N(x) + xN(y)x^{-1}$ $(x, y \in W)$. Hence there is a W-action on the set $\mathcal{P}(T)$ such that $wA = N(w) + wAw^{-1}$ $(w \in W,\ A \subseteq T)$.

Let $\mathcal{A}$ be the set of initial sections of reflection orders on T [**9**, (2.2) and (2.6)]. By [**9**, (2.7)], $wA \in \mathcal{A}$ for all $A \in \mathcal{A}$ and $w \in W$.

Recall that for $w \in W$, $N(w)$ is a finite set of cardinality $l(w)$. For $A \in \mathcal{A}$, and $v, w \in W$, define

$$l_A(v, w) = l(wv^{-1}) - 2\sharp[N(vw^{-1}) \cap vA] \in \mathbb{Z}$$

and set $l_A(w) = l_A(1, w)$. Note that $l_A(v, w) = l_{vA}(wv^{-1})$.

1.1 PROPOSITION. *For any* $x, y, z \in W$, $l_A(x,y) + l_A(y,z) = l_A(x,z)$.

PROOF. It will suffice to show that $l_A(x) + l_{xA}(y) = l_A(yx)$. This follows since

$$l_A(x) = \sharp(B) - 2\sharp(B \cap A)$$
$$l_{xA}(y) = \sharp(C) - 2\sharp(C \cap (A+B))$$

and

$$l_A(yx) = \sharp(B+C) - 2\sharp(A \cap (B+C))$$

where $B = N(x^{-1})$ and $C = x^{-1}N(y^{-1})x$.

1.2 PROPOSITION. *Let* $A \in \mathcal{A}$, $x \in W$, $t \in T$. *Then* $l_A(x, tx) > 0$ *iff* $t \notin xA$.

PROOF. Note that $t \notin xA$ iff $t \in txA$ and that $l_A(tx, x) = -l_A(x, tx)$ by (1.1). Hence it will suffice to show that $l_A(t) > 0$ if $t \in T \setminus A$. This assertion is readily checked if (W, S) is dihedral, and we reduce to this case.

Recall that if $W' = \langle W' \cap T \rangle$ is a reflection subgroup of W, then $\chi(W') = \{ t \in T \mid N(t) \cap W' = \{t\}\}$ is a set of Coxeter generators for W' [**7**, (3.3)] and that $A \cap W'$ is an initial section of a reflection order of the reflections $W' \cap T$ of W' [**9**, (2.4)(ii)]. Thus, for $w \in W'$, one may define $l_{A\cap W'}(w)$ in the Coxeter system $(W', \chi(W'))$. It is enough to prove that if $t \in T \setminus A$ then

$$1 + \sum_{W'} (l_{A\cap W'}(t) - 1) = l_A(t) \tag{1.2.1}$$

where the sum is over all maximal dihedral subgroups W' of W containing t [**8**, (3.2)]. But for such W',

$$l_{A\cap W'}(t) = \sharp(N(t) \cap W') - 2\sharp[(N(t^{-1}) \cap W') \cap (A \cap W')]$$

by [**7**, (3.3)(ii)]. Since every $t' \in T \setminus \{t\}$ is contained in a unique such W', (1.2.1) follows immediately.

1.3 DEFINITION. (cf. [**8**, §1]). For any $A \in \mathcal{A}$, let $\Omega_{(W,A)} = \Omega$ denote the directed graph with vertex set W and edge set $E_{(W,A)} = \{ (tw, w) \mid w \in W,\ t \in wA \}$. For any subset X of W, let $\Omega(X)$ denote the full subgraph of $\Omega_{(W,A)}$ on vertex set X.

It will sometimes be useful to regard $\Omega_{(W,A)}$ as an edge-labelled graph with edge-labelling $E_{(W,A)} \to T$ defined by $(x, y) \mapsto yx^{-1}$ $((x,y) \in E_{(W,A)})$.

1.4 PROPOSITION. *Fix* $A \in \mathcal{A}$. *Let* W' *be a reflection subgroup of* W, *and* $S' = \chi(W')$ *be its canonical set of Coxeter generaters. Then*

(1) $\Omega(W') = \Omega_{(W', W'\cap A)}$ *where the right hand side is the graph associated to the initial section* $W' \cap A$ *in* (W', S')

(2) *if* $x \in W$ *and* $N(x^{-1}) \cap W' = \emptyset$, *the map* $w \mapsto xw$, *for* $w \in W'$, *is an isomorphism of directed graphs* $\Omega(W') \to \Omega(xW')$

(3) *for any* $x \in W$, *the map* $v \mapsto vx^{-1}$ *gives an isomorphism of directed graphs* $\Omega_{(W,A)}(W'x) \to \Omega_{(W,xA)}(W')$ *and preserves the edge-labelling* $(v, w) \mapsto wv^{-1}$.

PROOF. We prove only (1); the proofs of (2), (3) are similar. Now both graphs have vertex set W'. Now the edge set of $\Omega(W')$ is

$$\begin{aligned}&\{\,(tw,w)\mid t\in wA,\ w\in W',\ tw\in W'\,\}\\ =&\{\,(tw,w)\mid t\in N(w)+wAw^{-1},\ w\in W',\ t\in W'\,\}\\ =&\{\,(tw,w)\mid t\in (N(w)\cap W')+w(A\cap W')w^{-1},\ w\in W'\,\}\end{aligned}$$

which is the edge set of $\Omega_{(W',W'\cap A)}$ by [**7**, (3.3)].

1.5 DEFINITION. Fix $A\in\mathcal{A}$. For $x,y\in W$ and $n\in\mathbb{N}$, let $C_n^A(x,y)$ be the set of paths of length n in $\Omega_{(W,A)}$ from x to y i.e. $C_n^A(x,y)=$

$$\{\,(x_0,\dots,x_n)\in W^{n+1}\mid (x_{i-1},x_i)\in E_{(W,A)}\ (i=1,\dots,n),\ x_0=x,\ x_n=y\,\}.$$

Note that by (1.1) and (1.2), $C_n^A(x,y)=\emptyset$ unless $l_A(x)\le l_A(y)$. Hence there is a partial order $\le_A$ on W such that $v\le_A w$ iff $C_n^A(v,w)\ne\emptyset$ for some $n\in\mathbb{N}$.

Thus, $v\le_A w$ $(v,w\in W)$ iff there exist $t_1,\dots,t_n\in T$ with $w=t_n\dots t_1v$ and $t_i\notin t_{i-1}\dots t_1vA$ for $i=1,\dots,n$.

1.6 REMARK. Note $\le_\emptyset$ is Bruhat order and for $A\in\mathcal{A}$, $\le_{T+A}$ is the reverse of the order $\le_A$.

For the remainder of this section, fix $A\in\mathcal{A}$ and write $\le$ for $\le_A$.

1.7 PROPOSITION. *If $v\le w$ then every maximal chain $v=v_0<v_1<\dots<v_n=w$ has length $n=l_A(v,w)$.*

PROOF. It is simple to check the claim if (W,S) is dihedral, and we reduce to this case.

It will suffice to show that if $l_A(x,tx)>1$ $(t\in T)$, then there exists $y\in W$ with $x<y<tx$. So suppose $l_A(x,tx)=l_{xA}(t)>1$. By (1.2.1), there is a maximal dihedral subgroup W' containing t such that $l_{A'}(t)=n>1$, where $A'=xA\cap W'$ is an initial section of $T\cap W'$. By the dihedral case, there exist $t_1,\dots,t_n\in W'\cap T$ such that $1<_{A'}t_1<_{A'}t_2t_1<_{A'}\dots<_{A'}t_n\dots t_1=t$ in W'. Then by (1.4), $x<t_1x<\dots<t_n\dots t_1x=tx$ as required.

1.8. Set $S'=\{\,t\in T\mid tA=\{t\}+A\,\}$ (cf. [**9**, (2.9)]), and let $W'=\langle S'\rangle$, $T'=\bigcup_{w\in W'}wS'w^{-1}$. It is easy to check that $w\mapsto wA+A$ $(w\in W')$ defines a map $N'\colon W'\to\mathcal{P}(T')$ satisfying $N'(s)=\{s\}$ $(s\in S')$ and $N'(xy)=N'(x)+xN'(y)x^{-1}$ $(x,y\in W')$. Hence by [**7**, §2], (W',S') is a Coxeter system (it is possible that $S'=\emptyset$).

For $x\in W$, define $\mathcal{L}_A(x)=\{\,s\in S\mid sx<x\,\}$ and $\mathcal{R}_A(x)=\{\,t\in S'\mid xt<x\,\}$. Note that if $s\in\mathcal{L}_A(x)$ then $l_A(sx,x)=1$ and that if $t\in\mathcal{R}_A(x)$ then $l_A(xt,x)=1$; the latter follows from the case (W,S) dihedral using (1.2.1).

Note $xtA=xA+\{xtx^{-1}\}$ $(x\in W,\ t\in S')$. It follows that if $s\in S$, $t\in S'$ and $x\in W$ satisfy $s\notin\mathcal{L}_A(x)$, $s\in\mathcal{L}_A(xt)$ then $t\notin\mathcal{R}_A(x)$ and $sx=xt$; this is a variant of the usual exchange condition for Coxeter systems. The following result is an analogue of the Z-property for Bruhat order.

1.9 PROPOSITION. *Let $x, y \in W$.*

(1) *If $s \in \mathcal{L}_A(x) \cap \mathcal{L}_A(y)$, then $x \le y \iff sx \le y \iff sx \le sy$.*
(2) *If $t \in \mathcal{R}_A(x) \cap \mathcal{R}_A(y)$, then $x \le y \iff xt \le y \iff xt \le yt$.*

PROOF. We prove only (2), the proof of (1) being similar. Note first that if $z \in W$, $t' \in T$ and $z < zt'$, then $zt < zt't$ for any $t \in S' \setminus \{t'\}$ (since $zt'z^{-1} \notin zA$ implies $zt'z^{-1} \notin zA + \{ztz^{-1}\} = ztA$).

Suppose $t \in \mathcal{R}_A(x) \cap \mathcal{R}_A(y)$. Then $xt \le x$, so $x \le y$ implies $xt \le y$. Conversely, suppose that $xt \le y$. Then there exist $n \in \mathbb{N}$ and $t_1, \dots, t_n \in T$ such that $y > yt_1 > yt_1t_2 > \dots > yt_1 \dots t_n = xt$. If all $t_i \neq t$, then $y > yt > yt_1t > yt_1t_2t > \dots > yt_1 \dots t_nt = x$. Otherwise, choose i maximal with $t_i = t$. Then

$$\begin{aligned} y > yt_1 > &yt_1t_2 > \dots > yt_1 \dots t_{i-1} \\ &= yt_1 \dots t_it > yt_1 \dots t_{i+1}t > \dots > yt_1 \dots t_nt = x. \end{aligned}$$

Hence $x \le y$ iff $xt \le y$. Similarly, $xt \le y$ iff $xt \le yt$.

1.10 EXAMPLE. In general, intervals $[x, y]$ in $\le_A$ need not resemble Bruhat intervals. For example, consider the "universal" Coxeter system (W, S) where $S = \{r, s, t\}$ and $W \cong \langle\, r, s, t \mid r^2 = s^2 = t^2 = 1 \,\rangle$. Let $A_1 = \cup_{n=1}^{\infty} N((rs)^n)$ and $A_2 = \cup_{n=1}^{\infty} N((rst)^n)$; one can show that $A_1, A_2 \in \mathcal{A}$. Now the interval $[1, rs]$ in $\le_{A_1}$ is of length 2 but has only one atom (namely s). Also, the interval $[1, rst]$ in $\le_{A_2}$ is of length 3 and is infinite; in fact, $1 < t'rst < rst$ for any $t' \in A_2$.

In the following sections, it will be shown that if the interval $[x, y]$ in $\le_A$ is finite and if every subinterval of length 2 has two atoms, then $[x, y]$ has many of the properties of Bruhat intervals.

2 Bruhat-like intervals in twisted orders

Throughout this section we fix $A \in \mathcal{A}$ and write $\le$ for $\le_A$. Define the relation $\longrightarrow$ on $W \times W$ to be the (reflexive) transitive closure of the relation $\longrightarrow'$ on $W \times W$ defined in (2.1) below (cf. (1.9)(1)).

2.1 DEFINITION. For $(x, y), (x', y') \in W \times W$, write $(x, y) \longrightarrow' (x', y')$ if there exists $r \in S$ satisfying one of (1)–(4) below

(1) $r \in \mathcal{L}_A(x),\ r \in \mathcal{L}_A(y),\ x' = rx,\ y' = ry$
(2) $r \notin \mathcal{L}_A(x),\ r \notin \mathcal{L}_A(y),\ x' = rx,\ y' = ry$
(3) $r \notin \mathcal{L}_A(x),\ r \in \mathcal{L}_A(y),\ x' = x,\ y' = ry$
(4) $r \notin \mathcal{L}_A(x),\ r \in \mathcal{L}_A(y),\ x' = rx,\ y' = y$

Thus, if $(x, y) \longrightarrow (x', y')$ then $x \le y$ can be deduced from $x' \le y'$ using (1.9)(1), without using transitivity of $\le$. Following are some technical properties of $\longrightarrow$.

2.2 LEMMA.

(1) *If* $(x,y) \longrightarrow' (x',y')$ *then* $l_A(x',y') \leq l_A(x,y)$, *and* $x \leq y$ *iff* $x' \leq y'$. *In particular,* $x \leq y$ *if* $(x,y) \longrightarrow (1,1)$.

(2) *If* $(x,y) \longrightarrow' (x',y')$ *then* $[x,y]$ *is finite iff* $[x',y']$ *is finite. In particular, if* $(x,y) \longrightarrow (1,1)$ *then* $[x,y]$ *is finite.*

(3) *If* $(x,y) \longrightarrow (1,1)$, *then for any* $z \in [x,y]$, $(x,z) \longrightarrow (1,1)$ *and* $(z,y) \longrightarrow (1,1)$.

(4) *If* $(x,y) \longrightarrow (1,1)$ *and* $l_A(x,y) = 2$, *then the interval* $[x,y]$ *has exactly two atoms.*

(5) *Suppose* $(x,y) \longrightarrow (1,1)$ *and that* $x < t_1x < t_2t_1x = y$ *where*$t_1, t_2 \in T$. *Let* W' *be the maximal dihedral reflection subgroup containing* $\langle t_1, t_2\rangle$. *Then either* $W' \cap xA$ *is finite or* $W' \cap (T + xA)$ *is finite.*

PROOF. Proofs of (1), (2) are immediate from (1.9)(1); for (2), one notes that if $(x,y) \longrightarrow' (x',y')$ then for some $r \in S$, $[x',y'] \subseteq [x,y] \cup r[x,y]$ and $[x,y] \subseteq [x',y'] \cup r[x',y']$. To prove (3), one shows that if $x \leq z \leq w \leq y$ and $(x,y) \longrightarrow' (x',y')$, then there exist z', w' with $x' \leq z' \leq w' \leq y'$ and either $(z,w) = (z',w')$ or $(z,w) \longrightarrow' (z',w')$; this is checked case by case using (1.9)(1). To prove (4), one first checks that if $(x,y) \longrightarrow' (x',y')$ where $l_A(x,y) = 2$, $l_A(x',y') = 1$ and $x' \leq y'$, then $[x,y] = \{x', rx', ry', y'\}$, for $r \in S$ as in (2.1). One then shows that if $(x,y) \longrightarrow' (x',y')$ where $x' \leq y'$, $l_A(x,y) = l_A(x',y') = 2$ and $[x',y']$ has two atoms, then $[x,y]$ has two atoms.

Finally, we prove (5). The assertion is true if $l_A(x,y) \leq 1$. We may assume inductively that the assertion is true for $[x'',y'']$ if $l_A(x'',y'') < l_A(x,y)$ and $(x'',y'') \longrightarrow (1,1)$, and also that the assertion is true for some $[x',y']$ where $(x,y) \longrightarrow' (x',y')$ and $(x',y') \longrightarrow (1,1)$. Choose $r \in S$ as in (2.1). If $r \notin \{t_1, t_2\}$, then (3) and induction prove that the assertion holds for $[rx, ry]$ and the chain $rx < rt_1x < rt_2t_1x = ry$, from which the desired conclusion follows. Now suppose for example that $t_2 = r$ (the case $t_1 = r$ is similar). Then $r \in \mathcal{L}_A(y)$. If $r \in \mathcal{L}_A(x)$, then $(x',y') = (rx, ry)$; the desired conclusion follows by applying the inductive hypothesis to the chain $x' < rx' < t_1rx' = y'$ in $[x',y']$. Finally, consider the case $t_2 = r$, $r \notin \mathcal{L}_A(x)$; let W' be the maximal dihedral subgroup containing $\langle r, t_1\rangle$, and let $A' = xA \cap W'$, an initial section of the reflections of W'. Then $1 \leq_{A'} t_1 \leq_{A'} rt_1$ and $1 \leq_{A'} r$ in the order $\leq_{A'}$ on W'. A simple computation for dihedral groups shows that either A' is finite or $(W' \cap T) \setminus A'$ is finite.

One also has the following corollary of the above proof.

2.3 COROLLARY. *Suppose that* $x \leq_A y$ *and that every closed subinterval* $[v,w]$ *of* $[x,y]$ *of length* 2 *has exactly two atoms. Then if* $(x,y) \longrightarrow (x',y')$, *every closed subinterval of length* 2 *of* $[x',y']$ *has exactly two atoms.*

PROOF. Suppose that $(x,y) \longrightarrow' (x',y')$. By (2.1), either $[x',y'] \subseteq [x,y]$ (in which case the result is trivial) or $(x',y') \longrightarrow' (x,y)$. Let $x' \leq v' \leq w' \leq y'$, $l_A(v',w') = 2$. By the proof of (2.2)(3), $(v',w') \longrightarrow (v,w)$ for some $x \leq v \leq w \leq$

y. Since $l_A(v,w)$ must be 1 or 2, the proof of (2.2)(4) shows that $[v',w']$ has two atoms.

2.4 LEMMA. *Let $x \le y$, $l_A(x,y) = 2$ and suppose that $[x,y]$ has (at least) two atoms. Let $\preceq$ be a fixed reflection order on T. Then there exist unique $v, w \in W$ with $x < v < y$, $x < w < y$ and $vx^{-1} \prec yv^{-1}$, $wx^{-1} \succ yw^{-1}$. Moreover, $vx^{-1} \prec wx^{-1}$ and $yw^{-1} \prec yv^{-1}$.*

PROOF. Choose $t_1, t_2 \in T$ so $x < t_1x < t_2t_1x = y$. Let W' be the maximal dihedral subgroup containing $\langle t_1, t_2\rangle$ and set $A' = xA \cap W'$, an initial section of the reflections of W'. Note that if $x < z < y$, then $xz^{-1}, zy^{-1} \in W' \cap T$ by [**8**, (3.1)]. Consider the interval $[1, t_1t_2]$ in the order $\le_{A'}$ on W'. It follows from (1.4)(3) that the map $W' \to W$ defined by $z \mapsto zx$ $(z \in W')$ restricts to an isomorphism of posets $\theta: [1, t_1t_2] \to [x,y]$ and that $\theta(z)\theta(z')^{-1} = zz'^{-1}$ for all $z, z' \in [1, t_1t_2]$. The lemma now follows by a simple computation in W'.

The following is a fundamental technical fact about the order $\le_A$.

2.5 PROPOSITION. *The following conditions are equivalent for $x \le y$ in W:*

(1) *the interval $[x,y]$ is finite and every closed subinterval of length two has two atoms*

(2) $(x,y) \longrightarrow (1,1)$

PROOF. It is clear that (2)$\Longrightarrow$(1) by (2.2)(2), (3), (4). Conversely, suppose that (1) holds. We prove first that if $x < y$, then there exists $t \in yA \setminus xA$. Let $n = l_A(x,y) \ge 1$.

Fix some reflection order $\preceq$ on T such that $T \setminus yA$ is an initial section of $\preceq$ i.e. yA is a final section of $\preceq$. Give T^n the lexicographic order induced by $\preceq$ on T [**9**, (4.2)] and for any maximal chain $m: y = y_0 > y_1 > \dots > y_n = x$ in $[x,y]$, associate an n-tuple $\lambda(m) = (y_0y_1^{-1}, y_1y_2^{-1}, \dots, y_{n-1}y_n^{-1})$ in T^n. Suppose that m is that maximal chain whose label $\lambda(m)$ is lexicographically first of all (finitely many) labels of maximal chains in $[x,y]$. Then $y_0y_1^{-1} \prec y_1y_2^{-1} \prec \dots \prec y_{n-1}y_n^{-1}$ by (2.3) (cf. the proof of [**9**,(4.3)]). Since $y_1 < y_0$ it follows that $y_0y_1^{-1} \in yA$; hence $y_{n-1}y_n^{-1} \in yA$ by choice of $\preceq$. But $y_{n-1}y_n^{-1} \notin xA$ since $y_n < y_{n-1}$. Hence we may take $t = y_{n-1}y_n^{-1}$.

Now we show that if $x \le y$ and (1) holds, then $(x,y) \longrightarrow (1,1)$. The result is clear if $l_A(x,y) = 0$. Suppose that $n = l_A(x,y) \ge 1$ and proceed by induction on n and then by induction on $m = \min\{\, l(t) \mid t \in yA \setminus xA \,\}$.

First suppose $m = 1$. Then there exists $r \in \mathcal{L}_A(y) \setminus \mathcal{L}_A(x)$. Note that $(x,y) \longrightarrow' (x, ry)$ and $l_A(x, ry) = l_A(x,y) - 1$; by (2.2)(2), (2.3) and induction, $(x, ry) \longrightarrow (1,1)$.

Now suppose $m > 1$; let $t \in yA \setminus xA$ with $l(t) = m$ and choose $r \in S$ with $l(rtr) = l(t) - 2$ ([**7**, (2.7)]). Note that $r \notin \mathcal{L}_A(y) \setminus \mathcal{L}_A(x)$ since $m > 1$; we claim also $r \notin \mathcal{L}_A(x) \setminus \mathcal{L}_A(y)$. For let $W' = \langle r, t\rangle$ have canonical Coxeter generators $\chi(W') = \{r, s\}$ $(s \in T)$. Note $t \ne s$ since $t \ne r$, $r \in N(t) \cap W'$; hence $l(s) < l(t)$ by [**7**, (3.10)]. Now if $r \in \mathcal{L}_A(x) \setminus \mathcal{L}_A(y)$, then $rx < x < tx$ and $ty < y < ry$

which imply $x < sx$ and $sy < y$ (for example, if $rx < x < tx$, then $r <_{A'} 1 <_{A'} t$ in W' where $A' = xA \cap W'$; a computation in W' shows $1 \leq_{A'} s$, hence $x < sx$). So $r \notin \mathcal{L}_A(x) \setminus \mathcal{L}_A(y)$.

It follows that $(x, y) \longrightarrow' (rx, ry)$. Now $l_A(rx, ry) = n$, and $rtr \in ryA \setminus rxA$ with $l(rtr) < m$; by (2.2)(2), (2.3) and induction $(rx, ry) \longrightarrow (1, 1)$. Hence $(x, y) \longrightarrow (1, 1)$ completing the proof.

2.6. Recall the definition of the graph $\Omega = \Omega_{(W,A)}$ in (1.3). We give some properties of the full subgraphs $\Omega([x, y])$ where $(x, y) \longrightarrow (1, 1)$.

First suppose that (W, S) is dihedral and either A is finite or $T \setminus A$ is finite. Then it is easy to see that if $l_A(x, y) = n \geq 0$, $\Omega([x, y])$ is isomorphic to the directed graph in [**8**, (1.2)] i.e. to the full subgraph of the Bruhat graph of (W, S) on any interval of length n. Denote this graph from [**8**, (1.2)] by Ω_n.

Now return to the case of general (W, S). Suppose that $(x, y) \longrightarrow (1, 1)$. It follows from (2.2)(5) and the above that the conditions on $\Omega_{(W,R)}([x, y])$ in [**8**, (3.4)] are also true of $\Omega_{(W,A)}([x, y])$. In particular, if the interval $[x, y]$ has two atoms (or coatoms) it is isomorphic to a Bruhat interval in a dihedral group. The following result generalizes [**8**, (3.3)].

2.7 PROPOSITION. *If $I = [x, y]$ where $(x, y) \longrightarrow (1, 1)$ then the isomorphism type of the directed graph $\Omega_{(W,A)}([x, y])$ is determined by the isomorphism type of the poset I.*

PROOF. We first show that if $z \in W$, $t \in T$, $l_A(z, tz) > 1$ and $(z, tz) \longrightarrow (1, 1)$, then there exists X' with $\{z, tz\} \subseteq X' \subseteq [z, tz]$ such that $\Omega_{(W,A)}(X') \cong \Omega_{2n+1}$ for some $n \geq 1$, where Ω_{2n+1} is as in (2.6).

Let $W_1, \ldots, W_p$ be the distinct maximal dihedral subgroups W' of W which satisfy $t \in W'$ and $l_{zA\cap W'}(t) > 1$ (see (1.2.1)). Write $l_{zA\cap W_i}(t) = 2k_i + 1$ ($k_i \geq 1$) and let $[1, t]_i = \{v \in W_i \mid 1 \leq_{zA\cap W_i} v \leq_{zA\cap W_i} t\}$ in W_i. By (1.4), the map $\theta_i \colon W_i \to W$ defined by $x \mapsto xz$ ($x \in W_i$) induces an isomorphism of directed graphs $\Omega_{(W_i, zA\cap W_i)}([1, t]_i) \to \Omega_{(W,A)}(X_i)$ where $X_i = \operatorname{Im}(\theta_i) \subseteq [z, tz]$. By (2.2)(5) and (2.6), it now follows that $\Omega_{(W,A)}(X_i)$ is isomorphic to Ω_{2k_i+1}, proving the first claim since $\{z, tz\} \subseteq X_i$.

Now suppose that $(x, y) \longrightarrow (1, 1)$ and there is a subset X' of $[x, y]$ (with $x, y \in X'$) so that the graph on vertex set X' and edge set $E_{(W,A)}(X') \cup \{(x, y)\}$ is isomorphic to Ω_{2n+1} for some $n \geq 1$. Using [**8**, (3.1)] one sees that $\langle vw^{-1} \mid v, w \in X' \rangle$ is a dihedral subgroup W' of W (cf. the proof of the claim (ii) in the proof of [**8**, (3.3)]); since yx^{-1} is a product of $2n + 1$ reflections in W', we have $yx^{-1} \in T$ and $(x, y) \in E_{(W,A)}(X')$.

One now sees that $\Omega_{(W,A)}([x, y])$ is determined by $[x, y]$ exactly as in the proof of [**8**, (3.3)].

One has the following corollary of the above proof.

2.8 COROLLARY. *Suppose that $(x, y) \longrightarrow (1, 1)$. If $yx^{-1} \in T$, then*

$$\sharp\{t \in T \mid x \leq yt \leq y\} \geq l_A(x, y).$$

PROOF. Note that by [**8**, (3.1)], one has $X_i \cap X_j = \{z, tz\}$ in the proof of (2.7) and that by (1.2.1), $l_A(z, tz) = 1 + \sum_{k=1}^{p} 2k_i$.

Since this paper was prepared, I have proved that (2.8) above remains true without the assumption $yx^{-1} \in T$. For reverse Bruhat intervals, this was a conjecture of Deodhar.

We conclude this section with some conditions on A which imply that $x \leq y$ iff $(x, y) \longrightarrow (1, 1)$ $(x, y \in W)$. Let $\mathcal{A}_0$ denote the set of initial sections A (i.e. elements of $\mathcal{A}$) such that for any dihedral reflection subgroup W' of W, either $W' \cap A$ is finite or $W' \cap (T \setminus A)$ is finite.

2.9 LEMMA. *The following conditions* (1), (2) *are equivalent*

(1) $A \in \mathcal{A}_0$ *and there exists* $z \in W$ *such that every interval* $[v, z]$ *is finite*
(2) *for* $x, y \in W$, $x \leq y$ *iff* $(x, y) \longrightarrow (1, 1)$.

PROOF. If (2) holds, then (1) holds by (2.5) and (2.2)(5). Conversely, suppose (1) holds. A simple induction on $l(w)$ using (1.9)(1) shows that every interval $[v, wz]$ is finite $(w \in W)$. Also, the proof of (2.4) shows that every closed interval of length two has two atoms, so (2) holds by (2.5).

2.10 REMARK. Let $A \in \mathcal{A}_0$ with $\{\, x \in W \mid x < z_0,\ l_A(x, z_0) = 1 \,\}$ finite for some $z_0 \in W$. Then it can be shown that $\{\, x \in W \mid x < z,\ l_A(x, z) = 1 \,\}$ is finite for all $z \in W$ and hence the conditions of (2.9) hold

2.11 PROPOSITION. *Let* $I_1 \subsetneq I_2 \subsetneq \ldots \subsetneq I_k \subseteq S$ *and set* $T_k = W_{i_k} \cap T$, *where* W_{I_k} *is the parabolic subgroup generated by* I_k. *Set* $A = T_1 + \ldots + T_k \subseteq T$. *Then* $A \in \mathcal{A}$, *and* A *satisfies the conditions of* (2.9).

PROOF. First we show $A \in \mathcal{A}$. This follows by repeated application of the following facts

(1) If $I \subseteq S$, there is a reflection order $\preceq$ on T such that $W_I \cap T$ is an initial section of $\preceq$ ([**9**, (2.3)])
(2) Let $\preceq'$ be any reflection order on the the reflections $W_I \cap T$ of W_I. Then the total order on T which (a) restricts to $\preceq'$ on $W_I \cap T$ (b) restricts to $\preceq$ on $T \setminus W_I$ and (c) satisfies $t \prec t'$ if $t \in W_I \cap T$, $t' \in T \setminus W_I$, is a reflection order (this follows easily from the definition of reflection orders).

To see that $A \in \mathcal{A}_0$, let W' be any dihedral subgroup of W and write $\chi(W') = \{r, s\}$. It is easily seen (e.g. by using the geometric realization of W [**7**, §4]) that there exists $K \subseteq S$ such that if $t \in (W' \cap T) \setminus \{r, s\}$ and $J \subseteq S$, then $t \in W_J$ iff $J \supseteq K$. Now $A \in \mathcal{A}_0$ is an immediate consequence.

Finally, we show that every interval $[v, 1]$ in the order $\leq_A$ is finite, proceeding by induction on k. If $k = 0$ then $A = \emptyset$ and $\leq_A$ is Bruhat order, for which the result is trivial since 1 is the minimum element. So suppose $k > 0$; let $v \in W$. Now $[v, 1]$ is empty unless $v \in W_{I_k}$. If $v \in W_{I_k}$, consider the order $\leq_{A'}$ on W_{I_k}, where $A' = T_1 + \ldots + T_{k-1}$ is an initial section of the reflections $W_{I_k} \cap T$ of W_{I_k}. The identity map is an order reversing bijection between the interval $[v, 1]$ in $(W, \leq_A)$ and the interval $[1, v]$ in $(W_{I_k}, \leq_{A'})$. Hence $[v, 1]$ is finite.

3 EL-labellings of Bruhat-like intervals

Throughout this section we fix $A \in \mathcal{A}$ and write $\leq$ for $\leq_A$. We let $\mathcal{R}$ denote the ring of Laurent polynomials $\mathcal{R} = \mathbb{Z}[q^{\frac{1}{2}}, q^{-\frac{1}{2}}]$ in an indeterminate $q^{\frac{1}{2}}$; also, let $a \mapsto \overline{a}$ be the ring involution of $\mathcal{R}$ with $\overline{q^{\frac{1}{2}}} = q^{-\frac{1}{2}}$ and set $\alpha = q^{\frac{1}{2}} - q^{-\frac{1}{2}}$. The main result of this section is the following

3.1 THEOREM. *Let $x, y \in W$ with $(x, y) \longrightarrow (1, 1)$. For any reflection order $\preceq$ on T, define $R_{\preceq}(x, y)$ by*

$$R_{\preceq}(x,y) = \sum_{n \in \mathbb{N}} \sum_{(t_1, \dots, t_n)} \overline{\alpha}^n$$

where the inner sum is over those $(t_1, \dots, t_n) \in T^n$ with $x < t_1 x < \dots < t_n \dots t_1 x = y$ and $t_1 \preceq t_2 \preceq \dots \preceq t_n$.

Then $R_{\preceq}(x, y)$ is independent of the choice of $\preceq$.

PROOF. Note that $l_A(x, y) \geq 0$ and that the assertion is trivial if $l_A(x, y) = 0$ since then $R_{\preceq}(x, y) = 1$. Hence suppose that $l_A(x, y) = n > 0$ and that the asertion is true for (x'', y'') if $(x'', y'') \longrightarrow (1, 1)$ and $l_A(x'', y'') < l_A(x, y)$. It will suffice to show that if $(x, y) \longrightarrow' (x', y')$ where $(x', y') \longrightarrow (1, 1)$ and the assertion holds for (x', y') then it holds for (x, y).

Assume that for some $s \in S$, we have $s \in \mathcal{L}_A(y)$, $y' = sy$ and either ($s \in \mathcal{L}_A(x)$, $x' = sx$) or ($s \notin \mathcal{L}_A(x)$, $x' = x$). (The other possibility is that for some $s \in S$, we have $s \notin \mathcal{L}_A(x)$, $x' = sx$ and either ($s \notin \mathcal{L}_A(y)$, $y' = sy$) or ($s \in \mathcal{L}_A(y)$, $y' = y$); the result for this second case may deduced from the case treated here by replacing A by $A + T$).

We will show there is a reflection order $\preceq_1$ such that

$$R_{\preceq}(x,y) = \begin{cases} R_{\preceq_1}(sx, sy) & \text{if } s \in \mathcal{L}_A(x) \\ R_{\preceq_1}(sx, sy) + \overline{\alpha} R_{\preceq_1}(x, sy) & \text{if } s \notin \mathcal{L}_A(x). \end{cases} \tag{3.1.1}$$

3.2. For a reflection order $\preceq$ on T, recall from [**9**, (2.5)] that the relation $\preceq''$ on T defined by $t \preceq'' t'$ (for $t, t' \in T$) if $t = s$ or ($t \neq s$, $s \prec t'$ and $t \preceq t'$) or ($t \neq s$, $t' \prec s$ and $sts \preceq st's$) is a reflection order on T; call $\preceq''$ the lower s-conjugate of $\prec$. Define the upper s-conjugate $\preceq'$ of $\preceq$ to be the reverse of the lower s-conjugate of the reverse of $\prec$. This is a reflection order such that $t \preceq' t'$ (for $t, t' \in T$) if $t' = s$ or ($t' \neq s$, $t \prec s$ and $t \preceq t'$) or ($t \neq s$, $s \prec t$ and $sts \preceq st's$). Observe that the upper s-conjugate of the lower s-conjugate of $\preceq$ is just the upper s-conjugate of $\preceq$.

3.3. Set $x'_0 = sx$ if $s \in \mathcal{L}_A(x)$ and $x'_0 = x$ if $s \notin \mathcal{L}_A(x)$. Note that if $v, w \in [x'_0, y]$, then the interval $[v, w]$ is finite (since $(x'_0, y) \longrightarrow (1, 1)$). Hence for $n \in \mathbb{N}$ we may define

$$X_n(v, w) = \{\, (x_0, \dots, x_n) \in C_n^A(v, w) \mid s \preceq x_0 x_1^{-1} \prec \dots \prec x_{n-1} x_n^{-1} \,\}$$

$$Y_n(v, w) = \{\, (y_0, \dots, y_n) \in C_n^A(v, w) \mid s \preceq y_0 y_1^{-1},\ y_0 y_1^{-1} \prec' \dots \prec' y_{n-1} y_n^{-1} \,\}$$

$$Z_n(v,w) = \{ (z_0, \dots, z_n) \in C_n^A(v,w) \mid z_0 z_1^{-1} \prec \dots \prec z_{n-1} z_n^{-1} \prec s \}.$$

Also, set $f_\prec(v,w) = \sum_{n\in\mathbb{N}} \sharp X_n(v,w)\,\overline{\alpha}^n$, $g_\prec(v,w) = \sum_{n\in\mathbb{N}} \sharp Y_n(v,w)\,\overline{\alpha}^n$ and

$$h_\prec(v,w) = \sum_{n\in\mathbb{N}} \sharp Z_n(v,w)\,\overline{\alpha}^n.$$

Note that

$$R_{\preceq}(v,w) = \sum_y h_\prec(v,y) f_\prec(y,w), \tag{3.3.1}$$

$$R_{\preceq'}(v,w) = \sum_y h_\prec(v,y) g_\prec(y,w). \tag{3.3.2}$$

3.4 LEMMA. *Let $v, w \in [x_0', y]$.*

(1) *If $s \in \mathcal{L}_A(v) \cap \mathcal{L}_A(w)$, then $f_\prec(v,w) = g_\prec(sv, sw)$.*

(2) *If $s \in \mathcal{L}_A(v) \cap \mathcal{L}_A(w)$, then*

$$g_\prec(v,w) - \overline{\alpha} g_\prec(v, sw) = f_\prec(sv, sw) - \overline{\alpha} f_\prec(v, sw).$$

(3) *If $s \in \mathcal{L}_A(w) \setminus \mathcal{L}_A(v)$, then $g_\prec(sv, sw) = f_\prec(v,w) - \overline{\alpha} f_\prec(sv, w)$.*

(4) *If $s \in \mathcal{L}_A(v) \setminus \mathcal{L}_A(w)$, then $f_\prec(v,w) = g_\prec(sv, sw) - \overline{\alpha} g_\prec(sv, w)$.*

PROOF. Define

$$K_n(v,w) = \{ (x_0, \dots, x_n) \in C_n^A(v,w) \mid s \prec x_0 x_1^{-1} \prec \dots \prec x_{n-1} x_n^{-1} \}$$

and $k_\prec(v,w) = \sum_{n\in\mathbb{N}} \sharp K_n(v,w)\,\overline{\alpha}^n$. Note that $X_N(v,w) = K_n(v,w)$ if $s \in \mathcal{L}_A(v)$ and

$$X_N(v,w) = K_n(v,w) \cup \{ (v, v_0, \dots, v_{n-1}) \mid (v_0, \dots, v_{n-1}) \in X_{n-1}(sv, w) \}$$

if $s \notin \mathcal{L}_A(v)$ where the union is of disjoint sets. Hence

$$k_\prec(v,w) = \begin{cases} f_\prec(v,w) & \text{if } s \in \mathcal{L}_A(v) \\ f_\prec(v,w) - \overline{\alpha} f_\prec(sv, w) & \text{if } s \notin \mathcal{L}_A(v). \end{cases}$$

Similarly, define $L_n(v,w) =$

$$\{ (x_0, \dots, x_n) \in C_n^A(v,w) \mid s \prec x_0 x_1^{-1},\ x_0 x_1^{-1} \prec' \dots \prec' x_{n-1} x_n^{-1} \prec' s \}$$

and $l_\prec(v,w) = \sum_{n\in\mathbb{N}} \sharp L_n(v,w)\,\overline{\alpha}^n$. Here,

$$l_\prec(v,w) = \begin{cases} g_\prec(v,w) & \text{if } s \notin \mathcal{L}_A(w) \\ g_\prec(v,w) - \overline{\alpha} g_\prec(v, sw) & \text{if } s \in \mathcal{L}_A(w). \end{cases}$$

Now if $x < tx$ ($x \in W$, $t \in T \setminus \{s\}$) then $sx < stx$. Hence by definition of $\preceq'$, the map $W^{n+1} \to W^{n+1}$ given by $(x_0, \dots, x_n) \mapsto (sx_0, \dots, sx_n)$ restricts to a bijection $K_n(v,w) \to L_n(sv, sw)$. Hence $k_\prec(v,w) = l_\prec(sv, sw)$ and the assertions of the lemma all follow immediately.

3.5. Let $\preceq'$, $\preceq''$ be as in (3.2). Note that by induction, $R_{\preceq}(v, sy) = R_{\preceq'}(v, sy)$ for all $v \in [x'_0, y]$. By induction on $l_A(v, sy)$, using (3.3.1), (3.3.2) we have $f_{\prec}(v, sy) = g_{\prec}(v, sy)$ for all $v \in [x'_0, y]$.

We now show that $f_{\prec}(v, y) = g_{\prec}(v, y)$ $(v \in [x'_0, y])$. Now if $s \notin \mathcal{L}_A(v)$,

$$\begin{aligned} g_{\prec}(v, y) &= \overline{\alpha} g_{\prec}(v, sy) + f_{\prec}(sv, sy) && \text{by (3.4) (4)} \\ &= \overline{\alpha} g_{\prec}(v, sy) + g_{\prec}(sv, sy) \\ &= \overline{\alpha} f_{\prec}(sv, y) + g_{\prec}(sv, sy) && \text{by (3.4) (1)} \\ &= f_{\prec}(v, y) && \text{by (3.4) (3).} \end{aligned}$$

If $s \in \mathcal{L}_A(v)$,

$$\begin{aligned} f_{\prec}(v, y) &= g_{\prec}(sv, sy) && \text{by (3.4) (1)} \\ &= f_{\prec}(sv, sy) \\ &= g_{\prec}(v, y) + \overline{\alpha}(f_{\prec}(v, sy) - g_{\prec}(v, sy)) && \text{by (3.4) (2)} \\ &= g_{\prec}(v, y). \end{aligned}$$

Using (3.3.1), (3.3.2) and induction on $l_A(v, y)$ we now find that $R_{\preceq}(v, y) = R_{\preceq'}(v, y)$. Hence $R_{\preceq}(x, y) = R_{\preceq'}(x, y)$ if $\preceq'$ is the upper s-conjugate of $\preceq$. Since $\preceq'$ is also the upper s-conjugate of $\preceq''$, we also have $R_{\preceq''}(x, y) = R_{\preceq'}(x, y)$. Note that s is the minimum element of T in the order $\prec''$ and that $R_{\preceq''}(x, y) = R_{\preceq}(x, y)$. Hence in order to prove (3.1.1), we may assume without loss of generality that s is the minimum element of T in the order $\preceq$. But then $f_{\prec}(x, y) = R_{\preceq}(x, y)$ and $g_{\prec}(x, y) = R_{\preceq'}(x, y)$. Hence by (3.4),

$$\begin{aligned} R_{\preceq}(x, y) &= f_{\prec}(x, y) \\ &= \begin{cases} g_{\prec}(sx, sy) & \text{if } s \in \mathcal{L}_A(x) \\ g_{\prec}(sx, sy) + \overline{\alpha} g_{\prec}(x, sy) & \text{if } s \notin \mathcal{L}_A(x) \end{cases} \\ &= \begin{cases} R_{\preceq'}(sx, sy) & \text{if } s \in \mathcal{L}_A(x) \\ R_{\preceq'}(sx, sy) + \overline{\alpha} R_{\preceq'}(x, sy) & \text{if } s \notin \mathcal{L}_A(x). \end{cases} \end{aligned}$$

Thus, (3.1.1) holds with $\preceq_1$ chosen to be $\preceq'$.

3.6 COROLLARY.

(1) *There is a unique family of polynomials $R_A(x, y)$, for $x, y \in W$ with $(x, y) \longrightarrow (1, 1)$, such that* (a), (b) *below hold*

(a) $R_A(1, 1) = 1$

(b) *If $s \in \mathcal{L}_A(y)$, then*

$$R_A(x, y) = \begin{cases} R_A(sx, sy) & \text{if } s \in \mathcal{L}_A(x) \\ \overline{\alpha} R_A(x, sy) + R_A(sx, sy) & \text{if } s \notin \mathcal{L}_A(x) \end{cases}$$

where it is understood that in case $s \notin \mathcal{L}_A(x)$ and $sx \not\leq sy$, the term $R_A(sx, sy)$ is zero.

(2) *If $(x, y) \longrightarrow (1, 1)$ then $R_A(x, y) = R_{\preceq}(x, y)$ for any reflection order $\preceq$ on T.*

(3) *If* $(x, y) \longrightarrow (1, 1)$, *then* $\sum_z R_A(x, z)\overline{R_A(z, y)} = \delta_{x,y}$.

PROOF. To prove (1) and (2), one sets $R_A(x, y) = R_{\preceq}(x, y)$ for some reflection order $\preceq$. This is independent of $\preceq$ by (3.1) and satisfies the relations in (1) by (3.1.1). The relations in (1) uniquely determine the $R_A(x, y)$ (for $(x, y) \longrightarrow (1, 1)$) by definition of $\longrightarrow$. To prove (3), note that $\sum_z R_{\preceq}(x, z)\overline{R_{\preceq'}(z, y)} = \delta_{x,y}$ by [**9**, (1.5)] where $\preceq'$ is the reverse reflection order of the order $\preceq$.

The following result is immediate from (3.6)(1).

3.7 COROLLARY. *If* $(x, y) \longrightarrow (1, 1)$, *then* $R_A(x, y)$ *is a monic polynomial of degree* $l_A(x, y)$ *in* $\overline{\alpha}$.

To conclude this section, we describe some consequences of (3.7) for the structure of the interval $[x, y]$. The results hold also for certain subposets $[x, y]_I^J$ $(I \subseteq J \subseteq S)$ which generalize the "descent classes" defined in [**2**].

3.8 DEFINITION. Fix $I \subseteq J \subseteq S$. Define $J' = S \setminus J$ and

$$D_I^J = \{\, y \in W \mid yA \cap W_{J'} = \emptyset,\ (T + yA) \cap W_I = \emptyset \,\}.$$

For $x, y \in D_I^J$ with $(x, y) \longrightarrow (1, 1)$, define $[x, y]_I^J = \{\, z \in D_I^J \mid x \leq z \leq y \,\}$.

Fix a reflection order $\preceq$ of which W_I is an final section and $W_{J'} \cap T$ is an initial section [**9**, 2.3].

3.9 PROPOSITION. *Let* $x, y \in D_I^J$ *with* $(x, y) \longrightarrow (1, 1)$. *Then*

(1) *Every maximal chain* $y = y_0 > y_1 > \ldots > y_n = x$ *of* $[x, y]_I^J$ *has length* $n = l_A(x, y)$.

(2) *The labelling* $(v, w) \mapsto wv^{-1}$, *for* $v, w \in [x, y]_I^J$ *with* $v \leq w$ *and* $l_A(v, w) = 1$, *is an EL-labelling of* $[x, y]_I^J$ *by* T.

PROOF. By (3.7), there is at most one maximal chain $y = y_0 > y_1 > \ldots > y_n = x$ of $[x, y]_I^J$ with $y_0y_1^{-1} \preceq \ldots \preceq y_{n-1}y_n^{-1}$. Now let $y = y_0 > y_1 > \ldots > y_n = x$ be that maximal chain of $[x, y]$ with label $(y_0y_1^{-1}, y_1y_2^{-1}, \ldots, y_{n-1}y_n^{-1})$ first of all such labels in the lexicographic ordering of T^n induced by $\preceq$. Then $y_0y_1^{-1} \preceq \ldots \preceq y_{n-1}y_n^{-1}$ as in the proof of (2.3). To complete the proof, it will suffice to show that $y_i \in D_I^J$ for $i = 0, \ldots, n$. The argument is similar to the proof of [**9**, (4.1)(ii)]; one shows that $y_iA \cap W_{J'} = \emptyset$ by descending induction on i (noting that if $y_iA \cap W_{J'} \neq \emptyset$ then $y_iA \cap J' \neq \emptyset$) and that $(T + y_iA) \cap W_I = \emptyset$ by induction on i.

The consequences of (3.9)(2) for the Möbius function, Stanley-Riesner ring and simplicial complex of the interval $[x, y]_I^J$ are described in [**1**]. Here, we merely mention the following

3.10 COROLLARY. *For* $x, y \in W$ *with* $l_A(x, y) = n \geq 2$ *and* $(x, y) \longrightarrow (1, 1)$, *the order complex associated to the open interval* $(x, y) = \{\, z \in W \mid x < z < y \,\}$ *is a combinatorial* $(n - 2)$*-sphere.*

Since this paper was submitted, I have proved that every (non-empty) finite interval in any order $\leq_A$ is CL-shellable. CL-shellability is slightly weaker than the EL-shellability of (3.9), but still implies that the order complex of any finite open interval in an order $\leq_A$ is either a combinatorial ball or a combinatorial sphere.

4 Modules for the Hecke algebra

As in §3, we fix $A \in \mathcal{A}$ and write $\leq$ for $\leq_A$. Recall the defintion of $\mathcal{R}$ and its involution $a \mapsto \bar{a}$ from §3. Let $\mathcal{H}$ denote the generic Hecke algebra of (W, S) over $\mathcal{R}$ (see [**14**]). Then $\mathcal{H}$ is an associative unital $\mathcal{R}$-algebra with free $\mathcal{R}$-basis $\{\tilde{T}_w\}_{w\in W}$ and multiplication determined by

$$\tilde{T}_r\tilde{T}_w = \begin{cases} \tilde{T}_{rw} & \text{if } l(rw) > l(w) \\ \tilde{T}_{rw} + \alpha\tilde{T}_w & \text{if } l(rw) < l(w). \end{cases}$$

Let $h \mapsto \overline{h}$ denote the ring involution $\sum_{w\in W} a_w\tilde{T}_w \mapsto \sum_{w\in W} \overline{a}_w\tilde{T}^{-1}_{w^{-1}}$ of $\mathcal{H}$.

Let $\mathcal{H}_A$ denote the set of formal $\mathcal{R}$-linear combinations $\sum_{w\in W} a_w\tilde{t}_w$ such that there exists $v \in W$ so $a_w = 0$ unless $w \leq v$.

4.1 PROPOSITION. *There is a left $\mathcal{H}$-module structure on $\mathcal{H}_A$ so that*

$$\tilde{T}_r(\sum_w a_w\tilde{t}_w) = \sum_w b_w\tilde{t}_w \quad \textit{where}$$

$$b_w = \begin{cases} a_{rw} & r \notin \mathcal{L}_A(w) \\ a_{rw} + \alpha a_w & r \in \mathcal{L}_A(w) \quad (r \in S). \end{cases}$$

PROOF. First motice that if X is any subset of W with a maximal element in $\leq$, then the set $\{1, r\}X$ has a maximal element (by (1.9)(1)) for any $r \in S$. This shows firstly that $\leq$ is directed (i.e. any finite subset of W has an upper bound in $\leq$; this is because it will be contained in some set $\{1, r_1\}\dots\{1, r_n\}$); hence $\mathcal{H}_A$ admits the obvious $\mathcal{R}$-module structure. Secondly, the above remark shows that the formal $\mathcal{R}$-linear combination $\sum_w b_w\tilde{t}_w$ is in $\mathcal{H}_A$.

For $r, s \in S$ and $n \in \mathbb{N}$, write $(\dots\tilde{T}_s\tilde{T}_r)_n$ for the product $(\dots\tilde{T}_r\tilde{T}_s\tilde{T}_r)$ with n factors. Now $\mathcal{H}$ is generated as unital $\mathcal{R}$-algebra by generators $\tilde{T}_r$, $r \in S$ subject to relations $\tilde{T}_r^2 = 1 + \alpha\tilde{T}_r$ and $(\dots\tilde{T}_s\tilde{T}_r)_n = (\dots\tilde{T}_r\tilde{T}_s)_n$ for distinct $r, s \in S$ with $n = \operatorname{ord}(rs)$ finite. It will suffice to show that $\tilde{T}_r^2\tilde{t}_w = (1 + \alpha\tilde{T}_r)\tilde{t}_w$ $(r \in S)$ and for distinct $r, s \in S$

$$(4.1.1) \qquad (\dots\tilde{T}_s\tilde{T}_r)_n\tilde{t}_w = (\dots\tilde{T}_r\tilde{T}_s)_n\tilde{t}_w, \quad \text{if } n = \operatorname{ord}(rs) < \infty.$$

More generally, let J be any subset of S such that W_J is finite and let $\mathcal{H}(W_J)$ denote the corresponding subalgebra of $\mathcal{H}$, spanned as $\mathcal{R}$-module by $\{\tilde{T}_w\}_{w\in W_J}$. Fix any $w \in W$; then $wA \cap W_J$ is a (finite) initial section of the reflections $W_J \cap T$ of W_J, so by [**9**, (2.11)], we have $wA \cap W_J = N(w')$ for some $w' \in W_J$. Let $x = w'^{-1}w$, and let $\mathcal{H}'_A(W_Jw)$ be the $\mathcal{R}$-submodule of $\mathcal{H}_A$ spanned as $\mathcal{R}$-module by $\{\tilde{t}_{vw}\}_{v\in W_J} = \{\tilde{t}_{vx}\}_{v\in W_J}$. Let θ denote the isomorphism $\theta\colon \mathcal{H}(W_J) \to \mathcal{H}'_A(W_Jw)$ of $\mathcal{R}$-modules defined by

$$\theta(\sum_{v\in W_J} a_v\tilde{T}_v) = \sum_{v\in W_J} a_v\tilde{t}_{vx}.$$

Now for any $v \in W_J$, one has $vxA \cap W_J = N(v)$ and hence $\mathcal{L}_A(vx) \cap J = vxA \cap J = \{ s \in J \mid l(sv) < l(v) \}$. It follows that for $h \in \mathcal{H}'_A(W_J w)$ and $r \in J$, one has

$$\tilde{T}_r h = \theta(\tilde{T}_r \theta^{-1}(h)).$$

The claim (4.1.1) is now immediate. The proof also shows that the left $\mathcal{H}(W_J)$-module $\mathcal{H}'_A(W_J w)$ is isomorphic to $\mathcal{H}(W_J)$ and that the poset $W_J w$ (in the order induced by $\leq$) is isomorphic to W_J (in Bruhat order), for finite W_J. Note also that $\mathcal{H} \cong \mathcal{H}_\emptyset$ as left $\mathcal{H}$-module.

Since this paper was submitted, I have noticed that the submodule of $\mathcal{H}_A$ spanned as $\mathcal{R}$-module by the $\tilde{t}_w$ is isomorphic to the left regular module for $\mathcal{H}$.

For the rest of this section we assume for notational simplicity that $A \in \mathcal{A}$ satisfies the conditions of (2.9) i.e. for $x, y \in W$, one has $x \leq y$ iff $(x, y) \longrightarrow (1, 1)$. Recall that this implies that every interval $[x, y]$ is finite.

4.2 PROPOSITION.

(1) *There is an $\mathcal{R}$-antilinear involution $k \mapsto \overline{k}$ of the $\mathcal{R}$-module $\mathcal{H}_A$ defined by*

$$\sum_w a_w \tilde{t}_w \mapsto \sum_w b_w \tilde{t}_w$$

where $b_w = \sum_{x \geq w} \overline{a_x} R_A(w, x)$ and $R_A(w, x)$ is as defined in (3.6).

(2) *For $h \in \mathcal{H}$ and $k \in \mathcal{H}_A$, one has $\overline{hk} = \overline{h}\,\overline{k}$.*

(3) *For any $w \in W$, there is a unique element $C'_{w,A}$ of $\mathcal{H}_A$ satisfying* (a)–(b) *below*

(a) $C'_{w,A} = \sum_v p_A(v, w) \tilde{t}_v$ *where $p_A(v, w) = 0$ if $v \not\leq w$, $p_A(w, w) = 1$ and $p_A(v, w) \in q^{-\frac{1}{2}}\mathbb{Z}[q^{-\frac{1}{2}}]$ if $v \neq w$*

(b) $\overline{C'_{w,A}} = C'_{w,A}$.

PROOF. The first assertion follows from (3.6)(3). To prove (2), it is sufficient to prove that $\overline{\tilde{T}_s \tilde{t}_w} = \overline{\tilde{T}_s}\ \overline{\tilde{t}_w}$ for $s \in S$, $w \in W$ and this follows readily from (3.6)(1). The proof of (3) is standard (see [**14**] or [**9**, (3.1) and (3.5)], for instance). An alternative proof of (4.2) may be based on (4.9)–(4.10) below.

4.3. Write $q^{l_A(v,w)/2} p_A(v, w) = P_A(v, w)$. Then one has $P_\emptyset(v, w) = P_{v,w}$ and $P_T(v, w) = Q_{w,v}$ where $P_{v,w}$ and $Q_{w,v}$ denote the Kazhdan-Lusztig polynomials and inverse Kazhdan-Lusztig polynomials [**14**] respectively. In general, if $v \leq w$ then $P_A(v, w)$ is a polyniomial in q (of degree at most $(l_A(v, w) - 1)/2$ if $v < w$) with constant term 1. In (4.4)–(4.8) we list some properties of the $P_A(v, w)$ and the $C'_{w,A}$ that will be needed later. The proofs are essentially identical with those of the corresponding facts from [**14**, §2] so we omit them.

Let $\mu_A(v, w)$ be the coefficient of $q^{-\frac{1}{2}}$ in $p_A(v, w)$ i.e. of $q^{(l_A(v,w)-1)/2}$ in $P_A(v, w)$. Define

$$\tilde{\mu}_A(v, w) = \begin{cases} \mu_A(v, w) & \text{if } v \leq w \\ \mu_A(w, v) & \text{otherwise.} \end{cases}$$

4.4 PROPOSITION. *If $w \in W$ and $r \in S \setminus \mathcal{L}_A(w)$ then*

$$C'_r C'_{w,A} = C'_{rw,A} + \sum_{v: r\in\mathcal{L}_A(v)} \mu_A(v,w) C'_{v,A}$$

4.5 COROLLARY. *If $w \in W$ and $r \in \mathcal{L}_A(w)$ then*

$$P_A(v,w) = q^c P_A(rv,rw) + q^{1-c} P_A(v,rw) - \sum_{x: r\in\mathcal{L}_A(x)} \mu_A(x,rw) q^{l_A(x,w)/2} P_A(v,x)$$

where c denotes 0 if $r \in \mathcal{L}_A(v)$ and 1 if $r \notin \mathcal{L}_A(w)$. Moreover, $P_A(1,1) = 1$.

4.6 COROLLARY. *If $v, w \in W$ and $r \in \mathcal{L}_A(w)$ then $P_A(v,w) = P_A(rv,w)$.*

4.7 COROLLARY. *If $v < w \in W$ and $r \in \mathcal{L}_A(w) \setminus \mathcal{L}_A(v)$ then $\mu_A(v,w) \neq 0$ iff $v = rw$. Moreover, $v = rw$ implies that $\mu_A(v,w) = 1$.*

4.8 COROLLARY. *If $w \in W$ and $s \in S$ then*

$$C'_s C'_{w,A} = \begin{cases} \sum_{v: s\in\mathcal{L}_A(v)} \tilde{\mu}_A(v,w) C'_{v,A} & \text{if } s \notin \mathcal{L}_A(w) \\ (q^{\frac{1}{2}} + q^{-\frac{1}{2}}) C'_{w,A} & \text{if } s \in \mathcal{L}_A(w). \end{cases}$$

Note that as in [**14**], the recurrence formula in (4.5) completely determines the $P_A(v,w)$ in general (by (2.5)).

In the next sections, we describe a generalization of [**9**,(3.3) and (3.7)]. This will be needed in order to obtain analogues of (4.4)–(4.8) for a right action of the Hecke algebra of (W', S') on $\mathcal{H}_A$, where (W', S') is as in (1.8).

4.9. Let $\mathcal{H}_A(W)$ denote the set of those functions $f_A: W \times \mathcal{A} \to \mathcal{R}$ satisfying the following conditions (1)–(3)

(1) there exists $v \in W$ such that $f_A(w,B) = 0$ for all $B \in \mathcal{A}$ unless $w \leq v$.
(2) for any $w \in W$, there is a finite subset T' of T such that $f_A(v,B_1) = f_A(v,B_2)$ if B_1, B_2 are both initial sections of the same reflection order $\preceq$ and $B_1 \cap T' = B_2 \cap T'$
(3) if $B \in \mathcal{A}$, $t \in T \setminus B$ and $tB = B + \{t\}$, then

$$f_A(v,tB) = \begin{cases} f_A(v,B) & \text{if } t \in vA \\ f_A(v,B) + \alpha f_A(tv,B) & \text{if } t \notin vA. \end{cases}$$

Define $\eta: \mathcal{H}_A(W) \to \mathcal{H}_A$ by $\eta_A(f_A) = \sum_{w\in W} f_A(w,\emptyset)\tilde{t}_w$.

4.10 THEOREM. *For $g_\emptyset \in \mathcal{H}_\emptyset(W)$ and $f_A \in \mathcal{H}_A(W)$, define $g_\emptyset f_A: W \times \mathcal{A} \to \mathcal{R}$ by $(g_\emptyset f_A)(w,B) = \sum_{y\in W} g_\emptyset(y,B) f_A(y^{-1}w, y^{-1}B)$.*

(1) *For $g_\emptyset, f_A$ as above, $g_\emptyset f_A \in \mathcal{H}_A(W)$. This makes $\mathcal{H}_\emptyset(W)$ into an associative $\mathcal{R}$-algebra and $\mathcal{H}_A(W)$ into a $\mathcal{H}_\emptyset(W)$-module.*
(2) *The map $\eta_A: \mathcal{H}_A(W) \to \mathcal{H}_A$ is bijective and $\eta_A(g_\emptyset f_A) = \eta_\emptyset(g_\emptyset)\eta_A(f_A)$. In particular, $\eta_\emptyset$ is an algebra isomorphism $\mathcal{H}_\emptyset(W) \to \mathcal{H}$, identifying $\tilde{t}_w \in \mathcal{H}_\emptyset$ with $\tilde{T}_w \in \mathcal{H}$. Regarding $\mathcal{H}_A(W)$ as a $\mathcal{H}$-module by means of $\eta_\emptyset^{-1}$, we have $\mathcal{H}_A(W) \cong \mathcal{H}_A$ as $\mathcal{H}$-modules.*

(3) *There is an involution* $f_A \mapsto \overline{f_A}$ *of* $\mathcal{H}_A(W)$ *defined by* $\overline{f_A}(w,B) = \overline{f_A(w,B+T)}$. *Moreover,* $\overline{(g_\emptyset f_A)} = \overline{g_\emptyset}\ \overline{f_A}$ *and* $\eta_A(\overline{f_A}) = \overline{\eta_A(f_A)}$.

PROOF. The proof is essentially the same as that of [**9**, (3.7)] so we omit it.

4.11 REMARK. Let $f_A \in \mathcal{H}_A(W)$. One sees by induction on $l(x)$ that

$$\tilde{T}_x \eta_A(f_A) = \sum_{v\in W} f_A(x^{-1}v, N(x^{-1}))\tilde{t}_v$$

for $x \in W$. Hence the values of the function f_A generalize the structure constants for the basis elements $\tilde{T}_x$ of $\mathcal{H}$ acting on the module $\mathcal{H}_A(W)$.

4.12. We now establish a right analogue of the result (4.10). Recall the Coxeter system (W', S') defined in (1.8), and let l' denote the length function on (W', S'). The generic Hecke algebra $\mathcal{H}'$ of (W', S') is free as an $\mathcal{R}$-module on $\mathcal{R}$-basis $\{\tilde{T}'_w\}_{w\in W'}$; the multiplication is determined by

$$\tilde{T}'_w \tilde{T}'_s = \begin{cases} \tilde{T}'_{ws} & \text{if } l'(ws) > l'(w) \\ \tilde{T}'_{ws} + \alpha \tilde{T}'_w & \text{if } l'(ws) < l'(w), \end{cases} \text{ for } w \in W' \ s \in S'.$$

There is an involution $\sum_{w\in W} a_w \tilde{T}'_w \mapsto \sum_{w\in W} \overline{a}_w \tilde{T}'^{-1}_{w^{-1}}$ of $\mathcal{H}'$ denoted by $h \mapsto \overline{h}$. An argument like the proof of (4.1) shows that there is a right $\mathcal{H}'$-module structure on $\mathcal{H}_A$ such that for $s \in S'$, $(\sum_w a_w \tilde{t}_w)\tilde{T}'_s = \sum_w b_w \tilde{t}_w$ where

$$b_w = \begin{cases} a_{ws} & s \notin \mathcal{R}_A(w) \\ a_{ws} + \alpha a_w & r \in \mathcal{R}_A(w). \end{cases}$$

In fact, $\mathcal{H}_A$ is a $(\mathcal{H}, \mathcal{H}')$-bimodule (by the analogue of the exchange condition in (1.8)). We now wish to prove that for $h \in \mathcal{H}_A$, $k \in \mathcal{H}'$ one has $\overline{hk} = \overline{h}\ \overline{k}$; this would follow by the same argument as for (4.2)(2) if one had a right analogue of (3.6)(1)(b). Such an analogue does hold (for arbitrary $A \in \mathcal{A}$, for an interval $[x, y]$ with $(x, y) \longrightarrow (1, 1)$), but this seems to be most easily shown by a variant of the following arguments (4.19)(1).

4.13. Let $\mathcal{H}(W')$ denote the set of those functions $h: W' \times \mathcal{A} \to \mathcal{R}$ satisfying the following conditions (1)–(3)

(1) there exists $v \in W'$ such that $h(w, B) = 0$ for all $B \in \mathcal{A}$ unless $w \leq' v$, where $\leq'$ denotes Bruhat order on (W', S').
(2) for any $w \in W'$, there exists a finite subset T' of T such that $h(w, B_1) = h(w, B_2)$ if B_1, B_2 are both initial sections of the same reflection order $\preceq$ and $B_1 \cap T' = B_2 \cap T'$
(3) if $B \in \mathcal{A}$, $t \in T \setminus (A + B)$ and $tB = B + \{t\}$, then

$$h(w, tB) = \begin{cases} h(w, B) + \alpha h(wt, B) & \text{if } t \in W' \text{ and } wt >' w \\ h(w, B) & \text{otherwise.} \end{cases}$$

Define a map $\eta': \mathcal{H}(W') \to \mathcal{H}'$ by $\eta'(h) = \sum_{z\in W'} h(z, A)\tilde{T}'_z$, for $h \in \mathcal{H}(W')$.

4.14 PROPOSITION. *For $h, k \in \mathcal{H}(W')$, define $hk: W' \times \mathcal{A} \to \mathcal{R}$ by*

$$(hk)(w, B) = \sum_{y \in W'} h(wy^{-1}, yB)k(y, B).$$

(1) *For $h, k \in \mathcal{H}(W')$, one has $hk \in \mathcal{H}(W')$. This product makes $\mathcal{H}(W')$ into an associative $\mathcal{R}$-algebra.*
(2) *The map $\eta': \mathcal{H}(W') \to \mathcal{H}'$ is an algebra isomorphism.*
(3) *The map $h \mapsto \overline{h}$ defined by $\overline{h}(w, B) = \overline{h(w, B + T)}$ is a ring involution of $\mathcal{H}(W')$ and $\eta'(\overline{h}) = \overline{\eta'(h)}$.*

PROOF. The proof is similar to that of [**9**, (3,3)] and so we merely indicate two key points. Firstly, for $r \in S'$ define $e'_r \in \mathcal{H}'(W)$ by

$$e'_r(w, B) = \begin{cases} 1 & w = r \\ \alpha & w = 1,\ r \in A + B \\ 0 & \text{otherwise} \end{cases}$$

for $w \in W'$, $B \in \mathcal{A}$. Then $\eta'(e'_r) = \tilde{T}'_r$.

Now suppose that $t \in T \setminus (A + B)$ and $tB = B + \{t\}$. For $f, g \in \mathcal{H}(W')$, one has by $(fg)(w, tB) = \sum_{y \in W'} f(wy^{-1}, yty^{-1}(yB))g(y, tB)$. Assume $t \in W'$. Observe that $yty^{-1} \in yB + A \iff yty^{-1} \in N(y) + yBy^{-1} + A \iff t \in N(y^{-1}) + y^{-1}Ay + A = N'(y^{-1})$ where by (1.8) and [**7**, (2.3)], $N'(y^{-1}) = y^{-1}A + A = \{t \in T' \mid l'(ty^{-1}) < l'(y^{-1})\}$ (T' denotes the set of reflections $W' \cap T$ of (W', S')). Hence $yty^{-1} \in yB + A$ iff $yt <' y$. Using this fact, one sees that essentially the same computations as in the proof of [**9**, (3.3)] show that fg satisfies (4.13)(3).

4.15 PROPOSITION. *For $f_A \in \mathcal{H}_A(W)$ and $h \in \mathcal{H}(W')$, define $f_A h: W \times \mathcal{A} \to \mathcal{R}$ by $(f_A h)(y, B) = \sum_{v \in W'} f_A(yv^{-1}, B)h(v, y^{-1}B)$.*

Then one has $f_A h \in \mathcal{H}_A(W)$. This makes $\mathcal{H}_A(W)$ into a right $\mathcal{H}(W')$-module. Moreover, $\eta_A(f_A h) = \eta_A(f_A)\eta'(h)$ and $\overline{(f_A h)} = \overline{f_A}\ \overline{h}$.

PROOF. Once again, we merely indicate an observation crucial to the computations of the proof. Let $y \in W$, $t \in T$. Note that $t \notin yA \iff y^{-1}ty \notin N(y^{-1}) + A$. Assume that $t \notin yA$. Then for any $v \in W'$,

$$\begin{aligned} t \notin yv^{-1} &\iff y^{-1}ty \notin N(y^{-1}) + v^{-1}A \\ &\iff y^{-1}ty \notin N(y^{-1}) + A + N'(v^{-1}) \\ &\iff y^{-1}ty \notin N'(v^{-1}). \end{aligned}$$

Similarly, if $t \in yA$ then $t \notin yv^{-1}A \iff y^{-1}ty \in N'(v^{-1})$.

4.16 COROLLARY. *For $h \in \mathcal{H}_A$ and $k \in \mathcal{H}'$ one has $\overline{hk} = \overline{h}\ \overline{k}$.*

From (4.16), it is easy to prove right analogues of all results (4.4)–(4.8); one could also define left, right, two-sided cells as in [**14**]. For our applications, we merely record the right-sided versions of (4.6) and (4.7).

4.17 COROLLARY. *If $v, w \in W$ and $s \in \mathcal{R}_A(w)$ then $P_A(v,w) = P_A(vs,w)$.*

4.18 COROLLARY. *If $v, w \in W$ and $s \in \mathcal{R}_A(w) \setminus \mathcal{R}_A(v)$, where $v < w$, then $\mu_A(v,w) \neq 0$ iff $v = ws$. Moreover, $v = ws$ implies that $\mu_A(v,w) = 1$.*

4.19 REMARKS. (1) Fix an arbitrary $A \in \mathcal{A}$. Consider any non-empty subset X of W which has the properties that $x, z \in X$, $x \leq z$ imply $(x,z) \to (1,1)$ and $y \in X$ for all $y \in W$ with $x \leq y \leq z$. The results (4.2)–(4.11) can be extended to the case X is closed under left multiplication by a parabolic subgroup W_J of (W,S) (one gets an action of the Hecke algebra of W_J), and similarly for (4.12)–(4.18).

(2) One expects the conjectures of [**9**, §3] to hold also in this more general context. For example, results of the next section show $P_A(v,w)$ has non-negative coefficients when (W,S) is crystallographic, $A = W_J \cap T$ and $v, w \in W^J$. It is possible to show that $P_A(v,w)$ has non-negative coefficients for any $A \in \mathcal{A}$, $v, w \in W$ with $(v,w) \longrightarrow (1,1)$ and $l_A(v,w) \leq 4$.

5 Relation to "parabolic" Kazhdan-Lusztig polynomials

In [**4**], Deodhar defined a family of polynomials $P^J_{\tau,\sigma}$ ($J \subseteq S$, $\tau, \sigma \in W^J$). If (W,S) is a crystallographic Coxeter system, the coefficients of $P^J_{\tau,\sigma}$ give dimensions of local intersection cohomology spaces for Schubert varieties in G/P, where G is a Kac-Moody group and P is the standard parabolic subgroup of G corresponding to W_J. The polynomials were also independently defined by L. Casian and D. Collingwood, who showed they determine multiplicities of highest weight irreducible modules for complex semisimple Lie algebras as composition factors of generalized Verma modules [**3**]. In this section, we will show that $P^J_{\tau,\sigma} = P_A(\tau,\sigma)$ ($\tau, \sigma \in W^J$) where $A = W_J \cap T$. It would be interesting to find a connection between the polynomials $P_A(v,w)$ for more general A and the geometry of G (for crystallographic W).

5.1. In this section, we fix $A = W_J \cap T$ where $J \subseteq S$. Let $\leq$ denote Bruhat order on W. Recall the order $\leq_A$ defined in (1.5). Note that in (1.8), $J \subseteq S'$ so $W_J \subseteq W'$.

Let $W^J = \{ w \in W \mid N(w^{-1}) \cap W_J = \emptyset \}$. Recall the following facts concerning W^J and W_J [**4**, (2.1)].

5.2 LEMMA.

(1) *Any $x \in W$ may be uniquely writtten in the form $x = \sigma w$ where $\sigma \in W^J$ and $w \in W_J$. Moreover, $l(\sigma w) = l(\sigma) + l(w)$.*

(2) *For $s \in S$ and $\sigma \in W^J$, exactly one of* (a)–(c) *below holds:*

   (a) *$s\sigma < \sigma$ and $s\sigma \in W^J$*

   (b) *$s\sigma > \sigma$ and $s\sigma \in W^J$*

   (c) *$s\sigma > \sigma$ and $s\sigma \notin W^J$. In this case, $s\sigma = \sigma r$ for a unique $r \in J$.*

We apply (5.2) to obtain some properties of $\leq_A$.

5.3 LEMMA.

(1) *For* $\sigma \in W^J$ *and* $v \in W_J$, *one has* $l_A(\sigma v) = l(\sigma) - l(v)$.
(2) *For* $\sigma \in W^J$ *and* $v \in W_J$, *one has* $\mathcal{L}_A(\sigma v) = \{ s \in S \mid s\sigma < \sigma \} \cup \{ s \in S \mid s\sigma = \sigma r \text{ for } r \in J \text{ with } rv > v \}$ *and* $\mathcal{R}_A(\sigma v) \cap J = \{ s \in J \mid vs > v \}$.
(3) *For* $\tau, \sigma \in W^J$, *one has* $\tau \leq \sigma$ *iff* $\tau \leq_A \sigma$.

PROOF. Properties (1) and (2) are immediate from the definitions and (5.2). For (3), proceed by induction on $l(\sigma)$. Note that if $\sigma = 1$, then $w \leq_A \sigma \iff w \in W_J$ $(w \in W)$. Hence the assertion is true for $\sigma = 1$. Suppose that $s \in S$ and the assertion holds for $s\sigma$, where $s\sigma < \sigma$.

Now if $s\tau < \tau$ then $\tau \leq \sigma \iff s\tau \leq s\sigma \iff s\tau \leq_A s\sigma \iff \tau \leq_A \sigma$ using (1.9)(1). Suppose that $s\tau > \tau$ and $s\tau \in W^J$. Using (1.9)(1) again, $\tau \leq \sigma$ iff $t \leq_A \sigma$. Finally, suppose that $s\tau > \tau$ and $s\tau \notin W^J$. Write $s\tau = \tau r$, $r \in J$. Then $\tau \leq \sigma \iff \tau \leq s\sigma \iff \tau \leq_A s\sigma$. Now $s\tau <_A \tau$ so $\tau \leq_A \sigma \iff s\tau \leq_A s\sigma$. But $r \in \mathcal{R}_A(s\sigma) \setminus \mathcal{R}_A(\tau r)$ so using (1.9)(2), $s\tau \leq_A s\sigma \iff \tau \leq_A s\sigma$.

5.4. Recall from [**4**, (3.9)] that the polynomials $P^J_{\tau,\sigma}$ $(\tau, \sigma \in W^J)$ (with $u = -1$) are uniquely determined by the properties $P^J_{1,1} = 1$, $P^J_{\tau,\sigma} = 0$ if $\tau \not\leq \sigma$ and

$$P^J_{\tau,\sigma} = \tilde{P} - \sum_{\substack{\theta \in W^J \\ s\theta < \theta \text{ or } s\theta \notin W^J}} \mu(\theta, s\sigma) q^{(l(\sigma)-l(\theta))/2} P^J_{\tau,\theta}$$

$(s \in S,\ s\sigma < \sigma)$ where $\mu(\tau, \sigma)$ is the coefficient of $q^{(l(\sigma)-l(\tau)-1)/2}$ in $P^J_{\tau,\sigma}$ and

$$\tilde{P} = \begin{cases} P^J_{s\tau,s\sigma} + qP^J_{\tau,s\sigma} & \text{if } s\tau < \tau \\ P^J_{\tau,s\sigma} + qP^J_{s\tau,s\sigma} & \text{if } s\tau > \tau \text{ and } s\tau \in W^J \\ (1+q)P^J_{\tau,s\sigma} & \text{if } s\tau > \tau \text{ and } s\tau \notin W^J. \end{cases}$$

5.5 PROPOSITION. *Let* $A = W_J \cap T$. *Then for* $\tau, \sigma \in W^J$, *the polynomials* $P^J_{\tau,\sigma}$ *(with* $u = -1$*) and* $P_A(\tau, \sigma)$ *are equal.*

PROOF. One has $P_A(\tau, \sigma) = 0$ unless $\tau \leq_A \sigma$ i.e. unless $\tau \leq \sigma$ by (5.3)(3), and $P_A(1,1) = 1$. Hence it will suffice to show that the recurrence formula of (5.4) is satisfied with $P_A(\tau, \sigma)$ replacing $P^J_{\tau,\sigma}$ and $\mu_A(\tau, \sigma)$ replacing $\mu(\tau, \sigma)$. Fix $\tau, \sigma \in W^J$ and $s \in S$ with $\tau \leq \sigma$ and $s\sigma < \sigma$.

Consider the recurrence formula (4.5) with $w = \sigma$, $v = \tau$ and $r = s$; note $s \in \mathcal{L}_A(\sigma)$. Suppose that $x \in W$, $s \in \mathcal{L}_A(x)$ and $\mu_A(x, s\sigma) \neq 0$. Write $x = \theta v$ with $\theta \in W^J$ and $v \in W_J$. If $v \neq 1$, choose $r \in J$ with $vr < v$. Then $r \in \mathcal{R}_A(s\sigma) \setminus \mathcal{R}_A(x)$ so $s\sigma = xr$ by (4.18). Now $xr >_A x$, so $\mathcal{L}_A(x) = S \cap xA \subseteq S \cap (xA \cup \{xrx^{-1}\}) = S \cap xrA = \mathcal{L}_A(xr)$, noting $rA = A + \{r\}$. Since $s \in \mathcal{L}_A(x)$, it follows that $s \in \mathcal{L}_A(xr) = \mathcal{L}_A(s\sigma)$ contrary to (5.3). Hence $v = 1$ and $x = \theta \in W^J$ (this is an argument from the elementary theory of cells in [**14**]). Now for $x = \theta \in W^J$, we have $l_A(x, \sigma) = l(\sigma) - l(\theta)$, and $s \in \mathcal{L}_A(x)$ iff $s\theta < \theta$ or

$s\theta \notin W^J$, by (5.3). Hence in (4.5),

$$\sum_{x: s\in\mathcal{L}_A(x)} \mu_A(x, s\sigma) q^{l_A(x,\sigma)/2} P_A(\tau, x)$$
$$= \sum_{\substack{\theta\in W^J \\ s\theta<\theta \text{ or } s\theta\notin W^J}} \mu_A(\theta, s\sigma) q^{(l(\sigma)-l(\theta))/2} P_A(\tau, \theta).$$

It remains to check that, letting $c = 0$ if $s \in \mathcal{L}_A(\tau)$ and $c = 1$ if $s \notin \mathcal{L}_A(\tau)$, one has $q^c P_A(s\tau, s\sigma) + q^{1-c} P_A(\tau, s\sigma) =$

$$\begin{cases} P_A(s\tau, s\sigma) + qP_A(\tau, s\sigma) & \text{if } s\tau < \tau \\ P_A(\tau, s\sigma) + qP_A(s\tau, s\sigma) & \text{if } s\tau > \tau \text{ and } s\tau \in W^J \\ (1+q)P_A(\tau, s\sigma) & \text{if } s\tau > \tau \text{ and } s\tau \notin W^J. \end{cases}$$

By (5.3), this is clear unless $s\tau > \tau$ and $s\tau \notin W^J$. In the latter case, one may write $s\tau = \tau r$ for some $r \in J$. Since $r \in \mathcal{R}_A(s\sigma) \setminus \mathcal{R}_A(\tau r)$ one has $P_A(\tau, s\sigma) = P_A(\tau r, s\sigma)$ by (4.17). Hence $(1+q)P_A(\tau, s\sigma) = P_A(s\tau, s\sigma) + qP_A(\tau, s\sigma)$ as required since $s \in \mathcal{L}_A(\tau)$ by (5.3). This completes the proof.

5.6 REMARK. The above result (5.5) generalizes [**4**, (3.4)] to the case when W_J is infinite. Note that $W_J \cap T$ plays the role of the "longest element" of W_J.

6 A Kazhdan-Lusztig conjecture

Let $\mathfrak{g}$ be a Kac-Moody Lie algebra associated to a symmetrizable generalized Cartan matrix. Let $\mathfrak{h}$ denote the Cartan subalgebra of $\mathfrak{g}$ and $\Pi \subseteq \mathfrak{h}^*$ be the set of simple roots. Denote the set of roots (respectively positive roots, real roots, positive real roots) by Δ (respectively $\Delta_+, \Delta^{\text{re}}, \Delta_+^{\text{re}}$).

Fix a (standard) non-degenerate invariant bilinear form on $\mathfrak{g}$ and let $(\cdot,\cdot)$ be the induced bilinear form on $\mathfrak{h}^*$. For $\alpha \in \Delta^{\text{re}}$, denote the reflection in α by $r_\alpha\colon \mathfrak{h}^* \to \mathfrak{h}^*$ (recall $r_\alpha(\lambda) = \lambda - \frac{2(\lambda,\alpha)}{(\alpha,\alpha)}\alpha$ $(\lambda \in \mathfrak{h}^*)$). Let $W = \langle r_\alpha \mid \alpha \in \Pi\rangle$ be the Weyl group of $\mathfrak{g}$ and $S = \{ r_\alpha \mid \alpha \in \Pi \}$ be the standard set of Coxeter generators of W. Maintain the notations T, N, χ etc. concerning (W, S) from previous sections. Let $\mathcal{O}$ be the BGG-category of $\mathfrak{g}$-modules [**6**, §3]. For any $\lambda \in \mathfrak{h}^*$, the Verma module M_λ with highest weight λ and its unique irreducible quotient $L(\lambda)$ both lie in $\mathcal{O}$. For any $M \in \mathcal{O}$ and $\lambda \in \mathfrak{h}^*$, $[M\colon L(\lambda)]$ denotes as usual the multiplicity of $L(\lambda)$ in M.

6.1 DEFINITION.

(1) Let Ω denote the directed graph with vertex set $\mathfrak{h}^*$ and edge set E consisting of those pairs $(\mu, \lambda) \in \mathfrak{h}^* \times \mathfrak{h}^*$ such that $\lambda = \mu - n\beta$ for some $\beta \in \Delta_+$ and non-zero $n \in \mathbb{N}$ with $2(\mu, \beta) = n(\beta, \beta)$.

(2) Let $\leq$ be the relation on $\mathfrak{h}^*$ such that $\mu \leq \lambda$ iff there is a path in Ω from μ to λ.

(3) Define $E^{\text{re}} = \{ (\mu, \lambda) \in E \mid \mu - \lambda \in \mathbb{Z}\beta \text{ for some } \beta \in \Delta_+^{\text{re}} \}$ and regard E^{re} as the edge set of a directed graph Ω^{re} on vertex set $\mathfrak{h}^*$. For $X \subseteq \mathfrak{h}^*$,

let $\Omega(X)$ and $\Omega^{\text{re}}(X)$ be the full subgraphs of Ω and Ω^{re} (respectively) on vertex set X.

6.2 REMARKS. (1) The definition of Ω and $\leq$ is suggested by a result of Kac-Kazhdan which generalizes to Kac-Moody algebras an earlier result of Bernstein, Gelfand and Gelfand. In the notation here, the result [**13**, Theorem 2] is that for $\mu, \lambda \in \mathfrak{h}^*$, one has $M[(\mu - \rho) : L(\lambda - \rho)] > 0$ iff $\mu \leq \lambda$ where $\rho \in \mathfrak{h}^*$ is an element satisfying $\rho(\alpha) = 1$ for all $\alpha \in \Pi$.

(2) If $\mu \leq \lambda$ then $\mu - \lambda \in \sum_{\alpha\in\Pi} \mathbb{N}\alpha$. It follows that $\leq$ is a locally finite partial order on $\mathfrak{h}^*$.

6.3. For the rest of this section, fix $\mu \in \mathfrak{h}^*$. Let $\Phi' = \{\,\alpha \in \Delta^{\text{re}} \mid \frac{2(\mu,\alpha)}{(\alpha,\alpha)} \in \mathbb{Z}\,\}$ and $W' = \langle\, r_\alpha \mid \alpha \in \Phi' \,\rangle$; then $\Phi' = \{\alpha \in \Delta^{\text{re}} \mid r_\alpha \in W'\,\}$. Set $\Phi'' = \{\alpha \in \Delta^{\text{re}} \mid (\alpha, \mu) = 0\,\} \subseteq \Phi'$ and $W'' = \langle\, r_\alpha \mid \alpha \in \Phi'' \,\rangle$. Set $S' = \chi(W')$ and $S'' = \chi(W'')$; note that $W' \cap T$ is the set of reflections of W'.

6.4 PROPOSITION. *Let $A = \{r_\alpha \mid \alpha \in \Phi',\ (\alpha, \mu) \leq 0\,\}$. Then*

(1) *A is an initial section of a reflection order of $W' \cap T$*
(2) *For any $s \in S''$, one has $sA = A + \{s\}$ in (W', S').*

PROOF. Fix $s \in S''$ (if $S'' \neq \emptyset$). Let V be the affine subspace of $\mathfrak{h}^*$ spanned by Π i.e. $V = \{\sum_{\alpha\in\Pi} c_\alpha \alpha \mid c_\alpha \in \Pi,\ \sum_{\alpha\in\Pi} c_\alpha = 1\,\}$. Let $\Psi^+ = \{\,\alpha \in V \mid \alpha = c\beta$ for some $\beta \in \Delta_+^{\text{re}},\ c > 0\,\}$. Note that the map $\Psi^+ \to T$ defined by $\alpha \mapsto r_\alpha$ ($\alpha \in \Psi^+$) is bijective. For each $c \in \mathbb{R}$, let $W_c = \langle\, r_\alpha \mid \alpha \in \Psi^+,\ (\alpha, \mu) = c\,\rangle$ and choose a reflection order $\preceq_c$ on the set $\{r_\alpha \mid \alpha \in \Psi^+,\ (\alpha, \mu) = c\,\}$ of reflections of the Coxeter system $(W_c, \chi(W_c))$ in such a way that if $S'' \neq 0$, then s is the last element of the order $\preceq_0$ on $W'' \cap T$ (see [**9**, (2.3)]). There is a unique reflection order $\preceq$ on T such that (1)–(2) below hold:

(1) $r_\alpha \prec r_\beta$ if $\alpha, \beta \in \Psi^+$ and $(\alpha, \mu) < (\beta, \mu)$.
(2) if $\alpha, \beta \in \Psi^+$, $(\alpha, \mu) = (\beta, \mu) = c$ and $r_\alpha \preceq_c r_\beta$, then $r_\alpha \preceq r_\beta$.

Now $A = \{\,t \in T \cap W' \mid t \preceq s\,\}$. Since $s \in W'$, both (1) and therosteritem2 follow using [**9**, (2.4)(ii) and (2.9)].

Recall the definition of the graph $\Omega_{(W',A)}$ on vertex set W' (1.3).

6.5 LEMMA. *Let $z \in W'$, $t \in W' \cap T$.*

(1) *If $t \in W''$, then $z\mu = zt\mu$*
(2) *Suppose $t \notin W''$. Then $(z, zt) \in E_{(W',A)}$ iff $(z\mu, zt\mu) \in E$.*

PROOF. The proof of (1) is trivial, so suppose $t \notin W''$, say $t = r_\alpha$ where $\alpha \in \Phi' \cap \Delta_+$ and $(\alpha, \mu) \neq 0$. Let $z(\alpha)^+$ denote that element of $\{\pm z(\alpha)\}$ which is in Δ^+. Then $(z, zt) \in E_{(W,A)}$ iff $t \notin N(z^{-1}) + A$ i.e. iff $(z(\alpha)^+, z(\mu)) > 0$. The last condition holds iff $(z\mu, r_{z(\alpha)} z\mu) \in E$ as required.

6.6 COROLLARY. *If $x, y \in W'$ and $x \leq_A y$ then $x\mu \leq y\mu$*

6.7 COROLLARY. *The order $\leq_A$ on W' is locally finite.*

PROOF. Note that if $t, t' \in W' \cap T$, $x \in W'$ and $(x\mu, xt\mu) = (x\mu, xt'\mu) \in E$ then $t = t'$. By (6.5), (6.6) and (6.2)(2), it follows that for $x, y \in W'$, the set $\{ t \in W' \cap T \mid t \notin W'',\ x <_A xt \leq_A y \}$ is finite. Also, for $t \in W'' \cap T$, we have $x <_A xt$ iff $t \in N(x^{-1})$ (since $W'' \cap T \subseteq A$). This shows that $\{ t \in W' \cap T \mid x <_A xt \leq_A y \}$ is finite, so $\leq_A$ is locally finite by (1.7).

The following lemma follows easily from W-invariance of Δ, Δ^{re} and implies that the order $\leq$ on $\mathfrak{h}^*$ satisfies an analogue of the Z-property (1.9).

6.8 LEMMA. *Let $(\mu, \lambda) \in E$ and $s \in S$.*

(1) *If $\lambda \neq s\mu$ then $(s\mu, s\lambda) \in E$.*
(2) *If $(\mu, \lambda) \notin E^{\mathrm{re}}$, then $\lambda \neq s\mu$ and $(s\mu, s\lambda) \notin E^{\mathrm{re}}$.*

Now recall the relations $\longrightarrow'$, $\longrightarrow$ on $W' \times W'$ defined in (2.1), with respect to the initial section A of $W' \cap T$.

6.9 LEMMA. *Let $x, y, x', y' \in W'$ with $x' \leq_A y'$ and $(x, y) \longrightarrow' (x', y')$. If $\Omega([x'\mu, y'\mu])$ is equal to $\Omega^{\mathrm{re}}([x'\mu, y'\mu])$ then $\Omega([x\mu, y\mu]) = \Omega^{\mathrm{re}}([x\mu, y\mu])$.*

PROOF. Choose $r \in \chi(W')$ as in (2.1) and argue by induction on $l(r)$. If $l(r) = 1$, then $r \in S$ and the conclusion follows easily from (6.8). Otherwise, choose $s \in S$ with $l(srs) = l(r) - 2$ and note $s \notin W'$. Note that if $\alpha \in \Pi$, $\nu \in \mathfrak{h}^*$ and $\frac{2(\nu,\alpha)}{(\alpha,\alpha)} \notin \mathbb{Z}$, then there does not exist any path $(\nu = \nu_0, \nu_1, \dots, \nu_n)$ in Ω with $\nu_n = r_\alpha(\nu_{n-1})$. It follows that the map $\mathfrak{h}^* \to \mathfrak{h}^*$ defined by $\lambda \mapsto s\lambda$ ($\lambda \in \mathfrak{h}^*$) induces an isomorphism $\Omega([x'\mu, y'\mu]) \to \Omega([sx'\mu, sy'\mu])$ and similarly with $[x'\mu, y'\mu]$ replaced by $[x\mu, y\mu]$ or Ω replaced by Ω^{re} (again using (6.8)). One has

$$sW's = \langle r_\alpha \mid \alpha \in \Delta^{\mathrm{re}}_+,\ \frac{2(s\mu,\alpha)}{(\alpha,\alpha)} \in \mathbb{Z} \rangle,$$

$$sAs = \{ r_\alpha \mid \alpha \in \Delta^{\mathrm{re}}_+,\ -\frac{2(s\mu,\alpha)}{(\alpha,\alpha)} \in \mathbb{N} \}.$$

Moreover, $sx's \leq_{sAs} sy's$ and $(sxs, sys) \longrightarrow' (sx's, sy's)$ in $sW's$, where one takes $srs \in \chi(sW's)$ to satisfy (2.1). Since $l(srs) < l(r)$, the conclusion follows by induction.

6.10 COROLLARY. *If $(x, y) \longrightarrow (1, 1)$ in $W' \times W'$, then*

$$\Omega([x\mu, y\mu]) = \Omega^{\mathrm{re}}([x\mu, y\mu]).$$

6.11 LEMMA. *Let $(x, y) \in W' \times W'$ with $(x, y) \longrightarrow (1, 1)$. Then there is a unique family of Laurent polynomials $\tilde{P}_A(z, y)$ ($z \in W'$, $x \leq_A z \leq_A y$) satisfying*

(1) $\tilde{P}_A(y, y) = 1$
(2) $\tilde{P}_A(z, y) \in q^{-\frac{1}{2}}\mathbb{Z}[q^{-\frac{1}{2}}]$ *if* $x \leq_A z <_A y$
(3) $\tilde{P}_A(z, y) = \sum_{v : z \leq_A v \leq_A y} R_A(z, v) \tilde{P}_A(v, y)$.

PROOF. This is immediate from (3.6) (3) and [**9**, (1.2)].

6.12 REMARKS. (1) Note that $R_A(z,v)$ and hence $\tilde{P}_A(z,y)$ is completely determined by the graph $\Omega_{(W',A)}([x,y])$ and its natural labelling by $W' \cap T$; in particular, they are computable.

(2) Suppose $s \in \mathcal{L}_A(y) \setminus \mathcal{L}_A(x)$. Then $\tilde{P}_A(sx,y) = q^{\frac{1}{2}}\tilde{P}_A(x,y)$; one proves this by applying (4.19)(1) to an interval closed under left multiplication by the parabolic subgroup $\{1,s\}$. Similarly, if $r \in \mathcal{R}_A(y) \setminus \mathcal{R}_A(x)$ then $\tilde{P}_A(xr,y) = q^{\frac{1}{2}}\tilde{P}_A(x,y)$.

The following result was obtained for Kazhdan-Lusztig polynomials by G. Lusztig, by V. Deodhar [**5**] and by O. Gabber and A. Joseph [**10**].

6.13 LEMMA. *Let $x, y \in W' \times W'$ with $(x,y) \longrightarrow (1,1)$. Then*

$$\left[\frac{d}{dq^{\frac{1}{2}}}\overline{\tilde{P}_A(x,y)}\right]_{q^{\frac{1}{2}}=1} = \left[\sum_{t\in T: x<_A tx \leq_A y} P_A(tx,y)\right]_{q^{\frac{1}{2}}=1}.$$

PROOF. Regard $R_A(x,y)$ as a polynomial in $\overline{\alpha} = q^{-\frac{1}{2}} - q^{\frac{1}{2}}$. By (3.6.2), the coefficient of $\overline{\alpha}$ (resp., 1) in $R_A(x,y)$ is non-zero iff $yx^{-1} \in T$ (resp., if $y = x$), in which case it is 1 (resp., 1). The result follows by dividing the identity

$$\tilde{P}_A(x,y) - \overline{\tilde{P}_A(x,y)} = \sum_{v: x<_A v \leq_A y} R_A(x,v)\overline{\tilde{P}_A(v,y)}$$

by $q^{-\frac{1}{2}} - q^{\frac{1}{2}}$ and setting $q^{\frac{1}{2}} = 1$.

6.14 REMARK. There is also a polynomial identity corresponding to (6.13). In order to avoid introducing notation specifically to state this, assume in this remark that $A = \emptyset$. Define $c'_w : W \times \mathcal{A} \to \mathcal{R}$ as in [**9**, (3.8)]; thus, $c'_w(x,\emptyset) = q^{-(l(w)-l(x))/2}P_{x,w}$. Let $\preceq$ be a fixed reflection order on T and for $t \in T$, let $B_t = \{\, t' \in T \mid t' \preceq t \,\}$. The identity is

$$\overline{c'_w(x,\emptyset)} - c'_w(x,\emptyset) = (q^{\frac{1}{2}} - q^{-\frac{1}{2}}) \sum_{t\in T: x<xt\leq w} c'_w(xt, B_t),$$

which is immediate from [**9**, (3.2.2)].

Together with (6.10) and (6.12)(2), the result (6.13) shows that the following conjecture is compatible with the Jantzen sum formula [**13**, (4.2)].

6.15 CONJECTURE. *Let $\mu \in \mathfrak{h}^*$ and define W', W'' as in* (6.3). *Let $x, y \in W'$ with $N(x^{-1}) \cap W'' = N(y^{-1}) \cap W'' = \emptyset$ and suppose that $(x,y) \longrightarrow (1,1)$ in W'* (cf. (2.1)). Let $M = M(x\mu - \rho)$ have Jantzen filtration $M = M_0 \supseteq M_1 \supseteq \ldots$ and set $L = L(y\mu - \rho)$. Then

$$\overline{\tilde{P}_A(x,y)} = \sum_i [M_i/M_{i+1} : L] q^{i/2}.$$

6.16 REMARKS. (1) Let $x, y \in W'$ with $(x,y) \longrightarrow (1,1)$ in W'. Write $x = x_1x_2$ ($x_1 \in W'$, $x_2 \in W''$, $N(x_1^{-1}) \cap W'' = \emptyset$) and similarly for y. Then $x\mu = x_1\mu$, $y\mu = y_1\mu$ and $(x_1,y_1) \longrightarrow (1,1)$.

(2) If $\mathfrak{g}$ is a complex semisimple Lie algebra, then (6.15) becomes the Kazhdan-Lusztig conjecture (in its strong form [**10**]) and has been proved (for instance) for all integral μ ([**11**], [**12**]).

REFERENCES

1. A. Björner and M. Wachs, *Bruhat order of Coxeter groups and shellability*, Adv. in Math. **43** (1982), 87–100.
2. A. Björner and M. Wachs, *Generalized quotients in Coxeter groups*, Trans. A. M. S. **308** (1988), 1–37.
3. L. G. Casian and D. H. Collingwood, *The Kazhdan-Lusztig conjecture for generalized Verma modules*, Math. Z. **195** (1987), 165–184.
4. V. Deodhar, *On some geometric aspects of Bruhat ordering II: The parabolic analogues of Kazhdan-Lusztig polynomials*, J. of Alg. **111** (1987), 483–506.
5. ______, *On the Kazhdan-Lusztig conjectures*, Indag. Math **85** (1982), 1–17.
6. V. Deodhar, O. Gabber and V. Kac, *Structure of some categories of representations with highest weight of infinite-dimensional Lie algebras*, Adv. in Math. **45** (1982), 92–116.
7. M. Dyer, *Reflection subgroups of Coxeter systems*, J. of Alg **135** (1990), 57–73.
8. ______, *On the Bruhat graph of a Coxeter system*, Comp. Math. **78** (1991), 185–191.
9. ______, *Hecke algebras and shellings of Bruhat intervals*, preprint.
10. O. Gabber and A. Joseph, *Towards the Kazhdan-Lusztig conjecture*, Ann. Scient. Éc. Norm. Sup. **14** (1981), 261–302.
11. R. Irving, *The socle filtration of a Verma module*, Ann. Scient Éc. Norm. Sup. **21** (1988), 47–65.
12. ______, *Singular blocks of the category $\mathcal{O}$*, Math. Z. **204** (1990), 209–224.
13. V. G. Kac and D. A. Kazhdan, *Structure of representations with highest weight of infinite-dimensional Lie algebras*, Adv. in Math. **34** (1979), 97–108.
14. D. Kazhdan and G. Lusztig, *Representations of Coxeter groups and Hecke algebras*, Invent. Math. **53** (1979), 165–184.

DEPARTMENT OF MATHEMATICS, UNIVERSITY OF NOTRE DAME, NOTRE DAME, INDIANA, 46556

E-mail address: dyer@cartan.math.nd.edu

Contemporary Mathematics
Volume **139**, 1992

On bases of irreducible representations of quantum GL_n

I. GROJNOWSKI AND G. LUSZTIG

ABSTRACT. We define a q-analog of the special bases of [**DK**] and relate them to the canonical bases of [**L1**].

Introduction

Let V be a vector space of dimension n over $\mathbf{C}$ and let d be an integer ≥ 1. On the tensor space $V^{\otimes d}$ we have natural commuting actions of the general linear group $GL(V)$ and the symmetric group S_d; this has been classically used by Schur and Weyl to transfer information about the representations of S_d to information about the representations of $GL(V)$. A similar idea has been used more recently, in [**DK**], to construct certain special bases of the irreducible representations of $GL(V)$ appearing in the tensor space, starting from some known bases for the irreducible representations of S_d. (It is known [**S**] that any irreducible representation of S_d can be realized in the top homology of the variety of Borel subgroups containing a certain unipotent element in GL_d. The irreducible components of that variety provide a basis for that representation.)

According to [**J**] (see also [**DJ**]) the tensor space $V^{\otimes d}$ has a $q-$analogue; the commuting actions of $GL(V)$ and S_d become commuting actions of a quantized enveloping algebra $\mathbf{U}$ and of the Iwahori-Hecke algebra H corresponding to S_d. Imitating the method of [**DK**] one can construct certain projective bases (=decompositions as direct sums of lines) for the irreducible representations of $\mathbf{U}$ appearing in the $q-$analogue of the tensor space, starting from the bases of the irreducible representations of H defined in [**KL1**] (or, rather, the corresponding dual bases). The resulting bases for the irreducible representations of $\mathbf{U}$

1991 *Mathematics Subject Classification.* 1980 Math Subject classification (1985 Revision) Primary 20G99.

This paper is in final form and will not be submitted for publication elsewhere are called *special bases.* It is not immediately clear that these special bases are independent of the choices made; this will follow from our results.

On the other hand, any irreducible representation of $\mathbf{U}$ appearing in the tensor space has a basis defined as in [**L1**]. We call it a *canonical basis.*

In this paper we give a relationship between special and canonical bases: we show that a special basis coincides with the projective basis associated to a canonical basis.

The proof of our result is based on the method of [**BLM**]. We repeat the sketch given at the end of [**L2**,§**11**], where our result was announced.

In [**BLM**], the algebra $\mathbf{U}$ has been realized in a geometric setting; namely certain big quotients of it were realized in terms of the geometry of relative positions of pairs of $n-$step filtrations of some large $\mathbf{C}^d$. This language is quite appropriate for the discussion of the canonical bases of [**L1**]. On the other hand, the Iwahori-Hecke algebra of S_d is realized in terms of the geometry of relative positions of pairs of complete flags in $\mathbf{C}^d$ and this language is quite appropriate for the discussion of the bases of [**KL1**]. To get a bridge between these two points of view, we realize the tensor space (and the action of $\mathbf{U}$ and H on it) in a similar geometric setting namely in terms of the geometry of relative positions of pairs consisting of an $n-$step filtration of $\mathbf{C}^d$ and a complete flag in $\mathbf{C}^d$. (Note that the number of such relative positions is $n^d = \dim(\mathbf{C}^n)^{\otimes d}$.) In this setting, the two points of view ([**KL1**] and [**L1**]) coexist and can be successfully compared.

1. The algebra U and the tensor space

1.1 Let v be an indeterminate. We fix an integer $n \geq 2$. Following Drinfeld and Jimbo, we define the quantized enveloping algebra $\mathbf{U}$ to be the $Q(v)-$ algebra with generators

$$E_{i,j} \quad (i,j \in [1,n], |i-j| = 1) \text{ and } O(\mathbf{j}) \quad (\mathbf{j} = (\mathbf{j}_1, \dots, \mathbf{j}_n) \in \mathbf{Z}^n)$$

and relations

$$O(\mathbf{j})O(\mathbf{j}') = O(\mathbf{j}+\mathbf{j}') \quad (\mathbf{j}, \mathbf{j}' \in \mathbf{Z}^n);$$

$$E_{i,j}^2 E_{k,l} - (v+v^{-1}) E_{i,j} E_{k,l} E_{i,j} + E_{k,l} E_{i,j}^2 = 0$$
if $i,j,k,l \in [1,n], \quad (i-j)(k-l) = 1, \quad (j-k)(i-l) = 0;$

$$E_{i,j}E_{k,l} = E_{k,l}E_{i,j} \text{ if } i,j,k,l \in [1,n], \quad (i-j)(k-l) = 1, \quad (j-k)(i-l) > 0;$$

$$E_{i,j}E_{j,i} - E_{j,i}E_{i,j} = (v-v^{-1})^{-1}(O_{i,j} - O_{j,i}) \text{ if } i,j \in [1,n], \quad |i-j| = 1$$

where $O_{i,j} = O(\mathbf{j})$ and $\mathbf{j}_h$ is 1 for $h = i$, is -1 for $h = j$ and is 0 for $h \neq i,j$;

$$O(\mathbf{j})E_{h,k} = v^{\mathbf{j}_h - \mathbf{j}_k} E_{h,k} O(\mathbf{j}), \text{ if } i,j \in [1,n], \quad |i-j| = 1, \quad \mathbf{j} \in \mathbf{Z}^n.$$

We denote by $\mathbf{U}^-$ the subalgebra of $\mathbf{U}$ generated by the elements $E_{i,j}$ with $i-j=1$.

1.2 $\mathbf{U}$ is a Hopf algebra with comultiplication Δ defined on the generators by

$\Delta(E_{i,j}) = E_{i,j} \otimes O_{i,j} + 1 \otimes E_{i,j}$ if $j=i+1$;

$\Delta(E_{i,j}) = E_{i,j} \otimes 1 + O_{i,j} \otimes E_{i,j}$ if $j=i-1$;

$\Delta(O(\mathbf{j})) = O(\mathbf{j}) \otimes O(\mathbf{j})$.

1.3 Let $\mathbf{V}$ be a $Q(v)$–vector space with a given basis $e_1, \ldots, e_n$. We shall regard $\mathbf{V}$ as a $\mathbf{U}$–module as follows:

$E_{i,j}e_j = e_i$ if $|i-j|=1$; $E_{i,j}e_h = 0$ if $|i-j|=1$ and $h \neq j$;

$O(\mathbf{j})e_h = v^{\mathbf{j}_h}e_h$.

1.4 Let V be a vector space of dimension d over a field $\mathbf{F}$. Let

$f = (0 = V_0 \subset V_1 \subset \cdots \subset V_m = V), f' = (0 = V'_0 \subset V'_1 \subset \cdots \subset V'_{m'} = V)$

be two filtrations of V. Let $c(f,f')$ be the $m \times m'$ matrix with entries

$$c_{ij} = \dim(V_{i-1} + (V_i \cap V'_j)) - \dim(V_{i-1} + (V_i \cap V'_{j-1})) \quad (1 \le i \le m, 1 \le j \le m').$$

Let X be the variety of all n–step filtrations $0 = V_0 \subset V_1 \subset \cdots \subset V_n = V$ of V and let X' be the variety of all complete flags $0 = V'_0 \subset V'_1 \subset \cdots \subset V'_d = V$ ($\dim V'_i = i$) in V. These are sets with a natural $GL(V)$–action. Hence $GL(V)$ acts naturally (diagonally) on $X \times X, X' \times X', X \times X'$. The map $(f,f') \to c(f,f')$ defines bijections

(a) from the set of $GL(V)$ orbits on $X \times X$ to the set Θ_d of $n \times n$ matrices with integer, ≥ 0 entries with sum of entries equal to d;

(b) from the set of $GL(V)$ orbits on $X' \times X'$ to the set of $d \times d$ matrices with entries $0, 1$ such that each row and each column contains exactly one entry 1 (which we identify with S_d in the obvious way);

(c) from the set of $GL(V)$ orbits on $X \times X'$ to the set of $n \times d$ matrices with entries $0, 1$ such that each column contains exactly one entry 1.

The last set will be identified with the set Π of sequences $i_1 i_2 \ldots i_d$ with each $i_k \in [1,n]$: to such a sequence corresponds the matrix whose (i_r, r) entry is 1 and the other entries are 0.

We shall write $\mathcal{O}_M$ for the $GL(V)$–orbit in $X \times X, X' \times X'$ or $X \times X'$ corresponding to M in Θ_d, S_d or Π.

1.5 Let q be a prime power. Let (M, M', M'') be in $\Theta_d \times \Theta_d \times \Theta_d$; we define an integer $g_{M,M',M'';q}$ as follows: we take $\mathbf{F} = F_q$, a finite field with q elements, we choose $(f_1, f_2) \in \mathcal{O}_{M''}$ and we let $g_{M,M',M'';q}$ be the number of $f \in X$ such that $(f_1, f) \in \mathcal{O}_M, \quad (f, f_2) \in \mathcal{O}_{M'}$.

Then $M \cdot M' = \sum_{M''} g_{M,M',M'';q} M''$ defines a (semisimple) Q–algebra structure on the Q-vector space with basis indexed by Θ_d.

1.6 Similarly, if (M, M', M'') is in $S_d \times S_d \times S_d$, we define an integer $g_{M,M',M'';q}$ as follows: we take $\mathbf{F} = F_q$, we choose $(f_1, f_2) \in \mathcal{O}_{M''}$ and we let $g_{M,M',M'';q}$ be the number of $f \in X'$ such that $(f_1, f) \in \mathcal{O}_M, \quad (f, f_2) \in \mathcal{O}_{M'}$.

Then $M \cdot M' = \sum_{M''} g_{M,M',M'';q} M''$ defines a (semisimple) Q–algebra structure on the Q-vector space with basis indexed by S_d.

1.7 Let (M, M', M'') be in $\Theta_d \times \Pi \times \Pi$. We define an integer $g_{M,M',M'';q}$ as follows: we take $\mathbf{F} = F_q$, a finite field with q elements, we choose $(f_1, f_2) \in \mathcal{O}_{M''}$ and we let $g_{M,M',M'';q}$ be the number of $f \in X$ such that $(f_1, f) \in \mathcal{O}_M$, $(f, f_2) \in \mathcal{O}_{M'}$.

Then $M \cdot M' = \sum_{M''} g_{M,M',M'';q} M''$ defines a left module structure over the algebra in 1.5 on the Q−vector space with basis indexed by Π.

1.8 Let (M, M', M'') be in $\Pi \times \Pi \times S_d$; we define an integer $g_{M,M',M'';q}$ as follows: we take $\mathbf{F} = F_q$, we choose $(f_1, f_2) \in \mathcal{O}_M$ and we let $g_{M,M',M'';q}$ be the number of $f \in X'$ such that $(f_1, f) \in \mathcal{O}'_M$, $(f, f_2) \in \mathcal{O}_{M''}$.

Then $M' \cdot M'' = \sum_M g_{M,M',M'';q} M$ defines a right module structure over the algebra in 1.6 on the Q−vector space with basis indexed by Π. This right module structure commutes with the left module structure in 1.7.

1.9 In each of the cases 1.5-1.8, there exist elements $g_{M,M',M''} \in \mathbf{Z}[v^2]$ which specialize to $g_{M,M',M'';q}$ for $v^2 = q$ (for all q as above).

These elements $g_{M,M',M''}$ are then the structure constants of a semisimple $Q(v)$−algebra $\mathbf{A}$ with $Q(v)$−basis e_M indexed by $M \in \Theta_d$ (in case 1.5); of a semisimple $Q(v)$−algebra H with $Q(v)$−basis e_M indexed by $M \in S_d$ (in case 1.6); and of an $(\mathbf{A}, H)$ bimodule structure on the $Q(v)$−vector space $\mathbf{T}$ with basis e_M indexed by $M \in \Pi$ (in cases 1.7, 1.8). (We shall also write T_w instead of e_M for $M = w \in S_d$.)

It is clear that the image of $\mathbf{A}$ (resp. of H) in the algebra of endomorphisms of $\mathbf{T}$ is the full centralizer of the image of H (resp. of $\mathbf{A}$). Hence if E is a simple right H-module which appears in the H−module $\mathbf{T}$, then $\tilde{E} = \mathrm{Hom}_H(E, \mathbf{T})$, with the $\mathbf{A}$−module structure inherited from from $\mathbf{T}$, is a simple $\mathbf{A}$−module.

Note that H is just the Iwahori-Hecke algebra corresponding to S_d and $\mathbf{A}$ is the algebra defined in [**BLM, 1.2**].

1.10 Assume now that $\mathbf{F}$ is algebraically closed. For M in Θ_d, S_d or Π we denote by IC_M the intersection cohomology complex of the closure of $\mathcal{O}_M$. Its i-th cohomology sheaf restricted to an orbit $\mathcal{O}_{M'}$ in the closure of $\mathcal{O}_M$ is a constant local system on $\mathcal{O}_{M'}$; the dimension of the stalks of this local system is denoted $n_{M',M;i}$. (We normalize IC_M so that $n_{M',M;0} = 1$.) We shall write $M' \le M$ whenever $\mathcal{O}_{M'}$ is contained in $\mathcal{O}_M$. Let $d(M) = \dim \mathcal{O}_M$. Let $r(M)$ be the dimension of the image of $\mathcal{O}_M$ under the first projection $X \times X \to X, X' \times X' \to X'$ or $X \times X' \to X$. We set $[M] = v^{-d(M)+r(M)} e_M$ and $\{M\} = \sum_{M' \le M} P_{M',M}[M']$ where $P_{M',M} = \sum_i n_{M',M;i} v^{i-d(M)+d(M')} \in \mathbf{Z}[v^{-1}]$ is 1 for $M' = M$ and has constant term 0 for $M' < M$. The elements $[M]$ form an $Q(v)$−basis of $\mathbf{A}, H$ or $\mathbf{T}$; the elements $\{M\}$ form another basis. We shall write C'_w instead of $\{M\}$ for $M = w \in S_d$, as in [**KL1, KL2**].

1.11 Following [**BLM**] we define a homomorphism $\gamma : \mathbf{U} \to \mathbf{A}$ of algebras by the following rules. The generator $E_{i,j}$ (where $|i - j| = 1$) is mapped to $\sum_M [M]$ where M runs through all elements of $M \in \Theta_d$ whose (i, j) entry is 1 and all other off-diagonal entries are 0; the generator $O(\mathbf{j})$ is mapped to the sum $\sum_M v^{m_1 \mathbf{j}_1 + \cdots + m_n \mathbf{j}_n}[M]$ where M runs through all diagonal matrices $M \in \Theta_d$

(with diagonal entries $m_1, \ldots, m_n$). The homomorphism γ is surjective, hence if $E, \tilde{E}$ are as in 1.9, the $\mathbf{A}$-module $\tilde{E}$ regarded as a $\mathbf{U}$-module via γ is still simple.

1.12 The H module structure on $\mathbf{T}$ is given as follows. For each $j \in [1, d-1]$ let $T_j = T_{s_j}$ where $s_j \in S_d$ is the transposition $(j, j+1)$. We have

$$e_{i_1 \ldots i_d} T_j = \begin{cases} e_{i_1 \ldots i_{j-1} i_{j+1} i_j i_{j+2} \ldots i_d} & \text{if } i_j < i_{j+1} \\ v^2 e_{i_1 \ldots i_j i_{j+1} \ldots i_d} & \text{if } i_j = i_{j+1} \\ v^2 e_{i_1 \ldots i_{j-1} i_{j+1} i_j i_{j+2} \ldots i_d} + (v^2 - 1) e_{i_1 \ldots i_j i_{j+1} \ldots i_d} & \text{if } i_j > i_{j+1}. \end{cases}$$

1.13 The $\mathbf{U}$-module structure on $\mathbf{T}$ is given as follows (see [**BLM, 2.3, 3.2**]):

$$E_{i,j} e_{i_1 \ldots i_d} = v^{-\sharp\{s | i_s = i\}} \sum_{p: i_p = j} v^{2\sharp\{s | (s-p)(j-i) > 0, i_s = i\}} e_{i_1 \ldots i_{p-1} i i_{p+1} \ldots i_d} \text{ for } |i-j| = 1$$

$$O(\mathbf{j}) e_{i_1 \ldots 1_d} = v^{\mathbf{j}_{i_1} + \cdots + \mathbf{j}_{i_d}} e_{i_1 \ldots i_d}.$$

We call $\mathbf{T}$ *the tensor space*. This is justified by the fact that the $\mathbf{U}$-module $\mathbf{T}$ is isomorphic to $\mathbf{V}^{\otimes d}$ (see 1.3), regarded as a $\mathbf{U}$-module via the comultiplication of $\mathbf{U}$; the isomorphism is

$$v^{\sharp\{(s,s') | s < s', i_s < i_{s'}\}} e_{i_1 \ldots i_d} \mapsto e_{i_1} \otimes \cdots \otimes e_{i_d}.$$

1.14 Let Z be the set of all functions $z : [1, n] \to \mathbf{N}$ such that $z(1) + \cdots + z(n) = d$. For any $z \in Z$, let Π_z be the set of all $i_1 \ldots i_d \in \Pi$ such that $z(k) = \sharp\{s \in [1, d] | i_s = k\}$ for $k = 1, \ldots, n$. We thus have a partition $\Pi = \cup_z \Pi_z$; the corresponding partition $X \times X' = \cup_z (X \times X')_z$ where $(X \times X')_z = \cup_{M \in \Pi_z} \mathcal{O}_M$ is precisely the partition of $X \times X'$ into connected components. Let $\mathbf{T}_z = \{y \in \mathbf{T} | O_{\mathbf{j}} y = v^{\mathbf{j}_1 z_1 + \ldots \mathbf{j}_n z_n} y, \forall \mathbf{j}\}$; then $\mathbf{T}_z$ is spanned by the $e_{i_1 \ldots i_d}$ with $i_1 \ldots i_d \in \Pi_z$ and we have $\mathbf{T} = \oplus_z \mathbf{T}_z$ (*weight decomposition*).

By 1.12, $\mathbf{T}_z$ is an H-submodule of $\mathbf{T}$.

1.15 Let $\Pi_{\min} = \{i_1 \ldots i_d \in \Pi | i_1 \leq \cdots \leq i_d\}$. If $M \in \Pi$, we have $M \in \Pi_{\min}$ if and only if $\mathcal{O}_M$ is closed in $X \times X'$. For any $z \in Z$, the intersection $\Pi_z \cap \Pi_{\min}$ consists of a single element; we denote it M_z.

Let S_z be the set of all $w \in S_d$ such that $\{M_z\} C'_w \in Q(v) M_z$. This is a standard parabolic subgroup of S_d, isomorphic to a product of symmetric groups $S_{z(1)} \times S_{z(2)} \times \ldots$; we denote by w_z the longest element in S_z.

Let D_z be the set of elements $\delta \in S_z$ such that $l(\delta) = l(w_z) + l(w_z \delta)$ (where l is the standard length function).

1.16 If $\delta \in D_z$, we have $\{M_z\} C'_\delta = \gamma_z \{M'\}$ for some $M' \in \Pi_z$ and some $\gamma_z \in Q(v)^*$; moreover, the elements $\{M_z\} C'_\delta$ (for various $\delta \in D_z$) form a basis of $\mathbf{T}_z$.

2. Canonical basis

2.1 Let **R** be a finite dimensional simple **U**−module with a given non-zero vector x such that $E_{i,i+1}x = 0$ for all $i \in [1, n-1]$. There exist integers $a_1 \leq a_2 \leq \ldots a_n$ such that $O(\mathbf{j})x = v^{\mathbf{j}_1 a_1 + \ldots \mathbf{j}_n a_n} x$ for all $\mathbf{j}$. (We say then that **R** is of type $a_1 \leq a_2 \leq \cdots \leq a_n$.)

Let B be the canonical basis of $\mathbf{U}^-$ (see [**L1**]). Let $B_{\mathbf{R}}$ be the set of all non-zero vectors of **R** which are of the form bx for some $b \in B$. It is known (see [**L1**]) that $B_{\mathbf{R}}$ is a basis of **R** as a $Q(v)$−vector space; it is called the *canonical basis* of **R**.

2.2 Let $z \in Z$ (see 1.14) be such that $z(1) \leq z(2) \leq \cdots \leq z(n)$. For any $b \in B$, the product $b\{M_z\}$ (for the **U**−module structure of **T**) is equal to $\{M\}$ for some $M \in \Pi$ or to zero.

The key point in the verification of this fact is the following one. One of the definitions of B is in terms of intersection cohomology of orbit closures arising from a quiver of type A with the standard orientation; on the other hand, the elements $\{M\}$ for $M \in \Pi$ are defined in terms of intersection cohomology of Schubert varieties. But these two kinds of singular varieties are locally isomorphic (see [**Z**]). Our assertion follows from this fact and from definitions.

3. Special basis

3.1 Let $\mathcal{P}_d$ be the set of all sequences of integers $\lambda = (\lambda_1 \geq \lambda_2 \geq \ldots)$ with $\lambda_i = 0$ for large i and $\lambda_1 + \lambda_2 + \cdots = d$. Each $\lambda \in \mathcal{P}_d$ determines in a standard way a simple right H−module E_λ (up to isomorphism); for example to $\lambda = (d, 0, 0, \ldots)$ corresponds the sign representation of H.

If $\lambda, \lambda' \in \mathcal{P}_d$, we write $\lambda \geq \lambda'$ if $\lambda_1 + \ldots \lambda_i \geq \lambda'_1 + \ldots \lambda'_i$ for all $i \geq 1$.

3.2 Let **P** be a right H−module of finite dimension over $Q(v)$. For any $\lambda \in \mathcal{P}_d$ we denote by $\mathbf{P}^{\leq\lambda}$ (resp. $\mathbf{P}^{<\lambda}$) the sum of all $E_{\lambda'}$−isotypical components of **P** for various λ' such that $\lambda' \leq \lambda$ (resp. $\lambda' < \lambda$). We denote $\mathbf{P}^\lambda = \mathbf{P}^{\leq\lambda}/\mathbf{P}^{<\lambda}$; this is a E_λ−isotypical right H−module.

3.3 The previous discussion applies in particular to the H−module $\mathbf{P} = \mathbf{T}$ or to its submodules $\mathbf{T}_z$. Hence, $\mathbf{T}^\lambda$ and $\mathbf{T}_z^\lambda$ are well defined; we have
$\mathbf{T}^\lambda = \oplus_z \mathbf{T}_z^\lambda$.

Now D_z (see 1.15) is a union of right cells of S_d (see [**KL1, 1.2**]). For any right cell $\Gamma \subset D_z$ we define $\mathbf{T}_z^\Gamma$ to be the subspace of $\mathbf{T}_z$ spanned by the elements $\{M_z\}C'_\delta$ $(\delta \in \Gamma)$. We then have $\mathbf{T}_z = \oplus_\Gamma \mathbf{T}_z^\Gamma$ where Γ runs over the set of right cells in D_z (see 1.16). For each Γ as above, there is a well defined $\lambda' \in \mathcal{P}_d$ such that $\mathbf{T}_z^\Gamma$ is contained in $\mathbf{T}_z^{\leq\lambda'}$ and the image of $\mathbf{T}_z^\Gamma$ in $\mathbf{T}_z^{\lambda'}$ is an irreducible H−submodule; we then write $\Gamma \in \lambda'$.

The subspace $\oplus_{\Gamma \subset D_z; \Gamma \in \lambda} \mathbf{T}_z^\Gamma$ of $\mathbf{T}_z^{\leq\lambda}$ is mapped under the canonical projection $\mathbf{T}_z^{\leq\lambda} \to \mathbf{T}_z^\lambda$ isomorphically onto $\mathbf{T}_z^\lambda$. Thus, we may identify
$\mathbf{T}_z^\lambda = \oplus_{\Gamma \subset D_z; \Gamma \in \lambda} \mathbf{T}_z^\Gamma$.

This is a direct sum decomposition as an H−module and each summand is irreducible. On the other hand, we have a direct sum decomposition $\mathbf{T}^\lambda = \oplus_z \mathbf{T}_z^\lambda$

as an H-module (see 1.14). Combining with the previous decomposition we obtain a direct sum decomposition $\mathbf{T}^\lambda = \oplus_{z,\Gamma\subset D_z;\Gamma\in\lambda}\mathbf{T}_z^\Gamma$; in this decomposition, each summand is an irreducible H-submodule.

3.4 We define $\tilde{E}_\lambda = \mathrm{Hom}_H(E_\lambda, \mathbf{T})$. By 1.9, this is a simple $\mathbf{A}$-module, provided that E_λ appears in $\mathbf{T}$, i.e. provided that $\lambda_i = 0$ for $i > n$. In that case, $\tilde{E}_\lambda$ is also an irreducible $\mathbf{U}$-module (see 1.10). It is clear that

$\mathrm{Hom}_H(E_\lambda, \mathbf{T}) = \mathrm{Hom}_H(E_\lambda, \mathbf{T}^{\leq\lambda}) = \mathrm{Hom}_H(E_\lambda, \mathbf{T}^\lambda)$

Using this and the decomposition at the end of 3.3, we obtain

$\mathrm{Hom}_H(E_\lambda, \mathbf{T}) = \oplus_{z,\Gamma\subset D_z;\Gamma\in\lambda}\mathrm{Hom}_H(E_\lambda, \mathbf{T}_z^\Gamma)$.

This gives a decomposition of $\tilde{E}_\lambda$ as a direct sum of lines, or, equivalently, a projective basis, called the *special basis*. (This construction is a q-analog of the construction in [**DK**]. If v is specialized to 1, the projective basis constructed above becomes the basis dual to the one constructed in [**DK**], provided that the irreducibility conjecture for the characteristic varieties of Schubert varieties in $GL_n(\mathbf{C})$ is valid.)

4. Comparison of bases

4.1 Let $\lambda \in \mathcal{P}_d$ be such that $\lambda_i = 0$ for $i > n$. Let $z \in Z$ be given by $z(i) = \lambda_{n+1-i}$ for $i \in [1, n]$. Let Γ be the right cell of S_d containing w_z. We have $\Gamma \subset D_z$ and $\Gamma \in \lambda$. We have $\{M_z\}C'_{w_z} = \gamma_z\{M_z\}$ for some $\gamma_z \in Q(v)^*$; hence we have $\{M_z\} \in \mathbf{T}^{\leq\lambda}$.

4.2 We show that $E_{i,j}\{M_z\} \in \mathbf{T}^{<\lambda}$ provided that $j - i = 1$. First note that $E_{i,j}\mathbf{T}_z \subset \mathbf{T}_{z'}$ where $z' \in Z$ is defined by $z'(i) = z(i) + 1$, $z'(j) = z(j) - 1$, $z'(h) = 0$ for $h \neq i, j$; this is an H-linear map.

Moreover, if $\lambda' \in \mathcal{P}_d$ is obtained by writing the components of z' in decreasing order, followed by zeros, then it is clear that $\lambda' < \lambda$. On the other hand, the H-module $\mathbf{T}_{z'}$ is a deformation of the representation of S_d induced by the unit representation of the subgroup $S_{\lambda'_1} \times S_{\lambda'_2} \times \ldots$. It follows that $\mathbf{T}_{z'} \subset \mathbf{T}^{<\lambda}$, and our assertion follows.

4.3 From 4.2 we see that the $\mathbf{U}$-submodule L_λ of $\mathbf{T}^\lambda$ generated by the image t_λ of $\{M_z\}$ under the canonical homomorphism $\mathbf{T}^{\leq\lambda} \to \mathbf{T}^\lambda$ is an irreducible $\mathbf{U}$-module of type $z_1 \leq z_2 \leq \cdots \leq z_n$ (see 2.1). On the other hand, the H-submodule $t_\lambda H$ of $\mathbf{T}^\lambda$ is isomorphic to E_λ; we choose an H-isomorphism $\phi : E_\lambda \to t_\lambda H$. Then the map $L_\lambda \to \tilde{E}_\lambda = \mathrm{Hom}_H(E_\lambda, \mathbf{T}^\lambda)$ given by $ut_\lambda \to u\phi$ is a well defined isomorphism. By 2.2, any element of the canonical basis of L_λ is equal to the image of an element $\{M\}$ (with $M \in \Pi$) under the canonical homomorphism $\mathbf{T}^{\leq\lambda} \to \mathbf{T}^\lambda$. Hence, by 1.16, it is equal to $\{M_z\}C'_\delta$ for some $z \in Z$ and some $\delta \in D_z$. Thus, it belongs to one of the lines in the special projective basis of L_λ. Thus, the projective basis associated to the canonical basis of L_λ is carried under $L_\lambda \to \tilde{E}_\lambda$ onto the special basis of $\tilde{E}_\lambda$.

4.4 As a consequence, if $\tilde{E}_\lambda$, $\tilde{E}_{\lambda'}$ are isomorphic as $\mathbf{U}$-modules, then their special bases correspond to each other under a $\mathbf{U}$-isomorphism.

References

[BLM] A. A. Beilinson, G. Lusztig and R. MacPherson, *A geometric setting for the quantum deformation of* GL_n, Duke Math. J. **62** (1990), 655-677.

[DK] C. De Concini and D. Kazhdan, *Special bases for* S_N *and* $GL(n)$, Isr. J. Math. **40** (1981), 275-290.

[DJ] R. Dipper and G. James, *q-tensor space and q-Weyl modules*, Trans. Amer. Math. Soc. (1991).

[J] M. Jimbo, *A* $q-$*analogue of* $U(\mathbf{gl}(N+1))$, *Hecke algebra and the Yang-Baxter equation*, Letters math. phys. **11** (1986), 247-252.

[KL1] D. Kazhdan and G. Lusztig, *Representations of Coxeter groups and Hecke algebras*, Invent. Math. **53** (1979), 165-184.

[KL2] D. Kazhdan and G. Lusztig, *Schubert varieties and Poincaré duality*, Proc Symp. Pure Math, **36** (1980), 185-203.

[L1] G. Lusztig, *Canonical bases arising from quantized enveloping algebras*, Jour. Amer. Math. Soc. **3** (1990), 447-498.

[L2] G. Lusztig, *Canonical bases arising from quantized enveloping algebras, II; in "Common trends in mathematics and quantum field theories", ed. T. Eguchi et. al.*, Progr. Theor. Phys. Suppl. **102** (1990), 175-201.

[S] T. A. Springer, *Trigonometric sums, Green functions of finite groups and representations of Weyl groups*, Invent. Math. **36** (1976), 173-207.

[Z] A. V. Zelevinskii, *Two remarks on graded nilpotent classes*, Russian Math. Surveys **40** (1985), 249-250.

Massachusetts Institute of Technology, Cambridge, MA 02139

Contemporary Mathematics
Volume **139**, 1992

Morita Equivalence of Primitive Factors of $U(\mathrm{sl}(2))$

TIMOTHY J. HODGES*

ABSTRACT. The primitive factors of the enveloping algebra $U(\mathbf{sl}(2,\mathbf{C})$ are classified up to Morita equivalence using the Hattori-Stallings map.

Let $U = U(\mathbf{sl}(2,\mathbf{C}))$ be the enveloping algebra of the Lie algebra $\mathbf{sl}(2,\mathbf{C})$. The minimal primitive ideals of U are of the form $(\Omega - \alpha)$ where $\alpha \in \mathbf{C}$ and Ω is the Casimir element. In [3] Dixmier proved that the algebras $B_\alpha = U/(\Omega - \alpha)$ are non-isomorphic as $\mathbf{C}$-algebras. On the other hand, Stafford showed in [9] that many of the B_α are Morita equivalent via "translation functors" and asked whether it was true that all the simple B_α are equivalent. In this note we answer this question by showing that two such algebras are equivalent only if there is a translation functor defining equivalence. In passing we give a new and very short proof of Dixmier's result. The essential idea is to look at the Hattori-Stallings traces of the rank one projective modules.

The parameterization given above for the B_α is an unnatural one to work with in this context. By Duflo's theorem each minimal primitive is the annihilator of a Verma module $M(\lambda)$ for $\lambda \in \mathbf{C}$ (here we are making the natural identification of $\mathbf{h}^*$ with $\mathbf{C}$ via $\nu \longmapsto \nu(H)$). For this and other basic information, we refer the reader to [4]. Define D_λ to be $U/\mathrm{Ann}_U M(\lambda)$. Recall that $\mathrm{Ann}_U M(\lambda) = (\Omega - (\lambda^2 - 1))$ and hence that $\mathrm{Ann}_U M(\lambda) = \mathrm{Ann}_U M(\mu)$ if and only if $\lambda = \pm\mu$. Thus Dixmier's theorem states in this context that $D_\lambda \simeq D_\mu$ if and only if $\mu = \pm\lambda$. On the other hand, Stafford showed in [9] that D_μ is Morita equivalent to $D_{\mu+1}$ whenever $\mu \neq 0, -1$. Thus for each fixed $\lambda \notin \mathbf{Z}$, the algebras $\{D_{n\pm\lambda} : n \in \mathbf{Z}\}$ are Morita equivalent. We show that these, together with the sets $\{D_n : 0 \neq n \in \mathbf{Z}\}$ and $\{D_0\}$ are the $\mathbf{C}$-linear Morita equivalence classes. (Recall that a $\mathbf{C}$-linear equivalence is one given by a $\mathbf{C}$-functor; that is, a functor F such that the maps from $\mathrm{Hom}(A, B)$ to $\mathrm{Hom}(FA, FB)$ are $\mathbf{C}$-linear maps.) Using global dimension we may dispose of the integer cases immediately. For it is shown in [9] that

$$\operatorname{gldim} D_\lambda = \begin{cases} \infty & \text{if } \lambda = 0 \\ 2 & \text{if } \lambda \in \mathbf{Z}\backslash\{0\} \\ 1 & \text{if } \lambda \in \mathbf{C}\backslash\mathbf{Z} \end{cases}$$

1991 *Mathematics subject classification*. Primary, 17B35.

This paper is in final form and no version of it will be submitted for publication elsewhere.
*This work was partially supported by a joint grant from the National Security Agency and the National Science Foundation (MDA 904-89-H2046).

Thus D_0 can be equivalent to no other D_λ. Similarly the D_n for $n \in \mathbf{Z}\backslash\{0\}$ are equivalent to each other but to no other D_λ.

Let us briefly review the definition of the Hattori-Stallings map that we will be using for the calculations below. Let R be a ring, let $K_0(R)$ be the Grothendieck group and let $H_0(R) = R/[R,R]$ be the trace group. Let P be a projective left R-module. Then there exist $f_1, \ldots, f_n \in P^* = \mathrm{Hom}_R(P,R)$ and $x_1, \ldots, x_n \in P$ such that $1 = \sum_{i=1}^n f_i \otimes x_i \in P^* \otimes_R P \simeq \mathrm{End}_R P$. The trace of P is defined to be the element $tr(P) = \sum_{i=1}^n f_i(x_i) + [R,R]$ of $H_0(R)$ and this element is independent of the choice of the f_i and x_i. Since tr is additive on direct sums it induces a map $Tr : K_0(R) \to H_0(R)$ given by $Tr([P]) = tr(P)$. See [1,5,10] for further details.

It is well-known and fairly straightforward to check that $D_\lambda = \mathbf{C}\cdot 1 \oplus [D_\lambda, D_\lambda]$ and hence the natural map $\pi : \mathbf{C} \to H_0(D_\lambda)$ is an isomorphism [3;1.10]. It is also known that $K_0(D_\lambda) \simeq \mathbf{Z} \oplus \mathbf{Z}$. Since most of the proofs of this result in the literature involve the theory of D-modules, we give here an outline of an elementary proof of this fact.

First recall the standard realization of D_λ as a subring of the Weyl algebra $A_1(\mathbf{C})$. Let p and q be generators of $A_1(\mathbf{C})$ such that $[p,q] = 1$. Define:

$$e = -p, \quad f = q(qp - \lambda + 1), \quad h = -2qp + \lambda - 1.$$

Then there is a well-defined map from $U(\mathbf{sl}(2,\mathbf{C}))$ to $A_1(\mathbf{C})$ sending the canonical generators E, F and H to e, f and h respectively. The kernel of this map is $(\Omega - (\lambda^2 - 1)) = \mathrm{Ann} M(\lambda)$. Thus D_λ may be identified with the subalgebra of $A_1(\mathbf{C})$ generated by e, f and h. The following proof was also outlined in [6; 2.5] where more details may be found concerning the K-theoretical ideas involved.

PROPOSITION 1. *If $\lambda \neq 0$, then $K_0(D_\lambda) \simeq \mathbf{Z} \oplus \mathbf{Z}$. Furthermore if $\lambda \notin \{0, -1, -2, \ldots\}$ then the classes of D_λ and the projective ideal $P_\lambda = D_\lambda p + D_\lambda(qp - \lambda)$ form a basis for $K_0(D_\lambda)$.*

PROOF. Assume that $\lambda \notin \{0, -1, -2, \ldots\}$. Then it can be shown that $A_1(\mathbf{C})$ is a flat epimorphic extension of D_λ; that is, it lies between D_λ and its quotient division ring and is flat as a right D_λ-module. Furthermore, the class of torsion modules is generated by $M(-\lambda) = D_\lambda/P_\lambda$. It then follows from Quillen's localization sequence [6] that the Grothendieck group $G_0(D_\lambda)$ of the category of finitely generated left D_λ-modules is a free abelian group with basis the classes $[D_\lambda]$ and $[M(-\lambda)]$. Because D_λ has finite global dimension in this situation, the Cartan map is an isomorphism and the result follows.

Let $Q(D_\lambda)$ be the quotient division algebra of D_λ, let $rk : K_0(D_\lambda) \to \mathbf{Z}$ be the usual rank function given by $rk([P]) = \mathrm{length}(Q(D_\lambda) \otimes_{D_\lambda} P)$ and define $Rk_0(D_\lambda) = \mathrm{Ker}(rk)$. From Proposition 1 it is clear that $Rk_0(D_\lambda) \simeq \mathbf{Z}$.

LEMMA 2. *Let ξ be a generator for $Rk_0(D_\lambda)$. Then $Tr(\xi) = \pm\lambda^{-1}$.*

PROOF. We know from the proposition that the generators of $Rk_0(D_\lambda)$ are $\pm([P_\lambda] - [D_\lambda])$. Recall that $P_\lambda = D_\lambda p + D_\lambda(qp - \lambda)$ and identify P_λ^* as $\{x \in Q(D_\lambda) \mid P_\lambda x \subset D_\lambda\}$. Since $qp + (-1)(qp - \lambda) = \lambda$, we have that in

$P_\lambda^* \otimes P_\lambda$,

$$1 = \lambda^{-1}(q \otimes p + (-1) \otimes (qp - \lambda))$$

Thus $Tr([P_\lambda]) = \lambda^{-1}(pq - (qp - \lambda)) = (\lambda + 1)\lambda^{-1}$. Hence, since $Tr([D_\lambda]) = 1$, it follows that $Tr(\xi) = \pm\lambda^{-1}$, as required.

THEOREM 3 [3]. *Let $\lambda, \mu \in \mathbf{C}$. Then D_λ is isomorphic to D_μ as a $\mathbf{C}$-algebra if and only if $\mu = \pm\lambda$.*

PROOF. Let $\phi : D_\lambda \to D_\mu$ be a $\mathbf{C}$-algebra isomorphism. Then ϕ induces maps $K_0(\phi)$ and $H_0(\phi)$ such that the following diagram commutes [1]:

$$\begin{array}{ccc} K_0(D_\lambda) & \xrightarrow{Tr} & H_0(D_\lambda) \\ \downarrow K_0(\phi) & & \downarrow H_0(\phi) \\ K_0(D_\mu) & \xrightarrow{Tr} & H_0(D_\mu) \end{array}$$

Furthermore, after we have identified $H_0(D_\lambda)$ and $H_0(D_\mu)$ with $\mathbf{C}$ in the manner described above, the induced map $H_0(\phi) : H_0(D_\lambda) \to H_0(D_\mu)$ is just the identity map. Let ξ be a generator of $Rk_0(D_\lambda)$ such that $Tr(\xi) = \lambda^{-1}$. Since rank commutes with $K_0(\phi)$, we must have that $K_0(\phi)(\xi)$ is a generator of $Rk_0(D_\lambda)$. But then

$$\lambda^{-1} = Tr(\xi) = Tr(K_0(\phi)(\xi)) = \pm\mu^{-1}.$$

Hence $\mu = \pm\lambda$, as required.

We now turn to the problem of Morita equivalence. Given a Morita equivalence between D_λ and D_μ, we get a commutative diagram similar to that used above in Theorem 3. The difference in this case is that the induced map on the H_0's is no longer the identity.

LEMMA 4. *Suppose that $F : D_\lambda$-Mod $\to D_\mu$-Mod is a $\mathbf{C}$-linear equivalence of categories. Then $H_0(F) : H_0(D_\lambda) \to H_0(D_\mu)$ is given by multiplication by $(\mu + n)\mu^{-1}$ for some integer n.*

PROOF. By standard Morita theory, F is given by $P \otimes_{D_\lambda}$ - for some projective generator P in Mod-D_λ with $\mathrm{End}_{D_\lambda}(P) = D_\mu$. Since D_λ and D_μ are both Noetherian domains, the rank of P must be one as both a right D_λ-module and as a left D_μ-module. Since F is $\mathbf{C}$-linear, $H_0(F)$ is a $\mathbf{C}$-linear map; hence it suffices to calculate $H_0(F)(1)$. Now the map $H_0(F)$ is given by sending the coset of the element $\sum f_i \otimes a_i \in P^* \otimes_{D_\mu} P \simeq D_\lambda$ to the coset of the element $\sum a_i \otimes f_i \in P \otimes_{D_\lambda} P^* \simeq D_\mu$ [2; p.48]. Hence, in particular, $H_0(F)(1) = Tr([{}_{D_\mu}P])$. Since ${}_{D_\mu}P$ is a rank one projective D_μ-module, it follows from Lemma 2 that $Tr([{}_{D_\mu}P] - [D_\mu]) = n\mu^{-1}$ for some $n \in \mathbf{Z}$. Hence $Tr([{}_{D_\mu}P]) = 1 + n\mu^{-1} = (\mu + n)\mu^{-1}$. Thus $H_0(F)$ is given by multiplication by $(\mu + n)\mu^{-1}$.

THEOREM 5. *Let λ and μ be elements of $\mathbf{C}\backslash\{0\}$. Then D_λ and D_μ are $\mathbf{C}$-linearly Morita equivalent if and only if $\mu = m \pm \lambda$ for some $m \in \mathbf{Z}$.*

PROOF. ($\Leftarrow$) This follows from [9; Cor. 3.3].

($\Rightarrow$) Let $F : D_\lambda$-Mod $\to D_\mu$-Mod be a $\mathbf{C}$-linear Morita equivalence. It is well-known (see for example [8]) that F induces the following commutative diagram (where the vertical maps are isomorphisms):

$$\begin{array}{ccc} K_0(D_\lambda) & \xrightarrow{Tr} & H_0(D_\lambda) \\ \downarrow K_0(F) & & \downarrow H_0(F) \\ K_0(D_\mu) & \xrightarrow{Tr} & H_0(D_\mu) \end{array}$$

Now $K_0(F)$ must preserve rank. Hence it maps $Rk_0(D_\lambda)$ isomorphically onto $Rk_0(D_\mu)$. Let $\eta = K_0(F)(\xi)$ be a generator of $Rk_0(D_\mu)$ with $Tr(\eta) = \mu^{-1}$. Then using the commutativity of the above diagram we obtain that

$$\mu^{-1} = Tr(\eta) = H_0(F)(Tr(\xi)) = \pm\lambda^{-1}(\mu + n)\mu^{-1}$$

for some $n \in \mathbf{Z}$. Hence $\mu + n = \pm\lambda$.

We have tried in this article to take a non-technical approach, avoiding in particular the theory of D-modules. However all of the ideas involved are special cases of general concepts applicable to the case of an arbitrary semisimple Lie algebra. One may make similar conjectures about isomorphism and Morita equivalence classes of primitive factors in this general situation. Preliminary calculations suggest that although this technique of looking at traces can go a long way towards distinguishing between these factors, the question cannot be completely solved using this approach.

References

[1] H. Bass, "Euler characteristic and characters of discrete groups", *Invent. Math.* **35**(1976), 155-196.

[2] P.M. Cohn, Morita Equivalence and Duality, *Queen Mary College Lecture Notes*, Queen Mary College, London, 1976.

[3] J. Dixmier, "Quotients simples de l'algèbre enveloppante de $\mathbf{sl}_2$", *J. Algebra* **24**(1973), 551-564.

[4] J. Dixmier, Enveloping Algebras, North Holland, Amsterdam 1977.

[5] A. Hattori, "Rank element of a projective module", *Nagoya Math. J.* **25**(1965), 113-120.

[6] T.J. Hodges, "K-Theory of Noetherian rings", *Seminaire d'algebre,* (M.-P. Malliavin, ed.) Lecture Notes in Mathematics 1404, Springer-Verlag, Berlin 1989.

[7] M. Karoubi, "Homologie cyclique et K-thèorie algèbrique I", *C.R. Acad. Sci. Paris, Serie I,* **297**(1983), 447-450.

[8] C. Kassel, "Charactére de Chern bivariant", *K-theory*, **3**(1989), 367-400.

[9] J.T. Stafford, "Homological properties of the enveloping algebra $U(\mathbf{sl}_2)$", *Math. Proc. Camb. Phil. Soc.* **91**(1982), 29-37.

[10] J. Stallings, "Centerless groups - an algebraic formulation of Gottlieb's theorem", *Topology* **4**(1965), 129-134.

DEPARTMENT OF MATHEMATICS, UNIVERSITY OF CINCINNATI, CINCINNATI, OH 45221-0025

E-mail: hodges@ucbeh.bitnet

Contemporary Mathematics
Volume 139, 1992

Degeneracy of Schubert Varieties

C. Huneke and V. Lakshmibai

Abstract: Let X be a Schubert variety in the flag variety G/B and let R(X) be the multigraded ring of X. In this paper, we prove the Cohen-Macaulayness of R(X) (for $G = SL_n$) using deformation techniques.

§1. Introduction

Let G be a semisimple, simply connected Chevalley group defined over a field k. Let T be a maximal k-split torus of G, B a Borel subgroup containing T. Let W be the Weyl group of G relative to T. Let Q be a parabolic subgroup of G containing B, say $Q = \bigcap_{i=1}^{r} P_i$, P_i being maximal parabolic subgroups. Let W(Q) be the Weyl group of Q. For $w \in W/W(Q)$, let $X(w) = \overline{BwQ}$ (mod Q), the Zariski closure of the Bruhat cell endowed with the canonical reduced scheme structure, be the Schubert subvariety ofG/Q associated to w. Let L_i be the ample generator of $Pic(G/P_i)$, and $R(w) = \bigoplus_{L \geq 0} H^0(X(w),L)$ where the sum on the R.H.S. runs over all positive line

---

1980 *Mathematics Subject Classification (1991 Revision),* Primary 20G05, 20G10, Secondary 14F05, 14M15.

This paper is in final form and will not appear elsewhere.

bundles $L = L_1^{a_1} \otimes \cdots \otimes L_r^{a_r}$, $a_i \in \mathbb{Z}^+$. The aim of this paper is the study of R(w) using deformation techniques. Of course, now much is known about R(w) (cf. [MR], [KR], [R], [RR]). In [MR], [KR], [R], [RR], the geometric properties of R(w) are obtained using Frobenius splitting. Our approach here uses the "standard monomials" basis for R(w) as constructed in [LMS], [LS]. In fact, our techniques could be applied more generally to the case of a variety X for which k[x] has algebra generators $\{x_\alpha, \alpha \in H\}$ (H being some finite indexing set) such that certain of the monomials in the x_α's form a k-basis for k[x] (see §3 for more details).

Our main result is

(*) R(w) is Cohen-Macaulay, G being SL(n).

The spirit of proving (*) consists in deforming R(w) into a simpler algebra, $R^{def}(w)$ (by successive flat deformations using the explicit basis of R(w) given by "standard monomials") and then proving $R^{def}(w)$ is Cohen-Macaulay by means of a "principal radical system" as in [EH].

While dealing with Schubert varieties in G/B, one faces the following two main difficulties: Firstly, Schubert subvarieties do not behave well in G/B. For instance, there can be two Schubert subvarieties $X(w_1)$

and $X(w_2)$, both of codimension one in some $X(w)$, with no Schubert subvariety $X(w')$ of codimension two in $X(w)$ appearing in $X(w_1) \cap X(w_2)$ (see Remark under Proposition 4.8). Secondly, and more seriously, the "standard monomials" basis of $H^0(G/B, L)$ $(L \geq 0)$ does not restrict to a basis for $H^0(X(w), L)$. To be more precise (cf. §2), there can be a standard monomial on G/B whose restriction to X(w) is nonzero but is no longer standard on X(w). This difficulty is overcome by using the idea of deformation. In [DL] the ring $R^{def}(w)$ for the case of G/P, G any type and P a maximal parabolic, was proved to be Cohen-Macaulay using some nice combinatorial topological properties of the simplicial complex associated to a partially ordered set arising out of the given Schubert variety X(w). In the present case (G = SL(n) and Q any parabolic) the same procedure does not work. To overcome this difficulty we use the idea behind principal radical systems as developed in [EH]. This procedure consists mainly of analyzing the hyperplane section of a given Schubert variety which in turn involves a detailed analysis of the unions and intersections of Schubert varieties. <u>Up to radical</u> (i.e., set-theoretically) these unions and intersections are completely determined by the structure of the Weyl group and can be analyzed through a study of the Bruhat order (cf. §2) of the Weyl group W. We carry through this analysis in §4. The set-theoretic equalities (involving unions and intersections of Schubert varieties) are concluded to be ideal-theoretic in $R^{def}(w)$. We achieve this latter statement by being able to realize $R^{def}(w)$ as a discrete algebra with straightening law in the sense of [E]. Finally, we are able to prove that $R^{def}(w)$ (and therefore R(w)) is Cohen-Macaulay by using the inductive technique of principal radical systems.

Historically, (*) for the case G = SL(n) and Q a maximal parabolic was proved by several authors (cf. [H],[K],[La],[Mu]). Then in [DL], (*) was proved for any semisimple G and Q, a maximal parabolic subgroup of classical type. (In fact one of the aims of this paper is to show how the techniques used in [DL] could be extended to the case of G/B.) Finally, in [R],[RR], the result was proved for any semisimple G and any parabolic subgroup Q. Notwithstanding the results of [R], [RR], one of our aims in this paper is to show that deformation techniques could be used for the study of Schubert varieties, and also of more general varieties (cf. $[L]_2$; see also the concluding remarks in §3).

The order of the paper is as follows: In §2, we

give definitions and some known lemmas to be used later, and prove a strong straightening condition which R(w) satisfies and which will allow us to deform R(w). In §3, we make this deformation in detail. In §4, we discuss the Weyl group and its Bruhat order, and prove the main propositions necessary to discuss the ideal theory of R(w). In §5, we apply the results of §4 to discuss the intersection and union of Schubert varieties in $R^{def}(SL(n)/B)$. We also analyze the hyperplane section of a given Schubert variety. Finally, in §6 we prove $R^{def}(w)$ is Cohen-Macaulay using induction and several well-known lemmas from homological algebra. The Cohen-Macaulayness of $R^{def}(w)$ lifts immediately to the Cohen-Macaulayness of R(w).

§2. **Preliminaries**

Let G = SL(n), B = the Borel subgroup of G consisting of upper triangular matrices, and T = maximal torus in G consisting of diagonal matrices. Let R = {roots of G relative to T} and S = {simple roots of R relative to B}. Let $P_1, \cdots, P_{n-1}$ be <u>any</u> ordering of the maximal parabolic subgroups of G containing B. Let W be the Weyl group of G with respect to T, $W(P_i)$ the Weyl group of P_i, and W^i the set of minimal representatives of $W(P_i)$ in W.

As in the introduction, if $w \in W$, we let $X(w)$ be the Schubert variety in the flag variety G/B defined to be the Zariski closure, $\overline{BwB}$ (mod B) of the orbit of w in G/B endowed with the canonical reduced scheme structure.

Recall the partial order on W; if $\tau_1, \tau_2 \in W$, then $\tau_1 \geq \tau_2$ if and only if $X(\tau_1) \supseteq X(\tau_2)$. (See §4 for more facts concerning this order.) This partial order restricts to any $W/W(Q)$, where Q is any parabolic containing B, in an obvious way. Given a Q such that $Q \supseteq B$, we may assume without loss of generality that

$$Q = \bigcap_{i=1}^{r} P_i .$$

Definition 2.1 (cf. [LS]). A Young diagram (or tableau) in W/W(Q) of type $\underline{m} = (m_1, \cdots, m_r)$, where $m_i \geq 0$, $1 \leq i \leq r$, is a sequence $\lambda = (\lambda_{ij})$ with $\lambda_{ij} \in W^i$, $1 \leq j \leq m_i$, $1 \leq i \geq r$.

Definition 2.2 (cf. [LS]). A Young diagram $\lambda = (\lambda_{ij})$ is said to be a Young diagram on a Schubert variety $X(\varphi) \subset G/Q$ if $\varphi_i \geq \lambda_{ij}$, for all $1 \leq i \leq r$, $1 \leq j \leq m_i$ where $X(\varphi_i)$ is the projection of $X(\varphi)$ under $G/Q \to G/P_i$.

Definition 2.3 (cf. [LS]). A Young diagram $\lambda = (\lambda_{ij})$ on $X(\varphi)$, $\varphi \in W/W(Q)$ is said to be standard on $X(\varphi)$ if there exists a sequence $\theta = (\theta_{ij})$ (which we call a defining sequence for λ) so that

1) θ_{ij} is in $W/W(Q)$, $1 \le i \le r$, $1 \le j \le m_i$,

2) each θ_{ij} is a lifting of λ_{ij} under $W/W(Q) \to W/W_i$, and

3) $X(\varphi) \ge X(\theta_{11}) \ge X(\theta_{12}) \ge \cdots \ge X(\theta_{1m_1}) \ge X(\theta_{21}) \ge \cdots \ge X(\theta_{rm_r})$ (in G/Q).

More generally, a Young diagram $\lambda = (\lambda_{ij})$ is said to be <u>standard on a union of Schubert varieties</u> $Z = \bigcup_{i=1}^{t} X(\varphi_i)$ in G/Q, if λ is standard on $X(\varphi_i)$, for some i , $1 \le i \le t$. If $m_t = 0$ for any t , $1 \le t \le r$, the family $\{\theta_{tj}\}$ $1 \le j \le m_t$ is understood to be empty.

As above, let L_i be the ample generator of $\mathrm{Pic}(G/P_i)$. One knows that the extremal weight vectors in $H^0(G/P_i, L_i)$ give a k-basis for $H^0(G/P_i, L_i)$, which we shall denote by $\{p_\tau, \tau \in W^i\}$. (These are simply the Plücker coordinates.) Given a Young diagram $\lambda = (\lambda_{ij})$, we shall set

$$p(\lambda) = \prod_{i=1}^{r} \prod_{j=1}^{m_i} p_{\lambda_{ij}}$$

(Note that $p(\lambda) \in H^0(G/Q, L_{\underline{m}})$, where $L_{\underline{m}} = L_1^{m_1} \otimes \cdots \otimes L_r^{m_r}$.) Such a monomial will be called <u>standard on Z</u>, if λ is standard on Z (cf. Definition 2.3).

In [LS], one of the main theorems is:

<u>**Theorem**</u>. Let G, Q, Z be as above. Then the standard monomials on Z of length $\underline{m} = (m_1, \cdots, m_r)$ form a basis of $H^0(Z, L_{\underline{m}})$, where $L_{\underline{m}} = L_1^{m_1} \otimes \cdots \otimes L_r^{m_r}$.

We next make some identifications which we will use throughout the rest of the paper. As $W(SL(n))$ is S_n, the reflections in W are precisely the elements (i,j) of S_n which switch the ith and j^{th} positions. Taking the ordering for maximal parabolics so that

$$P_i = \left\{ A \in G \;\middle|\; A = \begin{pmatrix} * & * \\ 0_{n-i\times i} & * \end{pmatrix} \right\},$$

$W(P_i)$ can be identified with the subgroup of W generated by the reflections $\{(j,j+1),\ j \neq i\}$. Then $W^i \simeq I_n(i) = \{(j_1,\cdots,j_i) \mid 1 \le j_1 < j_2 <\cdots< j_i \le n\}$. Further, if $\tau_1,\tau_2 \in I_n(i)$, say $\tau_1 = (j_1,\cdots,j_i)$ and $\tau_2 = (\ell_1,\cdots,\ell_i)$, then the induced order in W^i is the following: $\tau_1 \ge \tau_2$ if and only if $j_k \ge \ell_k$, $1 \le k \le i$.

In the sequel, if $\underline{j} = (j_1,\cdots,j_k)$ is a k-tuple of distinct integers we write $\underline{j}^{inc}$ for the k-tuple obtained by rearranging the elements of $\underline{j}$ in ascending order.

One of the great difficulties in studying a Schubert variety $X(w)$ in G/B, as opposed to the Schubert varieties in G/P is that a standard monomial on G/B which is non-zero on $X(w)$ <u>may not be standard on</u> $X(w)$. More precisely, suppose $\varphi \in W/W(Q)$, and $\varphi' \in W/W(Q)$. Let λ be a standard Young diagram on $X(\varphi)$. If $\varphi' \ge \varphi$, then λ remains a standard Young diagram on $X(\varphi')$. However, if $\varphi \ge \varphi'$, and if λ is a Young diagram on $X(\varphi')$, <u>it need not</u>

be standard. We illustrate this point with an example. Let $G = SL(3)$, and let φ be the element (312). The Young diagram $\lambda = ((2),(13))$ is standard on G/B, since if we take $\theta_{11} = (231)$, $\theta_{21} = (132)$, then $\theta_{11} \geq \theta_{21}$ while θ_{11} projects to (2) in G/P_1 and θ_{21} projects to (13) in G/P_2. In addition, λ is a Young diagram on φ, since $\varphi_1 = (3) \geq (2)$, while $\varphi_2 = (13) \geq (13)$. However, λ is not standard on $X(\varphi)$. For if λ were standard on $X(\varphi)$, there would exist a $\theta'_{11} \in W$, $\theta'_{21} \in W$ such that θ'_{11} projects to (2), θ'_{21} projects to (13), $\theta'_{11} \geq \theta'_{21}$ and $\varphi \geq \theta'_{11}$. The condition that θ'_{11} projects to (2) and $\varphi \geq \theta'_{11}$ forces $\theta'_{11} = (213)$, while the condition that θ'_{21} projects to (13) forces $\theta'_{21} = (132)$. However in this case, $\theta'_{11} \not\geq \theta'_{21}$. In [HL] we have characterized those Schubert varieties φ for which every standard monomial on G/B remains standard on $X(\varphi)$ as the Kempf varieties. (See $[K]_2$, $[L]_1$ for the definition of Kempf varieties.)

Remark 2.3′ (Corollary 11.2, [LMS]). Let $X(\varphi)$ be a Schubert variety in G/Q, $\varphi \in W/W(Q)$. Let λ be a standard Young diagram on $X(\varphi)$. Then there exists a unique maximal defining tableau θ^+ relative to $X(\varphi)$ for λ, i.e., θ^+ is a defining tableau and if θ is any other defining tableau for λ with respect to $X(\varphi)$ then $\varphi \geq \theta^+ \geq \theta$.

We will also need the existence of a minimal

defining tableau for λ, and in this case the minimal tableau is unique and does not depend on φ.

Not only do the standard monomials of type $\underline{m}$ on $Z = \cup X(\varphi_i)$ provide a basis for $H^0(Z, L^{\underline{m}})$, they also satisfy a type of "straightening law" which will be extremely important for us.

We first need to observe a property satisfied by the Plücker coordinates, namely, given τ, τ_1 in W^i, $\tau \leq \tau_1$ if and only if $p_\tau | X(\tau_1) \neq 0$.

Let $A = S_{m_1} \times \cdots \times S_{m_r}$ the product of the symmetric groups. If $\lambda = (\lambda_{ij})$ is a $m_1 + \cdots + m_r$-tuple of elements where $\lambda_{ij} \in W^i$, $1 \leq j \leq m_i$, $1 \leq i \leq r$, and $a \in A$, $a = (a_1, \cdots, a_r)$ then by λ^a we denote the expression which permutes the first m_1 entries according to a_1, the next m_2 according to a_2, etc.

Definition 2.4. Let $M = \{\lambda = (\lambda_{ij}), \lambda_{ij} \in W^i, 1 \leq j \leq m_i, 1 \leq i \leq r\}$. We define a partial order $\overset{L}{\geq}$ on M as follows: Let $\lambda = (\lambda_{ij})$, $\mu = (\mu_{ij}) \in M$. Let us write $\lambda = (\lambda_1, \cdots, \lambda_m)$, $\mu = (\mu_1, \cdots, \mu_m)$ where $m = m_1 + \cdots + m_r$. We say $\lambda \overset{L}{\geq} \mu$, if either $\lambda = \mu$ or there exists a t, $1 \leq t \leq m$ such that $\lambda_i = \mu_i$, $1 \leq i < t$ and $\lambda_t \underset{\neq}{>} \mu_t$ (note that for any i, $1 \leq i \leq m$, λ_i and μ_i both $\in W^P$ for some maximal parabolic P and $\geq$ on W^P denotes the Bruhat order).

Proposition 2.5. Let $Z = \bigcup_{i=1}^{\ell} X(\varphi_i)$ be a union of Schubert varieties in G/Q. Let $p(\lambda)$ be a nonzero, nonstandard monomial on Z of type $\underline{m} = (m_1, \cdots, m_r)$. Let

$$(*): \quad p(\lambda) = \sum_{i=1}^{N} a_i p(\tau_i) \ , \ a_i \in k^*,$$

be the expression for $p(\lambda)$ as a sum of standard monomials on Z. Then for every i, $\tau_i \overset{L}{\underset{\neq}{\geq}} \lambda^a$, $a \in A$ (A being as above).

Proof. We may assume $m_1 \neq 0$ (if $m_1 = 0$, $m_2 \neq 0$, then we may work with $Q' = \bigcap_{i=2}^{r} P_i$ etc.). Let us write $\tau_i = (\tau_{jk}^{(i)})$, where $\tau_{jk} \in W^j$, $1 \leq k \leq m_j$, $1 \leq j \leq r$. Among $\{\tau_{11}^{(i)}$, $1 \leq i \leq N\}$ choose a minimal one, which we may suppose to be $\tau_{11}^{(1)}$ (by reindexing the τ_i's if necessary). Let us denote $\tau_{11}^{(1)}$ by just τ_{11}. Let us index the τ_i's so that $\tau_1, \cdots, \tau_s$, are all the elements such that $\tau_{11}^{(i)} = \tau_{11}$, $1 \leq i \leq s$ (note, in particular that $\tau_{11}^{(i)} \not\leq \tau_{11}$, $s < i \leq N$). For each τ_i , $1 \leq i \leq s$, let us fix a defining sequence $\theta = (\theta_{jk}^{(i)})$ on Z (note that τ_i being standard on Z, is standard on some $X(\varphi_t)$ and we fix a defining sequence θ for τ_i on $X(\varphi_t)$ so that $X(\theta_{11}^{(i)}) \subseteq X(\varphi_t)$). Let $Z_1 = \bigcup_{i=1}^{s} X(\theta_{11}^{(i)})$. Let us now restrict (*) to Z_1. Then $p(\tau_j)|_{Z_1} \equiv 0$, $j > s$ (Note that under π: $G/Q \longrightarrow G/P_1$, $\pi(X(\theta_{11}^{(i)})) = X(\tau_{11})$, $1 \leq i \leq s$, and $\pi(X(\theta_{11}^{(i)}) \not\subseteq X(\tau_{11})$, s

$< i \le N$. Thus $p(\tau_j)|_{Z_1} \equiv 0$, $s < j \le N$.) Further, $p(\tau_j)$, $1 \le j \le s$ is standard on Z_1. By linear independence of standard monomials on Z_1, we have $p(\lambda)|_{Z_1} \not\equiv 0$ (note that $\sum_{i=1}^{s} a_i p(\tau_i)$ is a nonzero sum of standard monomials on Z_1). This implies in particular that $\lambda_{1j} \le \tau_{11}$, $1 \le j \le m_1$ (note that $\pi(Z_1) = X(\tau_{11})$, π being as above). If $\lambda_{1j} = \tau_{11}$ for some j, $1 \le j \le m_1$, then we cancel $p_{\tau_{11}}$ and use induction on m_1 and r, the reasoning being the same for the case of a maximal parabolic (the case $r=1$). (Note that $I(Z_1)$, ideal of $Z_1 = \bigcap_{i=1}^{s} I(X(\theta_{11}^{(i)}))$, since Z_1 is reduced. Hence if $p_{\tau_{11}} F \in I(Z_1)$, then $F \in I(Z_1)$, since $I(X(\theta_{11}^{(i)}))$ is prime and $p_{\tau_{11}} \notin I(X(\theta_{11}^{(i)}))$, $1 \le i \le s$.) Thus, we may assume that $\lambda_{1j} \lneq \tau_{11}$, for all $1 \le j \le m_1$. From this we obtain that for every τ_i such that $\tau_{11}^{(i)}$ is minimal in $\{\tau_{11}^{(j)}, 1 \le j \le N\}$, $\tau_i \overset{L}{\gneq} \lambda$. If for some i, $\tau_{11}^{(i)}$ is not minimal, then there exists a j such that $\tau_{11}^{(i)} \gneq \tau_{11}^{(j)}$ and $\tau_{11}^{(j)}$ is a minimal element in $\{\tau_{11}^{(\ell)}, 1 \le \ell \le N\}$. But then from the discussion above, we have $\tau_{11}^{(j)} \gneq \lambda_{1t}$, $1 \le t \le m_1$. Hence it follows that $\tau_{11}^{(i)} \gneq \lambda_{1t}$, $1 \le t \le m_1$. From this, we obtain that $\tau^{(i)} \overset{L}{\gneq} \lambda^a$. This completes the proof of Proposition 2.5.

<u>Note</u>: A relation as in Proposition 2.5 will be referred

to as a straightening relation on Z.

§3. Deformation

We preserve the notations of §2. We now introduce few more notations. Given X(w) in G/Q, let us denote its projection under π_i: $G/Q \longrightarrow G/P_i$, by $X(w^{(i)})$, $1 \le i \le r$. We set Y_w or just Y as

$$Y = \{\beta \in W^i,\ \beta \le w^{(i)}\ ,\ 1 \le i \le r\}$$

and

$$Y_i = \{\beta \in W^i,\ \beta \le w^{(i)}\}$$

For a monomial T of type $(m_1, \cdots, m_r)$, and $\alpha \in Y_i$, we set

$$T_\alpha = \{\beta \in Y_i \,|\, p_\beta \text{ occurs in } T, \text{ and } \beta \ge \alpha\}$$

and

$$T^\alpha = \{\beta \in T_\alpha \,|\, \beta \gneq \alpha\}.$$

We have (from §2)

Theorem 3.1. Let $R = \underset{\underline{m}}{\oplus} H^o(X(w), L_{\underline{m}})$, $\underline{m} \in \mathbb{Z}^+$. Then

1. R is a k-algebra with algebra generators consisting of

$$\{p_\beta,\ \beta \in W^i\ ,\ \beta \le w^{(i)}\ ,\ 1 \le i \le r\}.$$

2. Monomials in p_β's standard on X(w) form a k-basis of R.

3. In any straightening relation

$$\underset{L}{p(\lambda)} = \Sigma a_i p(\tau_i),\ a_i \in k^*$$

for every i, $\tau_i \gneq \lambda^a$, $a \in A$ (notations as in Proposition 2.5). We set $J_R^1 = \{\beta \in W^1 |$ there exists a straightening relation in R, and a monomial F on the R.H.S. of this relation such that p_β occurs more number of times in F than it does on the L.H.S. of this relation}.

Lemma 3.2. Fix a maximal element α in J_R^1. Let $I = (p_\alpha)$. Let $I_j = I^j$, $j \geq 1$. Then I_j has a basis consisting of standard monomials on X(w) involving p_α at least j times.

Proof. Let T be a nonzero monomial in I^j. Further let T be nonstandard on X(w). Let

$$(*)\ T = \Sigma a_i T_i\ ,\ a_i \in k^*$$

be the expression for T as a sum of standard monomials on X(w). For each i, we have (by the maximality assumption on α), $T^\alpha \geq T_i^\alpha$. This together with the lexicographic order condition (cf. Proposition 2.5) implies that $T_\alpha \subseteq (T_i)_\alpha$ (and $T^\alpha = T_i^\alpha$). Hence we obtain that for every i, $(T_i)_\alpha - T_i^\alpha \supseteq T_\alpha - T^\alpha$, and hence $\#((T_i)_\alpha - T_i^\alpha) \geq \#(T_\alpha - T^\alpha) \geq j$ (since $T \in I^j$). Thus, each T_i involves p_α at least j times.

Proposition 3.3. Let α be as in Lemma 3.2. Let

$\mathcal{R} = \cdots Rt^2 \oplus Rt \oplus R \oplus I_1 t^{-1} \oplus I_2 t^{-2} \oplus \cdots$

where t is an indeterminate. Then

(1) $\mathcal{R}$ is a k[t]-algebra with algebra generators given by $\{P_\beta,\ \beta \in Y\}$ where $P_\beta = p_\beta$, $\beta \neq \alpha$, and $P_\alpha = p_\alpha t^{-1}$.

(2) Standard monomials give a k[t]-basis for $\mathcal{R}$ (here by a standard monomial in $\mathcal{R}$, we mean a monomial of the form

$P(\Lambda) = \pi P_{\lambda_{ij}}$, where $\Lambda = (\lambda_{ij})$ is a standard Young diagram on X(w)).

(3) In any straightening relation

$$P(\Lambda) = \Sigma a_i P(\tau_i) \ , \ a_i \neq 0$$

where Λ is a nonstandard Young diagram on X(w) of type $(m_1, \cdots, m_r)$, we have, $\tau_i \overset{L}{\underset{\neq}{>}} \Lambda^a$, $a \in A$, for all i.

Proof. 1) is clear

2) Linear Independence: The $k[t,t^{-1}]$-algebra $R[t,t^{-1}] = \mathcal{R}[t^{-1}]$ has a $k[t,t^{-1}]$-basis given by standard monomials (by change of base). Since t is a unit in $k[t,t^{-1}]$, the standard monomials in the P_β's are also linearly independent over $k[t,t^{-1}]$ (note that $P_\beta = p_\beta$ or $p_\alpha t^{-1}$ according as $\beta \neq$ or $= \alpha$) and hence are linearly independent over k[t].

Generation: For $a \in R$. let n(a) be the integer r such that $a \in I_r$, $a \notin I_{r+1}$. Now given a monomial f in the p_β's, f determines an unique monomial F in the P_β's givn by $F = ft^{-n(f)}$ (and conversely). Further F is standard if and only if f is. Let now F be a non-standard monomial in $\mathcal{R}$. This implies that f is non-standard. Let

$$f = \sum_i a_i f_i, \ a_i \in k^*$$

be the straightening relation for f. This implies

$$Ft^{n(f)} = \sum_i a_i F_i t^{n(f_i)}, \ a_i \in k^*$$

where we have (by Lemma 3.2) $n(f_i) \geq n(f)$. Hence

$$F = \sum_i a_i F_i t^{m_i}, \quad a_i \in k^*$$

where $m_i \geq 0$. This proves the generation.

3) follows from Proposition 2.5.

Proposition 3.4. Let α be as in Lemma 3.2. Let $I = (p_\alpha)$ and let

$$\mathcal{R}_\alpha = R/I \oplus I/I^2 \oplus \cdots \oplus I^j/I^{j+1} \oplus \cdots \; (= gr_I R).$$

Then

1. $\mathcal{R}_\alpha \approx \mathcal{R}/(t)$

2. $\mathcal{R}_\alpha$ is a graded k-algebra with algebra generators consisting of $\{\bar{p}_\beta, \beta \in Y\}$, where $\bar{p}_\beta = p_\beta + I$, $\beta \neq \alpha$, and $\bar{p}_\alpha = p_\alpha + I^2$.

3. Standard monomials give a k-basis for $\mathcal{R}_\alpha$ (here, by a standard monomial in $\mathcal{R}_\alpha$, we mean a monomial of the form $\overline{p(\Lambda)} = \overline{\pi p_{\lambda_{ij}}}$, where $\Lambda = (\lambda_{ij})$ is a standard Young diagram on $X(w)$).

4. In any straightening relation

$$\overline{p(\Lambda)} = \sum a_i \overline{p(\tau_i)}, \quad a_i \in k^*$$

where Λ is a nonstandard Young diagram on $X(w)$ of type $(m_1, \cdots, m_r)$, we have $\tau_i \underset{\neq}{\overset{L}{>}} \Lambda^a$, $a \in A$, for all i.

5. $J^1_{\mathcal{R}_\alpha} \subseteq J^1(R) - \{\alpha\}$.

Proof. 1) The map $\theta: \mathcal{R}_\alpha \longrightarrow \mathcal{R}/(t)(=k[t]/(t) \otimes \mathcal{R})$, $\theta(\bar{p}_\beta) = 1 \otimes P_\beta$ defines an isomorphism. From this 1) follows.

2) and 3) follow from 1) and Proposition 3.3.

4) For any $a \in R$, as above let $n(a)$ = the integer ℓ such that $a \in I^{\ell}$, $a \notin I^{\ell+1}$. Let f be a monomial and let $n(f) = j$. Further, let f be not standard. Let $f = \Sigma a_i f_i$, $a_i \neq 0$. be the expression for f as a sum of standard monomials. We have, $n(f_i) \geq j$ (in view of Lemma 3.2). Hence we obtain, $n(f) \leq \min \{n(f_i)\}$.

Claim. $n(f) = \min_i \{n(f_i)\}$.

Proof of claim. If possible, let $n(f) \lneq n(f_i)$, for all i. Then reading $f = \Sigma a_i f_i$ modulo I^{j+1}, we obtain $f \equiv 0 \pmod{I^{j+1}}$. Hence, $n(f) \geq j+1$, which is not possible, since $n(f) = j$.

Now claim implies that

$$f = \Sigma a_i f_i \qquad (\text{in } I^j/I^{j+1})$$

where the summation on the R.H.S. runs over $\{i \mid n(f_i) = \text{ord } f\}$. Thus, if we write $f_i = p(\tau_i)$ and $f = p(\Lambda)$, then we obtain

$$\overline{p(\Lambda)} = \Sigma a_i \overline{p(\tau_i)}$$

where the summation runs over $\{i \mid n(p(\tau_i)) = n(p(\Lambda))\}$. The assertion in 4) follows from Proposition 2.5.

5) From our discussion above, we obtain that I^j/I^{j+1} has a basis consisting of $\{\overline{p(\Lambda)} \mid \Lambda \text{ is standard, and } n(p(\Lambda)) = j\}$. Thus in any straightening relation

$$\overline{p(\Lambda)} = \Sigma a_i \overline{p(\tau_i)} \ (\text{in } I^j/I^{j+1})$$

p_α occurs the same number of times in <u>all</u> of the terms (including $\overline{p(\Lambda)}$). From this 5) follows.

This completes the proof of Proposition 3.4.

Proposition 3.5. The k[t]-algebra $\mathcal{R}$ is a flat deformation whose special fiber (t=0) is $\mathcal{R}_\alpha$ and whose general fiber (t invertible) is R.

Proof. That $\mathcal{R}$ is a deformation with general fiber R is clear. That $\mathcal{R}_\alpha$ is the special fiber of $\mathcal{R}$ follows from Proposition 3.4, (1). That $\mathcal{R}$ is a flat deformation, (i.e., $\mathcal{R}$ is k[t]-free) follows from Proposition 3.3, (2).

<u>The algebra $R^{def}(w)$</u>: We first observe that in carrying out the above 1-step deformation for R, we have used only those properties of R as given by Theorem 3.1. Now the properties as given by Theorem 3.1 hold for $\mathcal{R}_\alpha$ also (in view of Proposition 3.4). Hence we can carry out a similar 1-step deformation for $\mathcal{R}_\alpha$ also. Thus proceeding we obtain

Theorem 3.6. There exists a sequence of graded k-algebras $R = R_{11}, R_{12}, \cdots, R_{1t_1}$ such that

1. There is a flat deformation $\mathcal{R}_{1s}$ whose special fiber is R_{1s+1} and whose general fiber is R_{1s}, $1 \le s \le t_1 - 1$.
2. R_{1t_1} has algebra generators $\{x_\gamma \mid \gamma \in W^j,\ \gamma \le w^{(j)},\ j = 1, \cdots, r\}$.

3. Standard monomials in x_j's give a k-basis for R_{1t_1}.

4. In any straightening relation

$$(*): \ x(\Lambda) = \sum_{\ell} b_\ell x(\tau_\ell), \ b_\ell \in k^*$$

where Λ is a non-standard Young diagram on X(w) of type $(m_1, \cdots, m_r)$, we have $\tau_\ell \overset{L}{\underset{\neq}{>}} \Lambda^a$, $a \in A$, for all ℓ (here, for a Young diagram $\theta = (\theta_{ij})$, $1 \le j \le m_i$, $1 \le i \le r$, on X(w) of type $(m_1, \cdots, m_r)$, $x(\theta) = \pi_{ij} x_{\theta_{ij}}$).

5. $J^1(R_{1t_1}) = \emptyset$.

Remark 3.7. In (*) above, let $\Lambda = (\Lambda_{ij})$, $\tau_\ell = (\alpha_{ij}^{(\ell)})$, $1 \le j \le m_i$, $1 \le i \le r$. Then (4) and (5) of Theorem 3.6 imply that $\pi_j \Lambda_{1j} = \pi_j \alpha_{1j}^{(\ell)}$, for all ℓ. We next work with J_S^2 (we have denoted R_{1t_1} by S for simplicity of notation), where $J_S^2 = \{\beta \in W^2 |$ there exists a straightening relation in S, and a monomial F on the R.H.S. of this relation such that x_β occurs more number of times in F than it does on the L.H.S. of this relation$\}$. We have (similar to Lemma 3.2)

Lemma 3.8. Fix a maximal element β in J_S^2. Let $I = (p_\beta)$, and $I_j = I^j$, $j \ge 1$. Then I_j has a basis consisting of standard monomials (in x_γ's) involving x_β at least j times.

Proof. Let T be a nonzero monomial in I^j. Further let

$$(*) \qquad T = \Sigma a_i T_i, \ a_i \in k^*$$

be the straightening relation for T. For each i, we have (by the maximality assumption on β), $T^{\beta} \geq T_i^{\beta}$ (notation as in the proof of Lemma 3.2). This together with the lexicographic condition (cf. Theorem 3.6 (4)) implies in view of Remark 3.7 that $T_{\beta} \leq (T_i)_{\beta}$ (and $T^{\beta} = T_i^{\beta}$). From this we conclude (as in Lemma 3.2) that I_j has a basis consisting of standard monomials in x_{γ}'s involving x_{β} at least j times.

Proceeding as above we arrive at $R_{21} = R_{1t_1}$, $R_{22}, \cdots, R_{2t_2}$ such that

1. There is a flat deformation $\mathcal{R}_{2s}$ whose special fiber is R_{2s+1} and whose general fiber is R_{2s}, $1 \leq s \leq t_2 - 1$.
2. R_{2t_2} has algebra generators
$$\{y_{\gamma} \mid \gamma \in W^j,\ \gamma \leq w^{(j)},\ j = 1, \cdots, r\}$$
3. Standard monomials in y_{γ}'s give a k-basis for R_{1t_1}
4. In any straightening relation
$$y(\Lambda) = \sum_{\ell} c_{\ell} y(\varphi_{\ell}),\quad c_{\ell} \in k^*,$$
where Λ is a non-standard Young diagram on X(w) of type $(m_1, \cdots, m_r)$, we have
$$\varphi_{\ell} \underset{\neq}{\geq} \Lambda^a,\ a \in A, \text{ for all } \ell.$$
5. $J^i(R_{2t_2}) = \emptyset$, i = 1,2.

Proceeding thus (working with $J^3(R_{2t_2})$ etc.) we obtain

Theorem 3.9. There exists a sequence of graded k-algebras $R = R_1, \cdots, R_t$ satisfying,

1) There is a flat deformation $\mathcal{R}_s$ whose special fiber is R_{s+1} and whose general fiber is R_s.

2) R_t has algebra generators $\{q_\gamma \mid \gamma \in \bigcup_{j=1}^{r} W^j$ and $w^{(j)} \geq \gamma\}$.

3) Standard monomials in the q_γ's give a k-basis for R_t.

4) $J^i(R_t) = \emptyset$, $i = 1, \cdots, r$. We denote R_t by $R^{def}(w)$.

We summarize what we have shown.

a) $R^{def}(w)$ has a basis consisting of standard monomials. Further, any nonstandard monomial is zero (in view of (4), Theorem 3.9).

b) If $R^{def}(w)$ is Cohen-Macaulay, then $R = R(w)$ is Cohen-Macaulay.

As a consequence of (a), we have

Proposition 3.10: $R^{def}(w)$ is defined by square-free monomials.

Proof. let x_α be an indeterminate over k, one for each $\alpha \in W^j$, $1 \leq j \leq r$, $\alpha \leq w^{(j)}$. Map $k[x_\alpha] = S$ onto $R^{def}(w)$ by sending x_α to q_α. Let I be the kernel. Denote the ideal generated by the set of square-free nonstandard monomials by J. Clearly $J \subseteq I$. On the other hand, if $F = p(X_\lambda)$ is a monomial which maps to zero, then λ is not standard, and so $p(X_\lambda)$ is divisible by some $p(X_\tau)$ where X_τ is a square-free nonstandard monomial.

In particular, $R^{def}(w)$ is an algebra with straightening law [E], in fact is a discrete algebra with straightening law, a fact we shall make use of in §5. This is a significant improvement, as $R(w)$ itself is not an algebra with straightening law in any obvious way.

Concluding Remarks. The above procedure suggests that the 1-step deformation can be carried out in a more general context as described below.

Let R be a k-algebra with algebra generators $\{x_\alpha, \alpha \in H\}$ (H being a finite indexing set), such that certain monomials in the x_α's give a k-basis for R. Let us call these as standard monomials. Let F be a non-standard monomial. Let $F = \sum_i a_i F_i$, $a_i \in k^*$ be the expression for F as a sum of standard monomials. We shall refer to such a relation as a straightening relation. Let

$$J_R = \left\{ x_\alpha \;\middle|\; \begin{array}{l} \text{there exists a straightening relation } F = \Sigma a_i F_i \\ \text{such that } x_\alpha \text{ is present in some } F_i \end{array} \right\}$$

$$(*)\begin{cases} \text{Suppose } x_\alpha \text{ in } J_R \text{ is such that the ideal } I^r, \text{ where } I = \\ (x_\alpha) \text{ has a basis consisting of standard monomials} \\ \text{involving } x_\alpha \text{ at least } r \text{ times.} \end{cases}$$

Then it is easily seen that the above 1-step deformation

can be carried out. Thus if R is such that (*) holds at each step, then the above 1-step deformation can be carried out successively, and we arrive (by successive flat deformations) at the "discrete algebra" R^{disc}, where R^{disc} has algebra generators $\{y_\alpha, \alpha \in H\}$ such that

1. Certain monomials (to be called standard monomials) in the y_α's form a k-basis for R.

2. Any non-standard monomial $\equiv 0$.

The simpler algebra R^{disc} may be analyzed either by studying the combinatorics of H or by other means (for example, the principal radical system). Examples of such R's are given by Hodge algebras (cf. [DEP]), the rings R(w) as discussed above, k[X], where X is the variety consisting of pairs of matrices (A,B) of size m×n and n×m respectively such that AB = BA = 0 (cf. $[L]_2$).

§4. The Weyl Group

In this section we present a detailed analysis of the Weyl group in so far as we will need this analysis to determine the unions and intersections of Schubert varieties. We recall some facts about the Bruhat order of the Weyl group W. As a group, W is generated by a set of reflections S, called <u>simple reflections</u> which satisfy,

1) $s^2 = 1$ if $s \in S$,

2) $(s_i s_j)^{p_{ij}} = 1$, $p_{ij} \geq 2$,

3) all other relations among the generators are implied by 1) and 2). If $w \in W$ is equal to $s_1 \cdots s_q$, $s_i \in S$, then we call the word $s_1 \cdots s_q$ in the alphabet S an expression for w. The length $\ell(w)$ of $w \in W$ is the least integer q for which an expression $w = s_1 s_2 \cdots s_q$ exists. Such an expression of minimal length is said to be reduced. In general we let T = the set of conjugates of S, i.e., $T = \{wsw^{-1} | w \in W, s \in S\}$. The elements of T are commonly called reflections.

Definition 4.1 (cf. [C]). If $w, w' \in W$ we say that w precedes w' in the Bruhat order, written $w \leq w'$, if there exist reflections $t_1, \cdots, t_m \in T$ such that $w' = wt_1 \cdots t_m$ and $\ell(wt_1 \cdots t_i) > \ell(wt_1 \cdots t_{i-1})$ for $i = 1, 2, \cdots, m$.

For $W = S_n$ we may take the set S, called the set of simple reflections to be $s(i\ i{+}1) = (123 \cdots (i-1)(i+1)i(i+2) \cdots n)$; the set of reflections T will then be the set of all $s(ij) = (1 \cdots \overset{i^{th}}{j} \cdots \overset{j^{th}}{i} \cdots n)$. On the right s(ij) acts on another element $w = (a_1 \cdots a_n)$ in W by switching a_i and a_j, while on the left s(ij) acts by switching the integers i and j.

If $w \geq w'$ in W and $\ell(w) = \ell(w') + 1$, then $w = w's$ for some $s \in T$. In addition dim X(w) is precisely $\ell(w)$.

If $w = (a_1, \cdots, a_n)$ we may compute $\ell(w)$ as follows. Let $t_i(w)$ = number of a_j such that $j > i$ and $a_j < a_i$. Then

$$\sum_{i=1}^{n} t_i(w) = \ell(w).$$

If $J \subseteq S$, let W_J be the subgroup generated by J in W. These correspond to the Weyl groups of parabolic subgroups. If $s_i = s(i\ i{+}1)$, and $J = S - \{s_i\}$ then $W_J = W(P_i)$, where P_i is the parabolic defined by

$$P_i = \left\{ A \in SL_n \;\middle|\; A = \begin{pmatrix} * & * \\ 0_{n-i\times i} & * \end{pmatrix} \right\}.$$

__Notation 4.2__. Let $J \subseteq S$. Then we set $W^J = \{w \in W \mid ws > w$ for all $s \in J\}$. If J corresponds to the parabolic Q, then we also will write W^Q for W^J. If w is in W, then w can be factored uniquely $w = uv$ with $u \in W^J$ and $v \in W_J$, and in this case $\ell(w) = \ell(u) + \ell(v)$. In particular if $u \in W^J$ then u is the unique member of its coset uW_J having minimal length. We will later need two lemmas.

__Lemma 4.3__ [BW]. Assume $w \in W^J$, $w' \in W$, $w' > w$ and $\ell(w') = \ell(w) + 1$. Then either $w' \in W^J$ or $w' = ws$ for some $s \in J$.

__Lemma 4.4__ [BW]. If $w, w' \in W^J$, $w < w'$, then all maximal chains $w' = u_0 > u_1 \cdots > u_r = w$ in W^J have the same length $r = \ell(w') - \ell(w)$.

We will later use a description of W^Q in terms of

the identification of W with S_n. Suppose $Q = P_{i_1} \cap \cdots \cap P_{i_k}$, $i_1 < \cdots < i_k$. Let $w = (a_1, \cdots, a_n)$ be an element in W. Then $w \in W^Q$ if and only if $a_1 < a_2 < \cdots < a_{i_1}$, $a_{i_1+1} < \cdots < a_{i_2}$, $a_{i_2+1} < \cdots < a_{i_3}$, $\cdots$, $a_{i_{k-1}+1} < a_{i_{k-1}+2} < \cdots < a_{i_k}$.

We observe that we are now dealing with a _fixed_ ordering of the parabolics; we may identify W/W_i with increasing i-tuples of integers from 1 to n.

Remark 4.5. If $s, t \in T$ then $(st)^p = 1$ for $p = 1,2$ or 3. If $s = s(ij)$, $t = s(k\ell)$, then $p = 1$ if and only if $k = i$, $j = \ell$, and $p = 2$ (i.e., s and t commute) if and only if $\{i, j\} \cap \{k, \ell\} = \emptyset$.

Lemma 4.6. Suppose $s_1, s_2, t_1, t_2 \in T$, $s_1 \neq s_2$ and

$$s_1 s_2 = t_1 t_2. \tag{1}$$

If any distinct two of these reflections commute, then all of them commute and $\langle s_1, s_2 \rangle = \langle t_1, t_2 \rangle$.

Proof. Since $s_i^2 = t_i^2 = 1$, the equation (1) is equivalent to either one of the following two equations:

$$s_1 t_1 = s_2 t_2. \tag{2}$$

$$t_1 s_1 = t_2 s_2. \tag{3}$$

If $s_1 s_2 = s_2 s_1$, then $(s_1 s_2)^2 = 1$ and so $(t_1 t_2)^2 = 1$. Thus $t_1 t_2 = t_2 t_1$ also. If in addition $(s_1 t_j)^2 = 1$, then it is clear that $(s_i t_j)^2 = 1$ for all i, j and so all of these reflections commute.

Therefore we may assume $(s_1t_1)^2 \neq 1$, in which case $(s_1t_1)^3 = 1$, that is $s_1t_1s_1t_1 = t_1s_1$. Using Equation (3) we find $s_1t_2s_2t_1 = t_1s_1$, while from (2) we obtain $s_1t_2s_1t_2 = t_1s_1$. As we are assuming $(s_1t_2)^2 \neq 1$, we have $(s_1t_2)^3 = 1$, and this together with the latter equation shows $t_1s_1 = t_2s_1$, or $t_1 = t_2$, This contradicts $s_1 \neq s_2$. Therefore if $s_1s_2 = s_2s_1$, then all the reflections commute.

Since Equation (1) is equivalent to (2) or (3), the above proof also proves the case when $s_it_i = t_is_i$. Finally let us assume that $s_1s_2 \neq s_2s_1$, $s_it_i \neq t_is_i$, and $s_1t_2 = t_2s_1$, say. We have, $(s_1s_2)^3 = 1$, i.e., $1 = s_1s_2s_1s_2s_1s_2 = t_1t_2s_1s_2s_1s_2$ (using (1)) $= t_1s_1t_2s_2s_1s_2$ (since $s_1t_2{=}t_2s_1$) $= t_1s_1t_1s_1t_1t_2$ (using (1) and (3)) $=$ s_1t_2 (note that $t_1s_1t_1 = s_1t_1s_1$). Hence we obtain that $s_1 = t_2$. This together with the hypothesis that two distinct elements of $\{s_i, t_i,\ i = 1,2\}$ commute, implies that $s_2t_1 = t_1s_2$. Proceeding as above, we have $1 = s_2s_1s_2s_1s_2s_1 = s_2t_1t_2s_1s_2s_1$ (using (1)) $= t_1s_2t_2s_1s_2s_1$ (since $s_2t_1 = t_1s_2$) $= t_1s_1$ (since $s_1 = t_2$). Hence we obtain $s_1 = t_1$. This implies $s_2 = t_2 = s_1 = t_1$. This completes the proof of the first statement.

Write $s_1 = (i_1, i_2)$, $s_2 = (j_1, j_2)$, $t_1 = (k_1, k_2)$ and

$t_2 = (\ell_1, \ell_2)$. By Remark 4.5, we see that $\{i_1, i_2, j_1, j_2, k_1, k_2, \ell_1, \ell_2\}$ are all distinct, unless some of s_1, s_2, t_1, t_2 are equal. However if these integers are all distinct, it is clear that $s_1 s_2 \neq t_1 t_2$. Since $s_1 \neq s_2$, s_1 must be equal to either t_1 or t_2, and then as $s_1 s_2 = t_1 t_2$ the other two also coincide.

This completes the proof of Lemma 4.6.

Lemma 4.7. Suppose $s_1 \neq s_2$ are in T, $t_1, t_2 \in T$ and $s_1 s_2 = t_1 t_2$. If $s_1 s_2 = s_2 s_1$, then $\langle s_1, s_2 \rangle = \langle t_1, t_2 \rangle$ and is a Klein four group. If $s_1 s_2 \neq s_2 s_1$, then $\langle s_1, s_2 \rangle$ is isomorphic to S_3 and $\langle s_1, s_2 \rangle = \langle t_1, t_2 \rangle$. In particular at most three of $\{s_1, s_2, t_1, t_2\}$ are distinct.

Proof. The first statement is immediate from Lemma 4.6. Let us then assume that $s_1 s_2 \neq s_2 s_1$. Our assumption implies (in view of Lemma 4.6) that no two of $\{s_1, s_2, t_1, t_2\}$ commute. That $\langle s_1, s_2 \rangle \approx S_3$ follows from the fact that two noncommuting reflections in S_n generate S_3. It remains to show that $\langle s_1, s_2 \rangle = \langle t_1, t_2 \rangle$. If two of $\{s_1, s_2, t_1, t_2\}$ coincide, then this result follows from the hypothesis that $s_1 s_2 = t_1 t_2$. We claim that all the four cannot be distinct. If possible, let us assume that all the four are distinct. Since no two of these four distinct reflections commute, we obtain that these four reflections are precisely of the form (i,j), (i,k),

(i,ℓ), (i,m). But then the hypothesis that $s_1s_2 = t_1t_2$ cannot hold. Hence the claim follows from this and so also the required result.

This completes the proof of Lemma 4.7.

Suppose $w = (a_1,\cdots,a_n)$ and $w_1 = (b_1,\cdots,b_n)$. If $w > w_1$ and $\ell(w) = \ell(w_1) + 1$, then (cf. [C]) $w_1 = ws_\alpha$, where s_α is the reflection corresponding to some root $\alpha = e_i - e_j (i<j)$. The condition $\ell(w_1) = \ell(w) - 1$ translates into the following conditions:

1. $a_i > a_j$, and
2. there does not exist a k, $i < k < j$, such that $a_j < a_k < a_i$.

The main result of this section is the following.

Proposition 4.8. Suppose w, w_1, w_2 are in W^Q, $w_1 \neq w_2$ and $\ell(w) - 1 = \ell(w_1) = \ell(w_2)$, $w > w_1$, $w > w_2$. Also suppose both w_1 and w_2 have strictly less projection on W/W_{i_1} than does w. Then the following statements hold.

1) There exists a $w' \in W$ such that $w_1,w_2 > w'$ and $\ell(w') = \ell(w) - 2$.

2) $w' \in W^Q$.

3) If $\theta \in W$ and $\theta < w_1$, $\theta < w_2$ then $\theta \leq w'$.

Remark. The assumption that both w_1 and w_2 have strictly less projection than w is crucial. For instance if $w = (3412)$, $w_1 = (1432)$, $w_2 = (3214)$, then there is no w' of codimension two in w which is less than both w_1 and w_2. Note that w_2 does not have a strictly less projection than w on W/W_1. On the other hand both w_1 and w_2 have strictly less projection than does w on W/W_2; however, if $Q = P_2 \cap P_3$, then $w_2 \notin W^Q$.

For another slightly different example, consider $Q = P_1 \cap P_3$. Then $w = (3142)$ is in W^Q, while both $w_1 = (1342)$ and $w_2 = (3124)$ are of codimension one in w and are both in W^Q. In addition, $w' = (1324)$ is the unique codimension two subvariety of w such that $w_1, w_2 \geq w'$. However, $w' \notin W^Q$. In this case both w_2 and w have the same projection.

__Proof__. Write $w = (a_1, \cdots, a_n)$, $w_1 = (b_1, \cdots, b_n)$, $w_2 = (c_1, \cdots, c_n)$. As $\ell(w_1) = \ell(w) - 1 = \ell(w_2)$, we may write $w_1 = ws_\alpha$, $w_2 = ws_\beta$ where s_α and s_β are reflections corresponding to the roots $\alpha = e_i - e_j$ and $\beta = e_k - e_\ell$. Without loss of generality, we may assume $i \leq k$. Since both w_1 and w_2 have strictly less projection than does w on W/W_{i_1}, we see that $i, k \leq i_1$ while $j, \ell > i_1$. Note that

$$b_r = \begin{cases} a_r & r \neq i, j \\ a_j & r = i \\ a_i & r = j \end{cases}$$

and

$$c_r = \begin{cases} a_r & r \neq k, \ell \\ a_k & r = \ell \\ a_\ell & r = k \ . \end{cases}$$

There are three cases to consider. We list them here.

<u>Case 1</u>. $i < k$, $j < \ell$

<u>Case 2</u>. $i < k$, $j > \ell$

<u>Case 3</u>. $i = k$, $j < \ell$

We note that the case $i < k$ and $j = \ell$ cannot occur. For in this case, we would have $a_j < a_i < a_k$. But then, $w_2 \notin W^Q$ since $c_k = a_j > c_i = a_i$.

Each case will require a separate argument.

<u>Case 1</u>. $i < k$, $j < \ell$. Then $s_\alpha s_\beta = s_\beta s_\alpha$, and we define $w' = ws_\alpha s_\beta = ws_\beta s_\alpha$. If $w' = (d_1, \cdots, d_n)$ then

$$d_r = \begin{cases} a_r & r \neq i, j, k, \ell \\ a_j & r = i \\ a_i & r = j \\ a_k & r = \ell \\ a_\ell & r = k \end{cases} \ .$$

We first show that $\ell(w') = \ell(w) - 2$ and that $w' < w_1$, $w' < w_2$. Since $w' = w_1 s_\beta$, we need only to show that there is no r, $k < r < \ell$ such that $b_\ell < b_r < b_k$. (Notice that $b_k > b_\ell$ since $b_k = a_k$, $b_\ell = a_\ell$ and $a_k > a_\ell$ as $w_2 = ws_\beta$ is of codimension one in w. Also, there is no r, $k < r < \ell$ such that $a_\ell < a_r < a_k$.) Since $b_r = a_r$ for $k < r < \ell$,

$r \neq j$, the only possible bad case is if $b_\ell < b_j < b_k$, i.e., if $a_\ell < a_i < a_k$.

However, $w_2 \in W^Q$ so that $c_i < c_k$. Since $c_i = a_i$, $c_k = a_\ell$, $a_i < a_\ell$, the case $a_\ell < a_i$ cannot occur. Thus $\ell(w') = \ell(w_1) - 1 = \ell(w) - 2$. It also easily follows that $w' < w_2$.

Next we must prove that $w' \in W^Q$.

Since $w_1, w_2 \in W^Q$, we see that

$$a_{i-1} < a_j < a_i$$

and

$$a_{k-1} < a_\ell < a_k.$$

Choose s, r such that

$$i_s + 1 \leq j \leq i_{s+1}$$

and

$$i_r + 1 \leq \ell \leq i_{r+1}.$$

Then in addition,

$$a_j < a_i < a_{j+1}$$

and

$$a_\ell < a_k < a_{\ell+1}.$$

(If no such s or r exists, then there are no conditions to be checked.)

Since $i < k$,

$$a_j < a_i \leq a_{k-1} < a_\ell < a_k$$

so that $d_1 < \cdots < d_{i_1}$. If $s \neq r$ the observations above prove that

$$d_{i_s+1} < \cdots < d_{i_{s+1}}$$

while

$$d_{i_r+1} < \cdots < d_{i_{r+1}}.$$

If $s = r$, then we need to prove that

$$d_{i_s+1} < \cdots < d_{i_{s+1}}$$

i.e., that

$$a_{i_j+1} < \cdots < a_{j-1} < a_i < a_{j+1} < \cdots < a_{\ell-1} < a_k < a_{\ell+1} <$$
$$\cdots < a_{i_s+1}.$$

This follows immediately from the above inequalities and from the fact that $w \in W^Q$. It follows that $w' \in W^Q$.

Lastly, we must show that w' has the property 3). If $(a_1, \cdots, a_k)$ are distinct integers, by $(a_1, \cdots, a_k)^{inc}$ we denote the unique sequence, $(a_{\pi 1}, \cdots, a_{\pi k})$ such that $a_{\pi 1} < \cdots < a_{\pi k}$. If $\alpha = (e_1, \cdots, e_n)$ and $\beta = (f_1, \cdots, f_n)$ are in W, recall that $\beta \le \alpha$ if and only if for all i, $1 \le i \le n$, $(f_1, \cdots, f_i)^{inc} \le (e_1, \cdots, e_i)^{inc}$ (where $\le$ is as in the Grassmanian).

Now let θ be as in 3). Set $\theta = (e_1, \cdots, e_n)$. If $\theta \not\le w'$, then we may choose m minimal such that

$$(e_1, \cdots, e_m)^{inc} \not\le (d_1, \cdots, d_m)^{inc}.$$

By assumption,

$$(e_1, \cdots, e_m)^{inc} \le (c_1, \cdots, c_m)^{inc}$$

and

$$(e_1, \cdots, e_m)^{inc} \le (b_1, \cdots, b_m)^{inc}.$$

If $(x_1, \cdots, x_m)$ and $(y_1, \cdots, y_m)$ are two sets of increasing integers, then we set $z = (z_1, \cdots, z_m) = \min\{(x_1, \cdots, x_m), (y_1, \cdots, y_m)\}$, where $z_i = \min\{x_i, y_i\}$. Clearly then,

$$(e_1, \cdots, e_m)^{inc} \le \min\{(c_1, \cdots, c_m)^{inc}, (b_1, \cdots, b_m)^{inc}\}.$$

We <u>claim</u>

$$\min\{(c_1, \cdots, c_m)^{inc}, (b_1, \cdots, b_m)^{inc}\} = (d_1, \cdots, d_m)^{inc}.$$

This clearly contradicts our minimality assumption on m and proves $\theta \le w'$.

If $m > \ell$, then $\{c_1, \cdots, c_m\} = \{b_1, \cdots, b_m\} = \{d_1, \cdots, d_m\}$ and our claim is obvious. If $m \le i_1$, then since $w_1, w_2 \in W^Q$, $c_1 < \cdots < c_m$ and $b_1 < \cdots < b_m$ and it is clear that

$$d_i = \min\{b_i, c_i\}.$$

If $i_1 < m < j$, then, $(b_1, \cdots, b_m) = (a_1, \cdots, a_{i-1}, a_j, a_i, \cdots, a_m)$ and $(c_1, \cdots, c_m) = (a_1, \cdots, a_{k-1}, a_\ell, a_k, \cdots, a_m)$, and $(d_1, \cdots, d_m) = (a_1, \cdots, a_{i-1}, a_j, a_{i+1}, \cdots, a_{k-1}, a_\ell, a_k, \cdots, a_m)$. To show $(d_1, \cdots, d_m)^{inc} = \min\{(b), (c)\}$, it is enough to show that the places of $b_i = a_j$ and $b_k = a_k$ in $(b_1, \cdots, b_m)^{inc}$ are the same respectively as those of $c_i = a_i$ and $c_k = a_\ell$ in

$(c_1,\cdots,c_m)^{inc}$. For, in this case since $a_\ell < a_k$ and $a_j < a_i$, $\min\{(b)),(c)\} = (d_1,\cdots,d_m)$. The place of b_i is equal to one plus the number of $b_r < b_i$ with $r \le m$. Thus to show b_i has the same place in $(b_1,\cdots,b_m)^{inc}$ as does c_i in $(c_1,\cdots,c_m)^{inc}$, it is enough to show the number of $b_r < a_j = b_i$ with $r \le m$ is the same as the number of $c_r < a_i = c_i$ with $r \le m$. We note if $a_r < a_i$ and $r \le m < j$, then $a_r < a_j$. If not, then there exists an $r < j$ such that $a_j < a_r < a_i$. Since if $r < i$, $a_r < a_j$ (as $w_1 \in W^Q$), we see that $i < r < j$. This contradicts the fact that $\ell(w_1) = \ell(w) - 1$. However, $\{b_r | r \le m \text{ and } r \ne i\} = \{a_r | r \le m,\ r \ne i\} = \{c_r | r \le m,\ r \ne i,k\} \cup \{a_k\}$. Since $a_j < a_i < a_\ell < a_k$, neither a_ℓ nor a_k have any effect upon the number of $a_r < a_j$ or $< a_i$. The proof for the places of b_k and c_k is similar. This proves our assertion.

If $j \le m < \ell$, then $\{b_1,\cdots,b_m\} = \{a_1,\cdots,a_m\}$, and $(c_1,\cdots,c_m) = (a_1,\cdots,a_i,\cdots,a_{k-1},a_\ell,a_{k+1},\cdots,a_m)$. Since $w_2 < w$, $(c_1,\cdots,c_m)^{inc} \le (a_1,\cdots,a_m)^{inc} = (b_1,\cdots,b_m)^{inc}$, so that the $\min\{(c)^{inc},(b)^{inc}\} = (c_1,\cdots,c_m)^{inc}$. However, $\{d_1,\cdots,d_m\} = \{c_1,\cdots,c_m\}$, and the conclusion is clear. This finishes the proof of Case 1.

<u>Case 2</u>. $i < k$, $j < \ell$. Again $s_\alpha s_\beta = s_\beta s_\alpha$ and we set $w' = ws_\alpha s_\beta = ws_\beta s_\alpha$. The proof of 1) and 2) in this case is similar to that of Case 1 and we leave the details to the

reader. Write $w' = (d_1, \cdots, d_n)$.

Let $\theta = (e_1, \cdots, e_n)$ be as in 3). We may assume $\theta \not\leq w'$, and choose m to be at least such that

$$(e_1, \cdots, e_m)^{\mathrm{inc}} \not\leq (d_1, \cdots, d_m)^{\mathrm{inc}}.$$

Again we make the claim that

$$(z_1, \cdots, z_m) = \min\{(c_1, \cdots, c_m)^{\mathrm{inc}}, (b_1, \cdots, b_m)^{\mathrm{inc}}\} = (d_1, \cdots, d_m)^{\mathrm{inc}}.$$

If $m \leq i_1$, this is again easy to see using the fact that $w_1, w_2 \in W^Q$. If $i_1 < m < \ell$, then $\{c_1, \cdots, c_m\} = \{a_1, \cdots, a_{k-1}, a_\ell, a_k, \cdots, a_m\}$ while $\{b_1, \cdots, b_m\} = \{a_1, \cdots, a_{i-1}, a_j, a_i, \cdots, a_m\}$ and $\{d_1, \cdots, d_m\} = \{a_1, \cdots, a_{i-1}, a_j, a_i, \cdots, a_{k-1}, a_\ell, a_k, \cdots, a_m\}$. As above it is enough to show the places of b_i and b_k in $(b_1, \cdots, b_m)^{\mathrm{inc}}$ are the same respectively as the places of c_i and c_k in $(c_1, \cdots, c_m)^{\mathrm{inc}}$. This follows as above.

If $\ell \leq m < j$, then $\{c_1, \cdots, c_m\} = \{a_1, \cdots, a_m\}$ and so $(z_1, \cdots, z_m) = (b_1, \cdots, b_m)^{\mathrm{inc}}$ as $w_1 < w$. However, $\{d_1, \cdots, d_m\} = \{b_1, \cdots, b_m\}$ in this case. Thus $(d_1, \cdots, d_m)^{\mathrm{inc}} = (z_1, \cdots, z_m)$ as was required. Finally, $m > j$ implies $\{d_1, \cdots, d_m\} = \{a_1, \cdots, a_m\} = \{b_1, \cdots, b_m\} = \{c_1, \cdots, c_m\}$ and our claim is obvious.

Case 3. $i = k$, $j < \ell$.

The order of a_i, a_j, a_ℓ is fixed. Since $\ell(w_1) = \ell(w_2) = \ell(w) - 1$, $a_j < a_i$ and $a_\ell < a_i$. On the other hand, if $a_\ell < a_j$, then $a_\ell < a_j < a_i$ and then $ws_\beta = w_2$ would not have length $\ell(w) - 1$. Hence,

$$a_j < a_\ell < a_i.$$

We define $w' = (d_1, \cdots, d_n)$ by

$$d_r = \begin{cases} a_r & r \neq i, j, \ell \\ a_j & r = i \\ a_\ell & r = j \\ a_i & r = \ell \end{cases} .$$

$w' = w_2 s_\alpha$, $w' = w_1(s_\alpha s_\beta s_\alpha)$. We note that $s_\alpha s_\beta \neq s_\beta s_\alpha$, consequently $(s_\alpha s_\beta)^3 = 1$ and $s_\alpha s_\beta s_\alpha = s_\beta s_\alpha s_\beta$.

We will show $\ell(w') = \ell(w_2) - 1$ and $w' < w_2$. For this it suffices to show that $c_j < c_i$, and if $i < m < j$ then c_m does not lie between c_i and c_j. However, $c_j = a_j$, $c_m = a_m$ for $i < m < j$ and $c_i = a_\ell$. By the discussion above, $a_j < a_\ell$ so that $c_j < c_i$. If $a_j < a_m < a_\ell$, then $a_j < a_m < a_i$ which contradicts the assumption that w_1 is not of codimension one in w. Thus, $w' < w_2$ and $\ell(w') = \ell(w) - 2$. Next observe that $w' < w_1$. For this, it is enough to observe that $b_\ell < b_j$. However, $b_i = a_i$ and $b_\ell = a_\ell$; since $a_\ell < a_i$, we see $b_\ell < b_j$.

Next, we need to prove $w' \in W^Q$. First assume there is an m with $i_m + 1 \leq j < \ell \leq i_{m+1}$. This cannot occur, for in this case w_1 is not in W^Q as $b_j < b_\ell$ since $b_j = a_i$, $b_\ell = a_\ell$. Thus such an m does not exist; it now easily follows that $w' \in W^Q$ from the fact that w_1 and w_2 are in W^Q.

Lastly we prove 3). Let $\theta = (e_1, \cdots, e_n)$ be as in 3). We have assumed that for any $1 \le m \le n$, $(e_1, \cdots, e_m)^{inc} \le$ both $(c_1, \cdots, c_m)^{inc}$ and $(b_1, \cdots, b_m)^{inc}$. However, it is trivial to check that,

$$(d_1, \cdots, d_m)^{inc} = \begin{cases} (b_1, \cdots, b_m)^{inc} & m < j \\ (c_1, \cdots, c_m)^{inc} & j \le m < \ell \\ (a_1, \cdots, a_m)^{inc} & m > \ell \end{cases}$$

Thus in all cases,

$$(e_1, \cdots, e_m)^{inc} \le (d_1, \cdots, d_m)^{inc}$$

which shows that $\theta \le w'$.

This completes the proof of Proposition 4.8.

§5. Ideal theoretic unions and intersections in $R^{def}(w)$.

In this section we apply our calculations of the previous section to understand the ideal theoretic union and intersection in the deformed ring $R^{def}(G/Q)$ of §3. The main result is Theorem 5.5. Accordingly, for the remainder of this section, we fix a parabolic $Q = P_{i_1} \cap \cdots \cap P_{i_k}$ with $i_1 < \cdots < i_k$. Schubert varieties in G/Q are indexed by $w \in W^Q$. If $w \in W^Q$, we let $R^{def}(w)$ be the ring R_t given by Theorem 3.9, for the variety $X(w)$ in G/Q. The k-algebra $R^{def}(w)$ has a basis consisting of standard monomials on X(w), and any nonstandard monomial is zero. For the variety G/Q itself, we abbreviate $R = R^{def}(G/Q)$. There is a surjection from R onto $R^{def}(w)$; we denote the kernel of this surjection by I(w), where I(w) is generated by the set of all square-free standard monomials on G/Q which are <u>not</u> standard on X(w). <u>We denote this set of square-free monomials by J(w)</u>. In particular we let H be the set of all square-free standard monomials on G/Q. We define a partial order on H as follows. If m and m′ are in H we say $m \geq m'$ if and only if m′ contains m as a submonomial. Of course, any monomial can be written uniquely in the form

$$m = m_1^{n_1} \cdots m_k^{n_k}$$

where

$$m_1 > \cdots > m_k.$$

With this partial order, R becomes an algebra with straightening law on H (cf. [E]). Indeed, any ring defined by square-free monomials is such an algebra [E]. We will use only the most basic properties of such an algebra in the discussion of this section. Although R has algebra generators indexed by the partially ordered set $\bigcup_{t=1}^{r} W^i t$, $R^{def}(w)$ is not an algebra with straightening law on the set of all $\alpha \in \bigcup_{t=1}^{r} W^i t$ such that $p_\alpha|_{X(w)} \neq 0$. It is possible to have two standard monomials $p_\tau p_\alpha$ and

$p_\alpha p_\beta$ on $X(w)$ without $p_\tau p_\beta$ being standard. In fact if $G = SL(3)$, and $\tau = (2)$, $\alpha = (1)$, $\beta = (13)$ with $w = (312)$ then this is the case.

Definition 5.1. A subset $J \subset H$ is said to be an ideal of H if whenever $a \in J$ and $\beta \le \alpha$, then $\beta \in J$ also. If J is an ideal of H, we let $\overline{J}$ denote the ideal of R generated by the monomials in J.

Remark. The set $J(w)$ is an ideal of H. For, $m \in J(w)$ if and only if m is nonstandard on $X(w)$; in particular if $m' \le m$, m' is also nonstandard on $X(w)$.

Lemma 5.2. Suppose R is an algebra with straightening law on H over A. If I and J are two ideals of H, then

1) $\overline{I} + \overline{J} = \overline{I \cup J}$,

2) $\overline{I \cap J} = \overline{I} \cap \overline{J}$.

Proof. See [E].

Lemma 5.3. Let $w_1, \cdots, w_k \in W^Q$.

1) $[I(w_1) \cap \cdots \cap I(w_{k-1})] + I(w_k) = [I(w_1) + I(w_k)] \cap \cdots \cap [I(w_{k-1}) + I(w_k)]$.

Assume $w, w_1, w_2 \in W^Q$ and w_1 and w_2 have strictly less projection on W/W_{i_1} than does w. Let w' be an element of W^Q as in Proposition 4.8. Then

2)
$$I(w_1) + I(w_2) = I(w').$$

Proof. Since $\overline{J(w_i)} = I(w_i)$ by Lemma 5.2, to prove 1) it is enough to prove

$$[J(w_1) \cap\cdots\cap J(w_{k-1})] \cup J(w_k) =$$
$$[J(w_1) \cup J(w_k)] \cap\cdots\cap [J(w_{k-1}) \cup J(w_k)].$$

However this equality is simply DeMorgan's law.

To prove 2) it suffices to show that

$$J(w_1) \cup J(w_2) = J(w').$$

Clearly $J(w_1) \cup J(w_2) \subset J(w')$, as any nonstandard monomial on $X(w_i)$, $i = 1,2$ remains nonstandard on $X(w')$.

Suppose conversely that m is a monomial which is standard on both $X(w_1)$ and $X(w_2)$. We must show m is standard on $X(w')$. Let λ be the Young diagram to which m corresponds. Let θ be the (unique) minimal defining diagram for λ (cf. Remark 2.3′) and let θ_1 be the first member of θ. Since m is standard on both $X(w_1)$ and $X(w_2)$, both $w_1 \geq \theta_1$ and $w_2 \geq \theta_1$. Hence Proposition 4.8, 3) implies $\theta_1 \leq w'$. It follows that λ is a standard Young diagram on $X(w')$, i.e., m is standard on $X(w')$. This completes the proof of Lemma 5.3.

In proving $R^{def}(w)$ is Cohen-Macaulay, it will be crucial to understand the representation of a hyperplane section of $X(w)$ as a union of Schubert varieties. Of course we actually do this in R.

Let $w \in W^Q$ and let $\tau = w^{(i_1)}$ (notation as in §2). Of course τ is the maximal element of the set Y_{i_1} (cf. §3).

Definition 5.4. Set $B(w) = \{w' \in W^Q | w \geq w', \ell(w) - 1 = \ell(w')$ and $p_\tau|_{X(w')} = 0\}$.

To say $p_\tau|_{X(w')} = 0$ is simply to say that the projection τ' of w' on W/W_{i_1} is strictly less than τ. Define $J = \langle p_\alpha | \alpha \in W^{i_1}, \alpha \leq \tau\rangle$.

Theorem 5.5. With notation as above

1) $(p_\tau, I(w)) = \bigcap_{w' \in B(w)} I(w') \cap (J + I(w))$ (*)

2) Let $w = (a_1, \cdots, a_n) \in W^Q$. If $a_{i_1+1} < a_{i_1}$, then $J + I(w) \supset I(w')$ for some $w' \in B(w)$.

Proof. 1). If $w'' \leq w$, then $I(w') \supseteq I(w)$. Since $p_\tau|_{X(w')} = 0$ if $w' \in B(w)$, $p_\tau \in I(w')$. Thus $(p_\tau, I(w)) = \bigcap_{w' \in B(w)} I(w') \cap (J + I(w))$. To prove the reverse inclusion, let F be a monomial in the right-hand intersection of (*). Since we may read (*) modulo $I(w)$, we may assume F is standard on $X(w)$, and we are working in $R(w) \overset{def}{=} R/I(w)$. As $F \in (J + I(w))/I(w)$, we may assume $F = \sum_{i=1}^{m} p_{\alpha_i} F_i$ where $\alpha_i \in W/W_{i_1}$ $\tau \geq \alpha_i$, and $p_{\alpha_i} F_i$ is standard on $X(w)$. We wish to show that $\alpha_i = \tau$. Choose α_i minimal among $\{\alpha_1, \cdots, \alpha_m\}$, and rearrange them so that $\alpha_i = \alpha_1$. It suffices to show $\alpha_1 = \tau$. This we prove, by proving the following

Lemma 5.6. Fix $w \in W^Q$. Let λ be a standard Young diagram on $X(w)$ and suppose $\lambda_1 \in W^{i_1}$ and λ_1 is strictly less than τ (= projection of w on W/w_{i_1}). Then λ is a standard Young diagram on some $X(w')$ where $w' \in B(w)$.

Assume we have proved this lemma. Then the proof of 1) of Theorem 5.5 follows easily. For, $p_{\alpha_1}F_1$ being standard on $X(w)$ corresponds to a standard Young diagram λ on $X(w)$ with $\lambda_1 = \alpha_1$. If $\lambda_1 < \tau$ then the lemma implies that λ remains standard on some $X(w')$ with $w' \in B(w)$. In particular, $p_{\alpha_1}F_1 (=F)$ is nonzero on $X(w')$. However by assumption, $F|_{X(w')} = 0$ for all such w'.

Proof of Lemma 5.6. Since λ is standard on $X(w)$, there is a defining diagram θ for λ such that $w \geq \theta_1$. (Notice $\theta_1 \in W^Q$.) Since θ_1 projects to λ_1 and $\lambda_1 < \tau$, we see $\theta_1 < w$. We induct on $\ell(w) - \ell(\theta_1)$ to prove the lemma. If $\ell(w) - \ell(\theta_1) = 1$, then $\theta_1 \in W^Q$ and λ is standard on $X(\tau_1)$.

Assume $\ell(w) - \ell(\theta_1) = 2$. By [BW] there are precisely two elements w_1 and w_2 in W such that $w > w_1, w_2 > \theta_1$. Write $w_1 = ws_1$, $w_2 = ws_2$, $\theta_1 = w_1t_1$, $\theta_1 = w_2t_2$. Then $s_1t_1 = s_2t_2$ and by Lemma 4.7, $\langle s_1, s_2\rangle = \langle t_1, t_2\rangle = \langle s_1, t_1\rangle = \langle s_2, t_2\rangle$. The facts that $w, \theta_1 \in W^Q$ imply that at least one of w_1, w_2 is in W^Q.

Case 1. Both $w_1, w_2 \in W^Q$. We need to show at least one of them has a different projection on W/W_{i_1} than does w.

If not, both $s_1, s_2 \in W_{i_1}$. In this case, $\langle t_1, t_2 \rangle = \langle s_1, s_2 \rangle \subseteq W_{i_1}$ and so $s_1 t_1 \in W_{i_1}$. Then $\theta_1 = w s_1 t_1$ has the same projection on W/W_{i_1} as does w. This is a contradiction.

Case 2. $w_1 \in W^Q$, $w_2 \notin W^Q$. Then by Lemma 4.3, $t_2 \in W(Q)$. If $s_1 \notin W_{i_1}$, then $w_1 \in W^Q$ and has strictly less projection on W/W_{i_1} than does w. So it is enough to prove $s_1 \notin W_{i_1}$. If possible, let $s_1 \in W_{i_1}$. If $s_1 s_2 = s_2 s_1$, then $\langle s_1, s_2 \rangle = \langle t_1, t_2 \rangle$ by Lemma 4.6 and so $s_1 = t_2 \in W(Q)$. But then $w_1 \notin W^Q$. Suppose $s_1 s_2 \neq s_2 s_1$. Since $t_2 \in W(Q)$, $t_1 = s_1 t_2 s_1 \in W_{i_1}$. But then $s_1 t_1 \in W_{i_1}$ and this is again a contradiction.

Finally assume $\ell(w) - \ell(\theta_1) > 2$. By Lemma 4.4, there is a $w_1 \in W^Q$ such that $w > w_1 > \theta_1$ and $\ell(w_1) = \ell(w) - 1$. If $w_1 \in B(w)$, then there is nothing to prove. If not, then w_1 projects to τ. By induction we may assume there is a w', $w' \in W^Q$, $w_1 > w' > \theta_1$, $\ell(w') = \ell(w_1) - 1$ such that $w' \in B(w_1)$. In this case $\ell(w) = \ell(w') + 2$ and $w \in W^Q$. Apply the case $\ell(w) - \ell(w') = 2$ to finish the proof of 1).

We now prove 2). Let d be the largest integer $i_1 + 1 \leq d \leq i_2$ such that $a_d < \max\{a_1, \cdots, a_{i_1}\}$. Then there is a k with $a_{k-1} < a_d < a_k$ (possibly $k = 1$, in which case we set $a_{k-1} = 0$). Consider the element $w' =$

$(a_1, \cdots, a_{k-1}, a_d, a_{k+1}, \cdots, a_{i_1}, \cdots, a_{d-1}, a_k, a_{d+1}, \cdots, a_n)$. Note that $w \geq w'$, and $\ell(w) - 1 = \ell(w')$ (since if $k + 1 \leq r \leq i_1$, $a_r > a_k$ and $a_r > a_d$, and if $i_1 + 1 \leq r \leq d - 1$, then by choice $a_r < a_d < a_k$). Further, the projection of w' on W/W_{i_1} is strictly less than τ. Thus $w' \in B(w)$. We claim $I(w) + J \supseteq I(w')$. Let $m \in J(w')$ and let $m \notin J(w)$. If $\lambda_1 \in W^{i_1}$ (equivalently, m is a monomial of multidegree $(m_{i_1}, \cdots, m_{i_k})$ with $m_{i_1} \neq 0$), then our assumptions imply that $\lambda_1 \leq \tau$ and hence $m \in J$. Let us then suppose that $m \in J(w')$, $m \notin J(w)$ and m is a monomial of multidegree $(m_{i_2}, \cdots, m_{i_k})$. Set $Q' = P_{i_2} \cap \cdots \cap P_{i_k}$. If we denote the projection of $X(w')$ under $\pi: G/Q \to G/Q'$ by $X(\theta)$, then our assumptions imply that m is not standard on $X(\theta)$. But this is a contradiction, since m is standard on $X(w)$ (note that $m \notin J(w)$) and $\pi(X(w)) = \pi(X(w')) = X(\theta)$. This completes the proof of 2) and hence also of Theorem 5.5.

§6. The arithmetic Cohen-Macaulayness of Schubert varieties in G/Q.

In this section we will prove,

Theorem 6.1. Let $Q = P_{i_1} \cap \cdots \cap P_{i_r}$ be any parabolic in $G = SL(n)$ containing B, and let $w \in W^Q$. Let $\underline{a} = (a_1, \cdots, a_r)$, $a_i \in \mathbb{Z}^+$, and $R(w) = \oplus_{\underline{a}} H^0(X(w), L_{i_1}^{a_i} \otimes \cdots \otimes L_{i_r}^{a_r})$.

Then $R(w)$ is Cohen-Macaulay.

As we observed in §3, to prove the above theorem it suffices to prove

Theorem 6.2. Let $Q = P_{i_1} \cap \cdots \cap P_{i_r}$ be any parabolic subgroup in $G = SL(n)$ containing B, and let $w \in W^Q$. Then the ring $R^{def}(w)$ is Cohen-Macaulay.

We will prove Theorem 6.2 by induction on the number r of maximal parabolics containing Q. We will also induct on $\dim R^{def}(w)$; to descend the Cohen-Macaulay property along the class of Schubert varieties we will use Theorem 5.5 and several elementary lemmas from homological algebra.

Lemma 6.3. Let $R = \bigoplus_{n\geq 0} R_n$ be a Noetherian graded ring with $R_0 = k$ and let I and J be two homogeneous ideals of R such that $\dim R/I = \dim R/J = d = \dim R/(I+J) + 1$. If R/I, R/J and $R/I+J$ are Cohen-Macaulay, then $R/I \cap J$ is Cohen-Macaulay of dimension d.

Proof. Let $M = R^+$ be the irrelevant ideal. It suffices to show (cf. [M]) that $(R/I \cap J)_M$ is Cohen-Macaulay. Consider the exact sequence

$$0 \to (R/I \cap J)_M \to (R/I)_M \oplus (R/J)_M \to (R/I + J)_M \to 0.$$

Taking local cohomology we find

$$\to H_M^{i-1}((R/I + J)_M) \to H_M^i((R/I \cap J)_M) \to$$
$$H_M^i((R/I)_M) \oplus H_M^i((R/J)_M) \cdots$$

The depth of a finitely generated R_M-module N is characterized by depth $N = \min\{i \mid H^i_M(N) \neq 0\}$. By assumption $H^i_M((R/I)_M) = H^i_M((R/J)_M) = 0$, for $i < d$, and $H^i_M((R/I + J)_M) = 0$, for $i < d - 1$. We conclude $H^i_M((R/I \cap J)_M) = 0$, for $i<d$. Hence,

$$d \le \operatorname{depth}(R/I \cap J)_M \le \dim(R/I \cap J)_M \le d$$

shows $(R/I \cap J)_M$ (and therefore $R/I \cap J$) is Cohen-Macaulay.

Remark. If R is as above and x is a homogeneous element of R which is not a zero-divisor, then R is Cohen-Macaulay if and only if R/xR is Cohen-Macaulay.

Lemma 6.4. Let $w \in W^Q$, $w = (a_1, \cdots, a_n)$.

1) If $w' \in B(w)$, then $\dim R^{def}(w') = \dim R^{def}(w) - 1$.

2) If $w_1, w_2 \in B(w)$ and w' is the element defined in Proposition 4.8, then

$$\dim R^{def}(w') = \dim R^{def}(w) - 2.$$

3) If $a_{i_1+1} < a_{i_1}$, then

$$\dim R^{def}(w)/J = \dim R^{def}(w) - 1.$$

Proof. To compute the dimension of $R^{def}(w)$, we note that $\dim R^{def}(w) = \dim R(w) = r + \ell(w)$, where $w \in W^Q$, $Q = P_{i_1} \cap \cdots \cap P_{i_r}$.

1) Since $\ell(w') = \ell(w) - 1$, 1) is clear.

2) As $w' \in W^Q$ by Proposition 4.8, and $\ell(w') = \ell(w) - 2$,

we obtain 2).

3) Observe that $R/(J + I(w)) = R^{def}(w)/J$ is isomorphic to $S^{def}(w)$ where S is the deformed ring of G/Q', $Q' = P_{i_2} \cap \cdots \cap P_{i_r}$. Hence

$$\dim R^{def}(w)/J = \dim S^{def}(w) = \ell(w) + r-1 = \dim R^{def}(w)-1.$$

To prove Theorem 6.2, we prove a more general result which will allow us to use induction.

Theorem 6.5. $Q = P_{i_1} \cap \cdots \cap P_{i_r}$ be any parabolic subgroup in $G = SL(n)$ containing B, and let $w \in W^Q$, $w_1, \cdots, w_k \in B(w)$. Let $\tau = w^{(i_1)}$, and let $J = I(w) + \langle p_\alpha | \alpha \in Y_{i_1} \rangle$.

Then,

1) $R^{def}(w)$ is Cohen-Macaulay.

2) $R/I(w_1) \cap \cdots \cap I(w_k)$ is Cohen-Macaulay.

3) If $w = (a_1, \cdots, a_n)$ and $a_{i_1} < a_{i_1+1}$, then $R/(I(w_1) \cap \cdots \cap I(w_k) \cap J)$ is Cohen-Macaulay.

Proof. We will prove all three statements by induction on r and on $\dim R^{def}(w)$. By this induction we may assume all three statements hold for any $w' \in W^{Q'}$, $Q' = P_{i_2} \cap \cdots \cap P_{i_r}$.

We prove 2) first, by induction on m. If $m = 1$, $R/I(w_1) \simeq R^{def}(w_1)$, and $\dim R^{def}(w_1) < \dim R^{def}(w)$. Hence by induction $R/I(w_1)(= R^{def}(w_1))$ is Cohen-Macaulay. Suppose $m > 1$. Let $K = I(w_2) \cap \cdots \cap I(w_k)$, and $d = \dim$

$R^{def}(w)$. By induction, R/K is Cohen-Macaulay of dimension $d - 1$. Also, $\dim R/I(w_1)$ is $d - 1$, and $R/I(w_1)$ is Cohen-Macaulay. By Lemma 6.3, to show $R/K \cap I(w_1)$ is Cohen-Macaulay, it suffices to show that $R/(K + I(w_1))$ is Cohen-Macaulay of dimension $d - 2$. By Lemma 5.3, $K + I(w_1) = (I(w_1) + I(w_2)) \cap \cdots \cap (I(w_1) + I(w_k))$. Since each pair w_1, w_j is in $B(w)$, Proposition 4.8 shows there is a unique $t_j \in W^Q$ such that $w_1 > t_j$, $w_j > t_j$, and $\ell(t_1) = \ell(w_1) - 1 = \ell(w_j) - 1$; further by Lemma 5.3, $I(w_1) + I(w_j) = I(t_j)$. Therefore, $K + I(w_1) = I(t_2) \cap \cdots \cap I(t_r)$, and $w_1 > t_j$, with $\ell(w_1) - 1 = \ell(t_j)$. If we can choose w_1 in such a way that each $t_j \in B(w_1)$, then by induction we may conclude that $R/K + I(w_1) = R/I(t_2) \cap \cdots \cap I(t_r)$ is Cohen-Macaulay of dimension $d - 2$. That such a w_1 can be chosen is proved in the following lemma.

Lemma 6.6. Let $w \in W^Q$, $w_1, \cdots, w_k \in B(w)$. Then we may choose w_1 (after rearranging $w_1, \cdots, w_k$) in such a way that $t_j \in B(w_1)$, where the notations are as above.

Proof. Write $w_j = ws_j$ and let $s_j = e_{k_j} - e_{\ell_j}$. Since $w_j \in B(w)$, $k_j \le i_1$ and $\ell_j > i_1$. Among the set $\{\ell_1, \cdots, \ell_k\}$ choose a maximal element. This maximal element, say ℓ_1, is unique. For if $\ell_1 = \ell_2$, then $s_1 = e_{k_1} - e_{\ell_1}$, $s_2 = e_{k_2} - e_{\ell_1}$. Since w covers w_1 and w_2, $a_{\ell_1} < a_{k_1}$ and $a_{\ell_1} < a_{k_2}$.

Suppose (without loss of generality) that $k_1 < k_2$. Since $w \in W^Q$, $a_{k_1} < a_{k_2}$. Then $a_{\ell_1} < a_{k_1} < a_{k_2}$, which implies $\ell(ws_2) < \ell(w) - 1$. Thus ℓ_1 is unique. We <u>claim</u> that $t_j \in B(w_1)$ for w_1 chosen in this manner. We may assume $j = 2$. We need to evaluate this, case by case. Write $s_1 = e_i - e_j$, $s_2 = e_k - e_\ell$. By assumption $j > \ell$.

<u>Case 1</u>. $i < k$. Since $j \neq \ell$. $s_1s_2 = s_2s_1$ and $t_2 = ws_1s_2 = w_1s_2$. Thus

$t_2 = (d_1, \cdots, d_n)$ where

$$d_r = \begin{cases} a_r & r \neq i, j, k, \ell \\ a_j & r = i \\ a_i & r = j \\ a_\ell & r = k \\ a_k & r = \ell \end{cases} .$$

The projection of t_2 on W/W_{i_1} is $(d_1, \cdots, d_{i_1})$ and this is strictly less than the projection of w_1. For, if $w_1 = (b_1, \cdots, b_n)$, then

$$b_r = \begin{cases} a_r & r \neq i, j \\ a_j & r = i \\ a_i & r = j \end{cases} .$$

Thus $d_s = b_s$ for $s \leq i_1$ unless $s = k$, in which case $a_\ell = d_k < b_k = a_k$.

<u>Case 2</u>. $i > k$. Again $s_1s_2 = s_2s_1$ and the same argument shows t_2 has strictly less projection on W/W_{i_1}.

Thus for Cases 1 and 2, any choice of w_1 would

suffice.

Case 3. $i = k$. Recall $j > \ell$. Then by Proposition 4.8, if $t_2 = (d_1, \cdots, d_n)$, then

$$d_r = \begin{cases} a_r & r \neq i, j, \ell \\ a_\ell & r = i \\ a_j & r = j \\ a_i & r = \ell \end{cases} .$$

Also, $a_\ell < a_j < a_i$. By choice of w_1, $w_1 = (b_1, \cdots, b_n)$ where

$$b_r = \begin{cases} a_r & r \neq i, j \\ a_j & r = i \\ a_i & r = j \end{cases} .$$

Thus, $d_r = b_r$ for $r \leq i_1$ unless $r = i$; in this case $a_\ell = d_i < b_i = a_j$, and so $t_2 \in B(w_1)$. This proves Lemma 6.6 and also finishes the proof of Theorem 6.5, part 2).

We now prove part 3 of Theorem 6.5). By Part 2), if $K = I(w_1) \cap \cdots \cap I(w_k)$, then R/K is Cohen-Macaulay of dimension $d - 1$. By Lemma 6.4, dim R/J is $d - 1$, and is Cohen Macaulay since $R/J \simeq R'/I'(w)$ where the "′" means we are working in G/Q'. Thus we apply our induction on the number of maximal parabolics containing Q. By Lemma 6.3, to prove $R/K \cap J$ is Cohen-Macaulay it suffices to prove $R/K + J$ is Cohen-Macaulay of dimension $d - 1$. However,

$$J + (I(w_1) \cap \cdots \cap I(w_k)) = (J + I(w_1)) \cap \cdots \cap (J + I(w_k)),$$

and

$$R/(J + I(w_\ell)) = R'/I(w_\ell).$$

Since $a_{i_1} < a_{i_1+1}$, $a_1 <\cdots< a_{i_1} < a_{i_1+1} <\cdots< a_{i_2}$ and hence $w \in W^{Q'}$. As each w_ℓ has strictly less projection than does w in G/P_{i_1}, we see that if $w_\ell = ws$, $s = e_i - e_j$, then $i \le i_1$ and $j > i_2$. For, if $j < i_2$, then $\ell(ws) \neq \ell(w) - 1$. When we apply s, we do not change the relative order of any of the segments $a_{i_m+1}, \cdots, a_{i_{m+1}}$. Thus $w_i \in W^{Q'}$ also. In addition since $j > i_2$ and w_ℓ has strictly less projection on W/W_{i_1} than w, it easily follows that w_ℓ has strictly less projection on W/W_{i_2} than does w. Thus $w_i \in B'(w)$. The induction now implies $R'/I'(w_1) \cap\cdots\cap I'(w_k) = R/(J + K)$ is Cohen-Macaulay of dimension d - 2. This proves part 3).

Finally we prove part 1). By Theorem 5.4,

$$(I(w), p_\tau) = \begin{cases} \bigcap_{w' \in B(w)} I(w') & , \text{ if } a_{i_1} > a_{i_1+1} \\ \bigcap_{w' \in B(w)} I(w') \cap J & , \text{ if } a_{i_1} < a_{i_1+1}. \end{cases}$$

In either case the proof of parts 2) and 3) show that

$$R/(I(w), p_\tau)$$

is Cohen-Macaulay. If p_τ is not a zero-divisor in $R/I(w)$, then clearly $R/I(w) = R^{\text{def}}(w)$ is Cohen-Macaulay.

However it is clear that since w projects to τ, if F is any standard mo#omial of X(w), with Young diagram λ, then $p_\tau F$ is also standard with diagram (τ,λ). This shows p_τ is not a zero-divisor in $R^{def}(w)$.

Corollary 6.7. Let $P_1,\cdots,P_n$ (where n = rank(G)) be any ordering of the maximal parabolics, $Q_i = P_i \cap\cdots\cap P_n$ and $w \in W/W(Q_i)$. Then R(w) is normal.

Proof. This follows at once from Serre's criterion of normality [S], the fact that R(w) is Cohen-Macaulay and the fact that the Schubert varieties are nonsingular in codimension 1 [C].

References

[BW] A Björner and M. Wachs, Bruhat order of Coxeter groups and shell-ability, Adv. Math. 43 (1982), 87-100.

[C] C.C. Chevalley, Sur les decompositions cellulaires des espaces G/B, non-published manuscript, 1958.

[DEP] C. DeConcini, D. Eisenbud and C. Procesi, Hodge Algebras, Asterisque 91 (1982).

[DL] C. DeConcini and V. Lakshmibai, Arithmetic Cohen-Macaulayness and arithmetic normality for Schubert varieties, Amer. J. Math. 103 (1981), 835-850.

[E] D. Eisenbud, Introduction to algebras with straightening laws, Ring Theory and Algebra III, Proceeding third Oklahoma conference, Marcel Dekker (1980).

[EH] J. Eagon and M. Hochster, Cohen-Macaulay rings, invariant theory, and the generic perfection of determinantal loci. Amer. J. Math. 93 (1971), 1020-1058.

[H] M. Hochster, Grassmanians and their Schubert subvarieties are arithmetically Cohen-Macaulay, J. Alg. 25 (1973), 40-57.

[HL] C. Huneke and V. Lakshmibai, A characterization of Kempf varieties in terms of standard monomials and the geometric consequences, J. Alg. 94 (1985), 52-105.

$[K]_1$ G.R. Kempf, Vanishing theorems for flag manifolds, Amer. J. Math. 98 (1976), 325-331.

$[K]_2$ G. Kempf, Schubert methods with an application to algebraic curves, Stichting Mathematisch Centrum, Amsterdam, 1971.

[KR] G. Kempf and A. Ramanathan, Multicones over Schubert varieties, Invent. Math. 87 (1987), 353-363.

$[L]_1$ V. Lakshmibai, Kempf varieties, J. Indian Math. Soc. 40 (1976), 299- 349.

$[L]_2$ ———————, The variety AB = BA = 0 (in preparation).

[LMS] V. Lakshmibai, C. Musili and C.S. Seshadri, Geometry of G/P-IV, Proc. Indian Acad. Sci. 88 (1979), 279-362.

[LS] V. Lakshmibai and C.S. Seshadri, Geometry of G/P-V, J. Alg. 100 (1986), 462-557.

[La] D. Laksov, The arithmetic Cohen-Macaulay character of Schubert schemes, Acta Mathematica 129 (1972), 1-9.

[M] J. Matijevic, Three local conditions on a graded ring, Trans. Amer. Math. Soc. 205 (1975), 275-284.

[MR] V.B. Mehta and A. Ramanathan, Frobenius splitting and cohomology vanishing for Schubert varieties, Ann. Math. 122, 1985, 27-40.

[Mu] C. Musili, Postulation formula for Schubert varieties, J. Indian Math. Soc. 36 (1972), 143-171.

[R] A. Ramanathan, Schubert varieties are arithmetically Cohen-Macaulay, Invent. Math. 80 (1985), 283-294.

[RR] S. Ramanan and A. Ramanathan, Projective normality of Flag varieties and Schubert varieties, Invent. Math. 79 (1985), 217-224.

[S] J.P. Serre, Algèbre local, Multiplicites, Lecture Notes in Mathematics 11, Springer-Verlag (1965).

[Se] C.S. Seshadri, Geometry of G/P-I (Standard monomial theory for a minuscule PJ; C.P. Ramanujan: A Tribute, 20.7 (Springer-Verlag) published for the Tata Institute of Fundamental Research, Bombay.

C. Huneke
Department of Mathematics
Purdue University
West Lafayette, Indiana

V. Lakshmibai
Department of Mathematics
Northeastern University
Boston, MA 02115

Contemporary Mathematics
Volume **139**, 1992

SINGULAR BLOCKS OF THE CATEGORY $\mathcal{O}, II$

RONALD S. IRVING

1. Introduction

The aim of this paper is to show that the structure of Verma modules with singular highest weights can be obtained via wall-crossing translation functors in much the same way that this is done in [**7**] for Verma modules with regular highest weights. Of course, one does not have the usual translation functors across walls for singular blocks of the category $\mathcal{O}$; what we do is identify such a block by the well-known equivalence of categories with a subcategory $\mathcal{HC}^{\#}$ of a block $\mathcal{HC}$ of Harish-Chandra modules for the corresponding complex group. We then study certain singular principal series modules in place of Verma modules, and the additional translation functors are available on the larger category $\mathcal{HC}$ of Harish-Chandra modules. These translation functors will take us out of the subcategory $\mathcal{HC}^{\#}$ equivalent to the singular block of $\mathcal{O}$, but we will still be able to make the desired arguments within $\mathcal{HC}$. In particular, we will find that singular Verma modules are rigid (their socle and radical filtrations coincide) and that the composition factors in layers of these filtrations are given by Kazhdan-Lusztig polynomials. In the regular case, this was the main result of [**7**]; in the singular case, this was proved in [**9**] by using deep results of Soergel to show that the regular result was preserved under translation to the wall. The main point of the approach in this paper is that both cases can be done uniformly and simultaneously in a way which does not involve reducing the singular case to the regular case, and which does not rely on the results of Soergel.

A secondary point is that this approach provides an interpretation of the Hecke algebra action on the appropriate filtered Grothendieck group associated to a singular block of $\mathcal{O}$. Given a Weyl group $\mathcal{W}$ with Hecke algebra $\mathcal{H}$, Deodhar in [**3**] associates to a parabolic subgroup $\mathcal{W}_S$ two different $\mathcal{H}$-modules (Hecke

---

1991 *Mathematics Subject Classification*. Primary 17B10, 22E47; Secondary 22E46.
This paper is in final form and no version of it will be submitted for publication elsewhere.

modules) with bases in bijection with $\mathcal{W}/\mathcal{W}_S$. He relates them by a map in [**4**]. One of these modules is the filtered Grothendieck group of a regular block of the category $\mathcal{O}_S$ associated to the parabolic subalgebra determined by S, with the action of $\mathcal{H}$ interpreted in terms of translation functors on the filtered category. It is clear from the results of [**9**] that the other Hecke module may be viewed as the filtered Grothendieck group of the singular block of $\mathcal{O}$ determined by S, but no interpretation of the $\mathcal{H}$-action is apparent.

We find in this paper that the translation functors introduced on the singular $\mathcal{O}$ block extend to functors on a filtered version of the category, and the induced action of the functors at the filtered character level coincides with the action of $\mathcal{H}$ on the Hecke module. Thus the two Hecke modules play analogous roles for two different categories associated to $\mathcal{W}_S$: a regular block of $\mathcal{O}_S$ and an S-singular block of $\mathcal{O}$. Two remarks should be made in this regard. First, recent work in the paper of Beilinson, Ginsburg, and Soergel shows a very strong relationship between these two categories [**15**]. Second, one important difference in the case of singular blocks of $\mathcal{O}$, in contrast to regular blocks of $\mathcal{O}$ or $\mathcal{O}_S$, is that translation on a Verma module (or more precisely on the corresponding principal series module) can produce a non-split extension of the Verma module by itself. Of course, no such extensions can exist within the category $\mathcal{O}$, or $\mathcal{HC}^{\#}$, which is why we must pass to the larger category $\mathcal{HC}$ in order to realize the translation functors.

There is one final point. From the perspective of Vogan's calculus for studying translation functors on categories of Harish-Chandra modules of real reductive groups ([**16**] and [**17**]), the fact that translation on a singular Verma module can produce a non-split extension of the module with itself corresponds to having real roots not in the τ-invariant. In fact, the Hecke modules we attach to singular blocks of $\mathcal{O}$ code three of the eight possibilities in the Vogan calculus: the two possibilities associated to complex roots, and the one just mentioned. Thus the methods of this paper will be applicable to regular blocks of Harish-Chandra modules for which the associated Hecke module of Lusztig and Vogan introduced in [**12**] and [**18**] has only these three possibilities. This includes blocks for complex groups, for which only the first two possibilities arise. In addition there are only three other cases: two infinite families of blocks and one of E_6 type. The most interesting of these is the infinite family which consists of the non-principal blocks for the groups $SU(n,n)$. For all of these blocks, the arguments of this paper carry over. This yields that the standard generalized principal series modules are rigid, with composition factors in layers given by appropriate inverse Lusztig-Vogan polynomials. These results follow as well via work of Casian [**2**]. The resemblance of these blocks of Harish-Chandra modules to singular blocks of the category $\mathcal{O}$ is worthy of note.

Work on this paper was done with partial support of the National Science Foundation and the National Security Agency. I profited from several helpful conversations with David Collingwood.

2. The Hecke module

Fix a Weyl group $\mathcal{W}$ with generating set of reflections B and let S be a fixed subset of B, generating the parabolic subgroup $\mathcal{W}_S$. Denote by $\mathcal{W}^S$ the collection of cosets $\mathcal{W}/\mathcal{W}_S$, which can be identified with the coset representatives of maximal length. Let us make this identification. Then $\mathcal{W}^S$ inherits an order $\leq$ from the Bruhat order of $\mathcal{W}$. Given $w \in \mathcal{W}$ and $s \in S$, there are three possibilities: $sw < w$ (in which case $sw \in \mathcal{W}^S$), $sw > w$ and $sw \in \mathcal{W}^S$, or $sw > w$ and $sw \notin \mathcal{W}^S$. For brevity, let us refer to the third case as $sw \notin \mathcal{W}^S$ and the second as $sw > w$ (so that $sw \in \mathcal{W}^S$ will be implicitly understood). Let us also note for future reference that $\mathcal{W}^S$ has a unique minimal element denoted by e and a unique longest element denoted by w^S.

With respect to our choices of $\mathcal{W}$ and S, let $\mathcal{M}^*$ be the free $\mathbb{Z}[q^{1/2}, q^{-1/2}]$-module with basis $\{m(w)^* : w \in \mathcal{W}^S\}$. Let $\mathcal{H}$ be the Hecke algebra associated to $\mathcal{W}$; it is a $\mathbb{Z}[q^{1/2}, q^{-1/2}]$-algebra with generators $\{T_s : s \in B\}$ and relations $(T_s + 1)(T_s - q) = 0$ for $s \in B$ and $T_s T_{s'} \ldots = T_{s'} T_s \ldots$ (n terms each) if ss' has order n in $\mathcal{W}$. We also introduce the elements $\{\theta_s : s \in S\}$, with $\theta_s = q^{-1/2}(T_s + 1)$.

Following Deodhar [**3**], we can make $\mathcal{M}^*$ into an $\mathcal{H}$-module in two ways:

$$\begin{aligned} T_s m(w)^* = & q^{1/2} m(sw)^* && \text{if } sw > w \\ & (q-1)m(w)^* + q^{1/2} m(sw)^* && \text{if } sw < w \\ & u.m(w)^* && \text{if } sw \notin \mathcal{W}^S, \end{aligned}$$

where u can be chosen to be -1 or q. (Here we have shifted each basis element of Deodhar by a suitable power of q.) The choice of -1 leads in the third case to $\theta_s m(w)^* = 0$. The resulting Hecke module is used in [**8**], with $\mathcal{W}^S$ replaced by ${}^S\mathcal{W}$, and the resulting $\mathcal{M}^*$ is shown to be the filtered Grothendieck group of regular blocks of the category $\mathcal{O}_S$ associated to the parabolic subalgebra defined by S. The action of θ_s is then related to the action of wall-crossing translation functors on a filtered version of $\mathcal{O}_S$. For the rest of this paper we will set $u = q$; thus in the third case we obtain $\theta_s m(w)^* = (q^{1/2} + q^{-1/2})m(w)^*$. (In the first two cases, we have $\theta_s m(w)^* = q^{-1/2}m(w)^* + m(sw)^*$ and $\theta_s m(w)^* = q^{1/2}m(w)^* + m(sw)^*$.)

Again following Deodhar, we find a second basis $\{l(w)^* : w \in \mathcal{W}^S\}$ related to the first by the appropriate Kazhdan-Lusztig and inverse Kazhdan-Lusztig polynomials:

$$\begin{aligned} l(w)^* &= \sum_{y, y \leq w} (-1)^{(\ell(w)-\ell(y))} q^{(\ell(w)-\ell(y))/2} \mathcal{P}^{\wedge}_{y,w}(q) m(y)^*, \\ m(w)^* &= \sum_{y, y \leq w} q^{(\ell(w)-\ell(y))/2} \mathcal{Q}^{\wedge}_{y,w}(q) l(y)^*. \end{aligned}$$

Here $\wedge$ denotes the automorphism on $\mathbb{Z}[q^{1/2}, q^{-1/2}]$ which replaces $q^{1/2}$ by $q^{-1/2}$. Recursive formulas for the polynomials $\mathcal{P}_{y,w}(q)$ are given in [**3**]. It follows from [**3**] that the inverse Kazhdan-Lusztig polynomials are inverse Kazhdan-Lusztig polynomials for $\mathcal{W}$; in other words, they are actually Kazhdan-Lusztig polynomials. Recall for y and w in $\mathcal{W}^S$ with $y < w$ that $\mu(y,w)$ denotes the coefficient in $\mathcal{P}_{y,w}(q)$ of $q^{(\ell(w)-\ell(y)-1)/2}$. It follows that μ coincides with the restriction to $\mathcal{W}^S$ of the usual μ-function defined via Kazhdan-Lusztig polynomials for $\mathcal{W}$.

The action of θ_s on the basis $\{l(w)^*\}$ can be calculated from the formulas for the polynomials, yielding:

$$\theta_s l(w)^* = \begin{cases} (q^{1/2}+q^{-1/2})l(w)^* + l(sw)^* + \sum_{y,y<w} \mu(y,w)l(y)^* & \text{if } sw > w \\ 0 & \text{if } sw < w \\ (q^{1/2}+q^{-1/2})l(w)^* + \sum_{y,y<w} \mu(y,w)l(y)^* & \text{if } sw \notin \mathcal{W}^S. \end{cases}$$

3. The categories of interest and translation functors

Let $\mathfrak{g}$ be a complex semisimple Lie algebra. Let $\mathfrak{b}$ be a Borel subalgebra containing a Cartan subalgebra $\mathfrak{h}$. With respect to these choices, the category $\mathcal{O}$ is defined as the category of finitely-generated $\mathfrak{g}$-modules which are $\mathfrak{h}$-semisimple and $\mathfrak{n}$-finite, where $\mathfrak{n}$ is the nilradical of $\mathfrak{b}$. Given $\lambda \in \mathfrak{h}^*$, we denote by $M(\lambda)$ and $L(\lambda)$ the Verma module of highest weight λ and its unique simple quotient. Let the Weyl group of $\mathfrak{g}$ act on $\mathfrak{h}^*$ by the dot action: $w.\lambda = w(\lambda+\rho)-\rho$, where ρ is the half-sum of the positive roots. Then λ is called dominant if it is maximal in its Weyl group orbit with respect to the usual ordering on $\mathfrak{h}^*$ and antidominant if it is minimal. Let μ be an antidominant weight and let $\mathcal{W}$ be the integral Weyl group associated to μ. It is the subgroup of the full Weyl group generated by the reflections s_α such that $s_\alpha.\mu \geq \mu$. Let χ_μ be the corresponding infinitesimal character on the center of the enveloping algebra of $\mathfrak{g}$. We denote by $\mathcal{O}_{\chi_\mu}$ the full subcategory of $\mathcal{O}$ consisting of modules with generalized infinitesimal character χ_μ and by $\mathcal{O}_\mu$ the full subcategory of $\mathcal{O}$ consisting of modules all of whose composition factors have the form $L(w.\mu)$ with $w \in \mathcal{W}$. Then $\mathcal{O}_\mu$ is a block of $\mathcal{O}$ and $\mathcal{O}$ is the direct sum of such blocks as μ varies over all antidominant weights. (For μ regular, $\mathcal{O}_{\chi_\mu}$ and $\mathcal{O}_\mu$ coincide, but in general $\mathcal{O}_{\chi_\mu}$ decomposes into a sum of $\mathcal{O}_\mu$'s, as μ varies over the antidominant weights with the same infinitesimal character.) Let B be the set of simple reflections for $\mathcal{W}$, let S be the subset $\{s_\alpha \in B : s_\alpha.\mu = \mu\}$, and let $\mathcal{W}^S$ be as in §2. Then $\{L(w.\mu) : w \in \mathcal{W}^S\}$ is a complete set of distinct simple modules of $\mathcal{O}_\mu$ up to isomorphism.

Let $\mathfrak{k}$ be a diagonal copy of $\mathfrak{g}$ in $\mathfrak{g} \times \mathfrak{g}$, as in §6.4 of [**10**]. A Harish-Chandra module for $\mathfrak{g}$ is a finitely-generated $\mathfrak{g} \times \mathfrak{g}$-module which as a $\mathfrak{k}$-module is a direct sum of finite-dimensional simple modules, each isomorphism type occuring only finitely often. We denote by $\mathcal{HC}$ the category of all Harish-Chandra modules. Given weights μ and λ in $\mathfrak{h}^*$ differing by an integral weight (so they have the same integral Weyl group $\mathcal{W}$), we may form the modules $\mathcal{L}(M(\lambda), M(\mu))$ and

$\mathcal{D}(M(\lambda), M(\mu))$. These are the $\mathfrak{k}$-finite parts of $\mathrm{Hom}_{\mathbb{C}}(M(\lambda), M(\mu))$ and $(M(\lambda) \otimes M(\mu))^*$, each of which carries a natural $\mathfrak{g} \times \mathfrak{g}$-module structure, and they are in $\mathcal{HC}$. The modules of the second type are called principal series modules. Each contains a single copy of the finite-dimensional $\mathfrak{k}$-module E whose highest weight is in the $\mathcal{W}$-orbit of $\lambda - \mu$; it thus has a distinguished composition factor, the one containing E, which we denote by $L(\lambda, \mu)$.

For the remainder of the paper, let us fix μ to be an antidominant weight and λ to be a dominant, regular weight. Let $\nu = (\mu, \lambda)$ and let χ_ν be the corresponding infinitesimal character on the center of the enveloping algebra of $\mathfrak{g} \times \mathfrak{g}$. The center has the form $Z \otimes Z$, so χ_ν can be decomposed as $\chi_\mu \otimes \chi_\lambda$. Let $\mathcal{HC}_{\chi_\nu}$ be the full subcategory of $\mathcal{HC}$ consisting of modules with generalized infinitesimal character χ_ν. Let $\mathcal{HC}^{\#}_{\chi_\nu}$ be the full subcategory of $\mathcal{HC}_{\chi_\nu}$ consisting of modules with the further requirement that $1 \otimes Z$ acts with genuine infinitesimal character $1 \otimes \chi_\lambda$. The relevance of Harish-Chandra modules to the study of the category $\mathcal{O}$ arises from a theorem of Bernstein and Gelfand [**1**]. (See also [**5**] and 6.27 of [**10**].) The theorem states that the functor $\mathcal{L}(M(\lambda), \text{—})$ from $\mathcal{O}_{\chi_\mu}$ to $\mathcal{HC}_{\chi_\nu}$ yields an equivalence of categories from $\mathcal{O}_{\chi_\mu}$ to $\mathcal{HC}^{\#}_{\chi_\nu}$. Let us denote by $\mathcal{HC}^{\#}_\nu$ the block of $\mathcal{HC}^{\#}_{\chi_\nu}$ corresponding to $\mathcal{O}_\mu$ under this equivalence and by $\mathcal{HC}_\nu$ the block of $\mathcal{HC}$ whose simple modules lie in $\mathcal{HC}^{\#}_\nu$.

Note that $\mathcal{HC}^{\#}_\nu$ contains the modules of the form $\mathcal{L}(M(\lambda), M(w.\mu))$, as w varies over $\mathcal{W}^S$. In fact, $\mathcal{L}(M(\lambda), M(w.\mu))$ is a principal series module, isomorphic to $\mathcal{D}(M(\mu), M(w^{-1}.\lambda))$. In the terminology of real groups, these are the standard principal series modules in $\mathcal{HC}_\nu$ for the simply-connected complex Lie group with Lie algebra $\mathfrak{g}$; thus, the category equivalence from $\mathcal{O}_\mu$ to $\mathcal{HC}^{\#}_\nu$ sends Verma modules to standard principal series modules. Also, $\{L(\lambda, w.\mu) : w \in \mathcal{W}^S\}$ is a complete set of distinct simple modules for $\mathcal{HC}_\nu$ and $\mathcal{HC}^{\#}_\nu$, up to isomorphism. In fact, for each $w \in \mathcal{W}^S$, the simple module $\mathcal{L}(\lambda, w.\mu)$ is isomorphic to $\mathcal{L}(M(\lambda), L(w.\mu))$ and occurs as the unique simple quotient of $\mathcal{L}(M(\lambda), M(w.\mu))$. (For these facts, see chapters 6 and 7 of [**10**], especially 6.9(2), 6.25, 6.26, 6.27, 6.29, and 7.23 .)

Just as there is a translation functor $T_\mu^{\mu'}$ from $\mathcal{O}_\mu$ to $\mathcal{O}_{\mu'}$ for a weight μ' with $\mu - \mu'$ integral, so too there are translation functors between blocks $\mathcal{HC}_\nu$ and $\mathcal{HC}_{\nu'}$. Specifically, if $\nu' = (\mu, \lambda')$ with $\lambda - \lambda'$ integral one can introduce a translation functor $R_\lambda^{\lambda'}$ from $\mathcal{HC}_\nu$ to $\mathcal{HC}_{\nu'}$; it acts on modules of the form $\mathcal{L}(M, N)$ by sending them to $\mathcal{L}(T_\lambda^{\lambda'} M, N)$. Similarly, if $\nu' = (\mu', \lambda)$, one can introduce a translation functor $S_\mu^{\mu'}$ which sends $\mathcal{L}(M, N)$ to $\mathcal{L}(M, T_\mu^{\mu'} N)$. (See §1.13 of [**5**], §3.1 of [**11**], and §6.33 of [**10**].)

If μ is regular, then for each simple reflection $s \in B$ the functor θ_s of translation across the s-wall is defined on $\mathcal{O}_\mu$. One chooses a weight μ' differing from μ by an integral weight with s the only simple reflection fixing μ'. Then $\theta_s = T_{\mu'}^{\mu} \circ T_\mu^{\mu'}$. Analogously, we define ${}^r\theta_s$ on $\mathcal{HC}_\nu$ as the composition $S_\mu^{\mu'} \circ S_{\mu'}^{\mu}$. Clearly $\mathcal{L}(M(\lambda), \text{—}) \circ \theta_s = {}^r\theta_s \circ \mathcal{L}(M(\lambda), \text{—})$; in particular, ${}^r\theta_s$ preserves the subcategory $\mathcal{HC}^{\#}_\nu$. Moreover, the category $\mathcal{HC}_\nu$ has an additional set of translation functors: for each s, choose λ' differing from λ by an integral weight and fixed only by s and define ${}^l\theta_s$ as $R^{\lambda}_{\lambda'} \circ R^{\lambda'}_{\lambda}$. In contrast to ${}^r\theta_s$, the functor ${}^l\theta_s$ will not preserve the subcategory $\mathcal{HC}^{\#}_{\nu}$.

If μ is not regular, then $\mathcal{O}_{\mu}$ no longer has translation functors across walls defined on it, and in parallel $\mathcal{HC}_{\nu}$ no longer has translation functors ${}^r\theta_s$. However, because λ is regular, one can still define translation functors ${}^l\theta_s$ on $\mathcal{HC}_{\nu}$ as above. The availability of these functors is the key to this paper, allowing us to mimic in $\mathcal{HC}_{\nu}$ arguments made with wall-crossing functors for regular blocks of $\mathcal{O}$. It is because these functors do not preserve $\mathcal{HC}^{\#}_{\nu}$ that we must work inside $\mathcal{HC}_{\nu}$ rather than in $\mathcal{O}_{\mu}$. Nevertheless any information we obtain about the standard principal series modules in $\mathcal{HC}_{\nu}$ can be transferred back to $\mathcal{O}_{\mu}$.

At this point let us reset our notation. The weights λ and μ are fixed already. For each $s \in B$ we will denote the functor ${}^l\theta_s$ on $\mathcal{HC}_{\nu}$ simply by θ_s. No ambiguity will arise. Also, let us denote $\mathcal{L}(M(\lambda), M(w.\mu))$ by $M(w)$ and $\mathcal{L}(M(\lambda), L(w.\mu))$ by $L(w)$. Any structural results on them will carry over to the corresponding highest weight modules $M(w.\mu)$ and $L(w.\mu)$. We begin with the basic facts on the behavior of θ_s on $M(w)$ and $L(w)$.

PROPOSITION 3.1. *Let $w \in \mathcal{W}^S$ and $s \in S$.*

(i) *If $sw > w$, there is a non-split exact sequence $0 \to M(sw) \to \theta_s M(w) \to M(w) \to 0$.*
(ii) *If $sw < w$, the embedding of $M(sw)$ in $M(w)$ yields an isomorphism of $\theta_s M(w)$ and $\theta_s M(sw)$. In particular, there is a non-split exact sequence $0 \to M(w) \to \theta_s M(w) \to M(sw) \to 0$.*
(iii) *If $sw \notin \mathcal{W}^S$, there is a non-split exact sequence $0 \to M(w) \to \theta_s M(w) \to M(w) \to 0$.*

PROOF. This follows from Proposition 3.2 of [**11**], combined with results of §2 of that paper. ◆

REMARKS. (1) Proposition 3.1 may be viewed as a special case of more general results of Vogan on translation functors and standard generalized principal series modules.

(2) The situation of case (iii) cannot occur in the category $\mathcal{O}_{\mu}$, since any self-extension of a Verma module splits. In particular, the module $\theta_s M(w)$ cannot lie in $\mathcal{HC}^{\#}_{\nu}$. However (iii) will arise whenever μ is singular. Specifically, if $s \in S$, then (iii) occurs for $\theta_s M(e)$, where e is the minimal element of $\mathcal{W}^S$. This is the reason why we must work in $\mathcal{HC}_{\nu}$ rather than $\mathcal{HC}^{\#}_{\nu}$ or $\mathcal{O}_{\mu}$.

PROPOSITION 3.2. *Let $w \in \mathcal{W}^S$ and let $s \in B$.*

(i) *If $sw > w$, then $\theta_s L(w)$ has simple socle $L(w)$, simple cap $L(w)$, and $rad\, \theta_s L(w)/soc\, \theta_s L(w)$ is a semisimple direct sum of $L(sw)$ and simple modules of the form $L(y)$ with $ys < y$ and $y < w$.*
(ii) *If $sw < w$, then $\theta_s L(w) = 0$.*

(iii) *If* $sw \notin \mathcal{W}^S$, *then* $\theta_s L(w)$ *has simple socle* $L(w)$, *simple cap* $L(w)$, *and* $rad\,\theta_s L(w)/soc\,\theta_s L(W)$ *is a semisimple direct sum, possibly* (0), *of simple modules of the form* $L(y)$ *with* $ys < y$ *and* $y < w$.

PROOF. First assume that μ is regular. Then all the statements besides the semisimplicity statements of (i) and (iii) are standard consequences of adjoint functor arguments, and the semisimplicity statements are what is called Vogan's conjecture. This is shown in [**17**] to be equivalent to the appropriate Kazhdan-Lusztig conjecture, which is proved in [**18**] for integral weights. To pass to the non-integral case, it is sufficient via the category equivalence $\mathcal{L}(M(\lambda), \_\_)$ to prove the Kazhdan-Lusztig conjecture for $\mathcal{O}_\mu$ in the non-integral case. But this follows from Theorem 11 of Soergel's paper [**14**].

Now assume that μ is not necessarily regular and let μ' be a regular weight differing from μ by an integral weight. Let $\nu' = (\lambda, \mu')$. Then we have the functor $S^\mu_{\mu'}$ from $\mathcal{HC}_{\nu'}$ to $\mathcal{HC}_\nu$ as well as wall-crossing functors θ_s on $\mathcal{HC}_\nu$ and ${}^l\theta_s$ on $\mathcal{HC}_{\nu'}$. They satisfy the relation

$$S^\mu_{\mu'} \circ {}^l\theta_s = \theta_s \circ S^\mu_{\mu'}.$$

Also, $S^\mu_{\mu'}$ is exact and takes the simple module $\mathcal{L}(M(\lambda), L(w.\mu'))$ to $L(w)$ if $w \in \mathcal{W}^S$ and to (0) otherwise. Combining these facts with the proposition's validity for $\mathcal{HC}_{\nu'}$ completes the proof. ♦

The situation of (iii) cannot occur in $\mathcal{O}_\mu$, just as with Proposition 3.1, but will occur in $\mathcal{HC}_\nu$ if μ is singular.

We turn to a result on Loewy length under translation. Recall that the *Loewy length* of a module M is the shortest possible length $s - r$ of a filtration $M = N^r \supset M^{r+1} \supset \ldots \supset M^S = 0$ of M having successive layers semisimple. Denote the Loewy length by $\ell\ell M$.

THEOREM 3.3. *Let* M *be a module in* $\mathcal{HC}_\nu$ *and let* $s \in B$. *Then* $\ell\ell\theta_s M \leq \ell\ell M + 2$.

PROOF. The analogous statement is proved for a regular block of $\mathcal{O}$ in the Appendix of [**7**]. The situation in $\mathcal{HC}_\nu$ is slightly different from that in $\mathcal{O}$ because it is possible for $\theta_s L(w)$ to be a non-split extension of $L(w)$ by itself, hence of Loewy length 2. (In particular this happens for $\theta_s L(e)$ if $s \in S$.) However, this does not affect the proof given in [**7**]. ♦

It follows, as was already proved in [**6**] for Verma modules, that $M(w)$ has Loewy length $\ell(w) + 1$. Let us also recall at this point the well-known fact that $M(w)$ has simple socle $L(e)$.

Using Theorem 3.3, one can copy (a corrected version of) the argument of [**8**] to show that in fact the translation functor θ_s extends to a functor on the filtered category $\mathcal{HC}^*_\nu$. Let us recall the relevant definitions. By a *filtered module* M^* we mean a module M in $\mathcal{HC}_\nu$ along with a filtration $M = M^r \supseteq M^{r+1} \supseteq \ldots \supseteq M^s = 0$ such that the layers M^i/M^{i+1} are semisimple. The filtered category $\mathcal{HC}^*_\nu$ has filtered modules as objects; a morphism from M^* to N^* is a module homomorphism $f : M \to N$ such that $f(M^i) \subseteq N^i$ for all i. A shift functor σ is defined on $\mathcal{H}C_\nu^*$, with $\sigma(M^*)$ having M as underlying module and $\sigma(M^*)^i = M^{i-1}$. Similarly one can define σ^t for any integer t. We need one more notion for filtered categories. Given a short exact sequence

$$0 \to K \underset{f}{\to} M \underset{g}{\to} N \to 0$$

in $\mathcal{H}C_\nu$ and a filtration M^* on M, the *induced filtrations* on K and N from M^* are given by $K^i = K \cap f^{-1}(M^i)$ and $N^i = g(M^i)$. Let us say that a sequence of maps

$$0 \to K^* \underset{f}{\to} M^* \underset{g}{\to} N^* \to 0$$

in $\mathcal{H}C_\nu^*$ is *exact* if the unfiltered version is exact and if K^* and N^* are the induced filtrations on K and N.

Define the Grothendieck group $Gr(\mathcal{H}C_\nu^*)$ of $\mathcal{H}C_\nu^*$, or the *filtered Grothendieck group*, to be the free $\mathbb{Z}$-module with basis $\{[M^*] : M^* \text{ a filtered module}\}$ modulo the submodule generated by elements of the form $[M^*] - [N^*]$ if $M^* \cong N^*$ and of the form $[M^*] - [K^*] - [N^*]$ for exact sequences as above. Given $w \in \mathcal{W}^S$, let $L(w)^*$ be the simple module $L(w)$ concentrated at $0 : L(w)^0 = L(w)$ and $L(w)^1 = 0$. Then $Gr(\mathcal{H}C_\nu^*)$ is a free $\mathbb{Z}$-module with basis $\{[\sigma^t L(w)^*] : w \in \mathcal{W}^S, t \in \mathbb{Z}\}$. We will identify $Gr(\mathcal{H}C_\nu^*)$ with the Hecke module $\mathcal{M}^*$ as free $\mathbb{Z}[q^{1/2}, q^{-1/2}]$-modules, with $[\sigma^t L(w)^*]$ corresponding to $q^{t/2} l(w)^*$. Thus, for a filtered module M^*, we have $[M^*] = \Sigma_{i\in\mathbb{Z}}\Sigma_{w\in\mathcal{W}^S}(M^i/M^{i+1} : L(w))q^{i/2}l(w)^*$. Under this identification, the action of $\mathcal{H}$ on $\mathcal{M}^*$ carries over to $Gr(\mathcal{H}C_\nu^*)$.

If the translation functor θ_s is to be extended to filtered modules in a way compatible with the action of θ_s on $Gr(\mathcal{H}C_\nu^*)$, in view of the formulas at the end of §2, we are led to expect $\theta_s L(w)^*$ to have $\sigma^{-1}L(w)^*$ and $\sigma L(w)^*$ at the top and bottom (assuming $sw \not< w$), and to have all other composition factors at layer 0. Proposition 3.2 shows that we can give $\theta_s L(w)^*$ such a filtration. If $\theta_s L(w)$ is just an extension of $L(w)$ by itself, for instance if $w = e$ and $s \in S$, then the filtration we wish to put on $\theta_s L(w)^*$ has an intermediate layer equal to 0. This phenomenon is the only difference from what happens in the filtered category $\mathcal{O}_S^*$. Despite this additional subtlety, one can follow the arguments of [**8**] to obtain the following result.

THEOREM 3.4. *For each $s \in B$ there is a functor θ_s defined on $\mathcal{H}C_\nu^*$ such that its restriction to $\mathcal{H}C_\nu$ is the usual translation functor θ_s and such that on a filtered module M^* one has $[\theta_s M^*] = \theta_s[M^*]$.*

PROOF. Given a filtered module M^* , we define the filtration $(\theta_s M)^*$ as follows: Let π_i be the composition of the canonical surjections of $\theta_s(M^i)$ to $\theta_s(M^i/M^{i+1})$ and of $\theta_s(M^i/M^{i+1})$ to its largest semisimple quotient. Let κ_{i-1} be the composition of the natural map of M^{i-1} to $\theta_s(M^{i-1})$ (arising via the adjoint functors of translation to and from the s-wall) and the map of $\theta_s(M^{i-1})$ into $\theta_s(M^i)$. Set $(\theta_s M)^i = Ker\pi_i + Im\kappa_{i-1}$. (Compare the diagram on page 77 of [8].) It is straightforward to check that $(\theta_s M)^i/(\theta_s M)^{i+1}$ is semisimple, using Theorem 3.3, and that θ_s is functorial. ♦

4. Rigidity and structure of standard principal series and Verma modules

Recall the radical filtration $\{rad^i M\}$ and the socle filtration $\{soc^i M\}$ of a finite length module M, from [7] for instance. The indexing of the socle filtration is opposite to our convention; let us call the filtration M^* given by $M^i = soc^{t-i} M$ the *standard socle filtration*, where t is the length of the socle filtration (the Loewy length). More generally, let us call any filtration *standard* if the indexing is arranged so that $M^0 = M$ and $M^1 \neq M$. Thus the indexing in the definition of the radical filtration is already standard. We call a filtration M^* on a module M a *Loewy filtration* if it has minimal possible length among filtrations with semisimple layers. Thus a standard Loewy filtration has the same length as and is intermediate between the radical and standard socle filtrations. We call M *rigid* if it has a unique Loewy filtration up to shifts; equivalently, the radical and standard socle filtrations coincide.

Following [7], we now introduce a local version of rigidity. Let us say a simple module L is *rigidly placed* in a module M if there are non-negative integers $a_0, a_1, \ldots$ such that for any standard Loewy filtration M^* of M and any integer $i \geq 0$ we have $(M^i/M^{i+1} : L) = a_i$. Notice that M is rigid if and only if every simple module is rigidly placed in M.

We wish to prove the rigidity of the standard principal series modules $M(w)$. Because of the possibility of case (iii) in Proposition 3.1, we cannot directly apply the argument of [7]. In fact, the analogue of Lemma 2.2.1 of [7] is false in the singular case. (The analogue would state that if $sw > w$ and $L(y)$ is rigidly placed in $M(w)$ and $M(sw)$, then $L(y)$ is rigidly placed in $\theta_s M(w)$. But this fails for $y = e$ if $s \in S$, as will follow from Corollary 4.3.) What we use instead as a substitute is the following result. It is the reason we have chosen to present the argument in the language of filtered categories, which was not used in [7].

PROPOSITION 4.1. *Let $w \in W$ and $s \in B$. Let $M(w)^*$ be a standard Loewy filtration on $M(w)$.*

(i) *Let $sw > w$. Then the filtrations induced on $M(w)$ and $M(sw)$ from $\theta_s(M(w)^*)$ by the short exact sequence $0 \to M(sw) \to \theta_s M(w) \to M(w) \to 0$ are Loewy filtrations .*

(ii) *Let $sw \notin \mathcal{W}^S$. Then the filtrations induced on $M(w)$ from $\theta_s(M(w)^*)$ by the short exact sequence $0 \to M(w) \to \theta_s M(w) \to M(w) \to 0$ are Loewy filtrations.*

PROOF. Assume $sw > w$. The module $M(w)$ has Loewy length $\ell(w)+1$ with simple socle $L(e)$. Let $t = \ell(w)$. Then for any standard Loewy filtration $M(w)^*$ we have $M(w)^t = L(e)$ and $M(w)^0/M(w)^1 = L(w)$. Moreover this accounts for all appearances of $L(w)$ and $L(e)$ in $M(w)$. Thus Proposition 3.2 and Theorem

3.4 imply that $\theta_s(M(w)^*)$ has $L(w)$ in layers 1 and -1, $L(sw)$ in layer 0, and $L(e)$ in layers $t-1$ and $t+1$. It follows that the induced filtration on $M(sw)$ satisfies $M(sw)^0/M(sw)^1 = L(sw)$ and $M(sw)^{t+1} = L(e)$. Since $M(sw)$ has Loewy length $\ell(sw)+1$, the filtration on $M(sw)$ is a Loewy filtration. Similarly, we find that the induced filtration on $M(w)$ satisfies $M(w)^{-1}/M(w)^0 = L(w)$ and $M(w)^{t-1} = L(e)$, so it is a Loewy filtration. This proves (i); the proof of (ii) is essentially the same. ◆

THEOREM 4.2. *Let $w \in \mathcal{W}^S$. Then $M(w)$ is rigid.*

PROOF. It suffices to prove for each in $y \in \mathcal{W}^S$ that $L(y)$ is rigidly placed in $M(w)$ for all $w \in \mathcal{W}^S$. If this is not the case, choose y maximal with $L(y)$ not rigidly placed in some $M(w)$. If $y = w^S$, then $w = w^S$. But the only appearance of $L(w^S)$ in $M(w^S)$ is as the unique simple top layer of any Loewy filtration, so $L(w^S)$ must be rigidly placed. Thus we may assume that $y < w^S$. Hence there exists $s \in B$ with $sy > y$. We will obtain a contradiction to maximality by showing that $L(sy)$ is not rigidly placed in one of $M(w)$ or $M(sw)$, thereby proving the Theorem.

By assumption, there are standard Loewy filtrations M^* and N^* of $M(w)$ and an integer i with $(M^i/M^{i+1} : L(y)) \neq (N^i/N^{i+1} : L(y))$. By Propositions 3.1 and 3.2, there is one appearance of $L(sy)$ in $\theta_s M(w)$ for each appearance of $L(y)$ in $M(w)$ and additional appearances in $\theta_s M(w)$ for each $L(z)$ in $M(w)$ with $\mu(sy,z) \neq 0$. These z's all satisfy $y < z$, so by maximality of y they are rigidly placed in $M(w)$. By Theorem 3.4, $L(sy)$ occurs in $\theta_s(M^*)$ or $\theta_s(N^*)$ in the same layers with the same multiplicities that $L(y)$ or the $L(z)$'s occurred. Since the $L(z)$'s are rigidly placed in $M(w)$ but $L(y)$ is not, we obtain two filtrations $\theta_s(M^*)$ and $\theta_s(N^*)$ on the module $\theta_s M(w)$ satisfying $(\theta_s(M^*)^i : L(sy)) \neq (\theta_s(N^*)^i : L(sy))$.

Suppose $sw > w$ (or $sw \notin \mathcal{W}^S$). By Proposition 4.1, the filtrations induced on $M(sw)$ and $M(w)$ (or $M(w)$ and $M(w)$) by $\theta_s(M^*)$ and $\theta_s(N^*)$ are Loewy filtrations. By maximality of y, the simple module $L(sy)$ must be rigidly placed in both these principal series modules. Using Theorem 3.4 again, we deduce that $(\theta_s(M^*)^i : L(sy)) = (\theta_s(N^*)^i : L(sy))$, a contradiction. This proves the Theorem. ◆

NOTATION. For $w \in \mathcal{W}^S$, let $M(w)^*$ be the radical filtration of $M(w)$. By Theorem 4.2, it is the unique standard Loewy filtration on $M(w)$.

COROLLARY 4.3. *Let $w \in \mathcal{W}^S$.*

(i) *If $sw > w$, then there is a short exact sequence $0 \to M(sw)^* \to \theta_s M(w)^* \to \sigma^{-1}M(w)^* \to 0$. In particular, $\theta_s[M(w)^*] = [M(sw)^*] + q^{-1/2}[M(w)^*]$.*

(ii) *If $sw \notin \mathcal{W}^S$, then there is a short exact sequence $0 \to \sigma M(w)^* \to \theta_s M(w)^* \to \sigma^{-1}M(w)^* \to 0$. In particular, $\theta_s[M(w)^*] = (q^{1/2} + q^{-1/2})[M(w)^*]$.*

PROOF. For (i), the filtrations induced from $\theta_s M(w)^*$ on $M(sw)$ and $M(w)$ are Loewy filtrations by Proposition 4.1. Theorem 4.2 then implies that these filtrations are shifts of the radical filtration. Since $L(w)$ occurs with multiplicity 1 in $M(w)$ and $M(sw)$, these shifts are determined by the layers in which $L(w)$ occurs. But Theorem 3.4 yields that $L(w)$ appears in $\theta_s M(w)^*$ in layers -1 and 1, from which (i) follows. Part (ii) is proved by the same argument. ♦

COROLLARY 4.4. *Let y, $w \in \mathcal{W}^S$ and let $M(w)^*$ be the standard radical filtration of $M(w)$.*

(i) $[M(w)^*] = m(w)^*$.

(ii) *If $y \leq w$, then*

$$\sum_i (rad^i M(w)/rad^{i+1} M(w) : L(y)) q^{(\ell(w)-\ell(y)-i)/2} = \mathcal{Q}_{y,w}(q).$$

PROOF. Part (ii) is a re-interpretation of (i). Part (i) follows immediately from Corollary 4.3 and Theorem 3.4, using the formulas in §2 for θ_s on the basis $\{m(w)^*\}$ of the Hecke module. ♦

COROLLARY 4.5. *Let y and w be elements of $\mathcal{W}^S$ with $y \leq w$. Then*

$$\dim Ext^1_{\mathcal{O}}(L(y.\mu), L(w.\mu)) = \dim Ext^1_{\mathcal{O}}(L(w.\mu), L(y.\mu)) = \mu(y, w).$$

PROOF. The first equality follows from the duality on $\mathcal{O}$. For the second, we use that $\dim Ext^1_{\mathcal{O}}(L(w.\mu), L(y.\mu)) = (rad^1 M(w.\mu)/rad^2 M(w.\mu) : L(y.\mu))$. This follows using the fact that any non-split extension of $L(y.\mu)$ by $L(w.\mu)$, being a highest weight module, is a homomorphic image of $M(w.\mu)$. The equality then follows from Corollary 4.4. ♦

REMARKS. (1) Theorem 4.2 and Corollary 4.4 are proved in [**9**] as consequences of the fact that $\dim Ext^1_{\mathcal{O}}(L(y.\lambda), L(w.\lambda)) = \dim Ext^1_{\mathcal{O}}(L(y.\mu), L(w.\mu))$ for a regular antidominant weight λ and a singular antidominant weight μ. This is essentially Corollary 4.5, and is proved in [**9**] using Soergel's calculation of $\dim Ext^i_{\mathcal{O}}(M(y.\mu), L(w.\mu))$, a deep result. The approach of this paper is more elementary. Moreover, with Corollary 4.5 in hand, we can obtain all the other results of [**9**] using the proofs given in that paper. In particular, we obtain a description of the radical filtrations of projective modules in $\mathcal{O}_\mu$ and obtain some results on filtered versions of the translation functors T^μ_λ and T^λ_μ between regular and singular blocks of $\mathcal{O}$.

(2) It would be of interest to find a more elementary approach to Soergel's theorem calculating $\dim Ext^i_{\mathcal{O}}(M(y.\mu), L(w.\mu))$, along the lines of this paper. If μ is regular, Vogan has shown how to use his semisimplicity conjecture and translation functors to obtain the identification of $\dim Ext^i_{\mathcal{O}}(M(y.\mu), L(w.\mu))$ with coefficients of the Kazhdan-Lusztig polynomials [**17**]. It is tempting, using translation functors for singular μ, to mimic his argument, but one can't do this directly because of the failure of translation to preserve $\mathcal{O}_\mu$ or $\mathcal{HC}^\#_\nu$.

REFERENCES

1. J. N. Bernstein and S. I. Gelfand, *Tensor products of finite and infinite dimensional representations of semisimple Lie algebras*, Comp. Math. 41 (1980), 245–285.
2. L. Casian, *Graded characters of induced representations for real reductive Lie groups I*, J. Algebra 123 (1989), 289–326.
3. V. V. Deodhar, *On some geometric aspects of Bruhat orderings II. The parabolic analogue of Kazhdan-Lusztig polynomials*, J. Algebra 111 (1987), 483–506.
4. V. V. Deodhar, *Duality in parabolic set up for questions in Kazhdan-Lusztig theory*, preprint.
5. O. Gabber and A. Joseph, *On the Bernstein-Gelfand-Gelfand resolution and the Dulfo sum formula*, Comp. Math. 43 (1981), 107–131.
6. R. S. Irving, *Projective modules in the category $\mathcal{O}_S$: Loewy series*, Trans. Amer. Math. Soc. 291 (1985), 733-754.
7. R. S. Irving, *The socle filtration of a Verma module*, Ann. scient. Éc. Norm. Sup. 21 (1988), 47-65.
8. R. S. Irving, *A filtered category $\mathcal{O}_S$ and applications*, Memoirs Amer. Math. Soc. 419 (1990), (See also Errata to appear.).
9. R. S. Irving, *Singular blocks of the category $\mathcal{O}$*, Math. Z. 204 (1990), 209-224.
10. J. Jantzen, *Einhüllende Algebren halbeinfacher Lie-Algebren*, Springer-Verlag, Berlin, Heidelberg, New York, Tokyo, 1983.
11. A. Joseph, *The Enright functor on the Bernstein-Gelfand-Gelfand category $\mathcal{O}$*, Invent. Math. 67 (1982), 423–445.
12. G. Lusztig and D. Vogan, *Singularities of closures of K orbits on flag manifolds*, Invent. Math. 71 (1983), 365–379.
13. W. Soergel, *$\mathfrak{n}$-cohomology of simple highest weight modules on walls and purity*, Invent. Math. 98 (1989), 565–580.
14. W. Soergel, *Kategorie $\mathcal{O}$, perverse Garben und Moduln Øber den Koinvarianten zur Weylgruppe*, Jour. Amer. Math. Soc. 3 (1990), 421–445.
15. A. Beilinson, V. Ginsburg, and W. Soergel, *Koszul duality patterns in representation theory (preliminary version)*, preprint (1991).
16. D. Vogan, *Representations of Real Reductive Lie Groups, Progress in Math. 15*, Birkhäuser, Boston, 1981.
17. D. Vogan, *Irreducible characters of semisimple Lie groups II: The Kazhdan-Lusztig conjectures*, Duke Math. J. 46 (1979), 805–859.
18. D. Vogan, *Irreducible characters of semisimple Lie groups III: Proof of the Kazhdan-Lusztig conjecture in the integral case*, Inv. Math. 71 (1983), 381-417.

DEPARTMENT OF MATHEMATICS, UNIVERSITY OF WASHINGTON, SEATTLE, WASHINGTON 98195

E-mail: irving@math.washington.edu

Contemporary Mathematics
Volume **139**, 1992

A Geometric Realization of Minimal $\mathfrak{k}$-type of Harish-Chandra Modules for Complex S.S. Groups [1]

Bertram Kostant and **Shrawan Kumar**

0. Introduction

Let G be a semi-simple connected simply-connected complex algebraic group (viewed as a real Lie group), with a fixed Borel subgroup B, a complex maximal torus $T \subset B$, and a maximal compact subgroup K. Let $\mathfrak{g}, \mathfrak{b}, \mathfrak{h}, \mathfrak{k}$ be their (real) Lie algebras (respectively). In this paper we will be concerned with irreducible $(\mathfrak{g}^{\mathbb{C}}, \mathfrak{k}^{\mathbb{C}})$-modules (also called Harish-Chandra modules), where $\mathfrak{g}^{\mathbb{C}} := \mathfrak{g} \otimes_{\mathbb{R}} \mathbb{C}$ (similarly $\mathfrak{k}^{\mathbb{C}}$) is the complexified Lie algebra. Since the Lie algebra pair $(\mathfrak{g}^{\mathbb{C}}, \mathfrak{k}^{\mathbb{C}})$ is canonically isomorphic (as complex Lie algebras) with the pair $(\tilde{\mathfrak{g}}, \Delta(\mathfrak{g}))$ (cf. § 1.1) (where $\tilde{\mathfrak{g}} := \mathfrak{g} \oplus \mathfrak{g}$ is the direct sum Lie algebra, $\Delta(\mathfrak{g})$ is the diagonal subalgebra and, G being a complex group, $\mathfrak{g}$ has the canonical complex structure), we can equivalently consider $(\tilde{\mathfrak{g}}, \Delta(\mathfrak{g}))$-modules. The infinitesimal character of an irreducible $(\tilde{\mathfrak{g}}, \Delta(\mathfrak{g}))$-module is represented by a pair (λ, μ) of dominant (with respect to $\mathfrak{b}$) elements in $\mathfrak{h}^* := \mathrm{Hom}_{\mathbb{C}}(\mathfrak{h}, \mathbb{C})$. In this paper we will only consider irreducible $(\tilde{\mathfrak{g}}, \Delta(\mathfrak{g}))$-modules with integral infinitesimal character (i.e. λ and μ are integral weights).

Let us assume that λ and μ as above are both, in addition, regular. We replace λ (resp. μ) by $\lambda+\rho$ (resp. $\mu+\rho$), where λ and μ are dominant (integral) weights. (The main body of the paper does not have this restriction but we put it here just as a simplifying assumption.) Now it is known (cf. [D] or [BB]; see Theorem 2.2 in this paper) that the Weyl group W (associated to G) parametrizes bijectively the irreducible $(\tilde{\mathfrak{g}}, \Delta(\mathfrak{g}))$-modules with infinitesimal character $(\lambda+\rho, \mu+\rho)$. Let us denote the irreducible $(\tilde{\mathfrak{g}}, \Delta(\mathfrak{g}))$-module thus associated to $w \in W$ by $N_w = N_w(\lambda+\rho, \mu+\rho)$. It is further known that the minimal $\Delta(\mathfrak{g})$-type of N_w is $V(\overline{\mu_w - \lambda})$, where $\mu_w := -w(\mu+\rho) - \rho, V(\overline{\mu_w - \lambda})$ is the (finite dimensional) irreducible G-module with highest weight $\overline{\mu_w - \lambda}$ and, for any $\beta \in \mathfrak{h}^*, \overline{\beta}$ denotes the unique dominant element in the W-orbit of β.

[1] 1991 Mathematics Subject Classification. Primary 22E47.

This paper is in final form and no version of it will be submitted for publication elsewhere.

On the other hand, for any $w \in W$, there is a certain distinguished irreducible $\Delta(\mathfrak{g})$-subquotient E_w (which is isomorphic with $V(\overline{\mu_w - \lambda})$ as a $\mathfrak{g}$-module) of the tensor product $\tilde{\mathfrak{g}}$ -module $V(\lambda+\rho)^* \otimes V(\mu+\rho)^*$ (cf. [Ku1; § 2]), where $V(\lambda+\rho)^*$ is the dual $\mathfrak{g}$-module. In particular, observe that the minimal $\Delta(\mathfrak{g})$-type of N_w coincides with E_w. The aim of this paper is to explain this coincidence in terms of a 'natural' geometrical construction, which we now describe :

By Beilinson-Bernstein (cf. Theorem 2.2), the module $N_w(\lambda+\rho, \mu+\rho)$ is realized as the space of global sections $H^0(\widetilde{G/B}, \tilde{\mathcal{F}}_w \otimes \mathcal{L}(\lambda \otimes \mu))$, where $\tilde{\mathcal{F}}_w$ is a certain $\mathcal{D}_{\widetilde{G/B}}$-module on the product flag variety $\widetilde{G/B} := G/B \times G/B$, and $\mathcal{L}(\lambda \otimes \mu)$ is a homogeneous line bundle (cf. §§1.3 and 2.1). The module $H^0(\widetilde{G/B}, \tilde{\mathcal{F}}_w \otimes \mathcal{L}(\lambda \otimes \mu))$ embeds as a submodule of the local cohomology module $H^{\ell}_{\tilde{X}_w/\partial\tilde{X}_w}(\widetilde{G/B}, \mathcal{L}(\lambda \otimes \mu))$ (cf. Lemmas 2.3 and 2.4); where $\ell := \dim_{\mathbb{C}} G/B - \ell(w), \tilde{X}_w := \overline{G(e,w)} \subset \widetilde{G/B}$, and $\partial\tilde{X}_w := \tilde{X}_w \setminus G(e,w)$. Now define a Kunneth map (got by taking the tensor product) ψ_w :

$$H^0(\widetilde{G/B}, \mathcal{L}) \otimes H^{\ell}_{\tilde{X}_w/\partial\tilde{X}_w}(\widetilde{G/B}, \mathcal{L}(-\rho \otimes -\rho)) \to H^{\ell}_{\tilde{X}_w/\partial\tilde{X}_w}(\widetilde{G/B}, \mathcal{L}(\lambda \otimes \mu)),$$

where $\mathcal{L} := \mathcal{L}(\lambda+\rho \otimes \mu+\rho)$. We further show (cf. Corollary 2.12) that the module $H^{\ell}_{\tilde{X}_w/\partial\tilde{X}_w}(\widetilde{G/B}, \mathcal{L}(-\rho \otimes -\rho))$ contains a unique $\Delta(\mathfrak{g})$-invariant ϑ. (Even though we do not need, it is the unique irreducible $(\tilde{\mathfrak{g}}, \Delta(\mathfrak{g}))$-module with infinitesimal character $(0,0)$.) We next prove (cf. Lemma 2.14) that the restricted map

$$\psi_w^{\vartheta} : H^0(\widetilde{G/B}, \mathcal{L}) \to H^{\ell}_{\tilde{X}_w/\partial\tilde{X}_w}(\widetilde{G/B}, \mathcal{L}(\lambda \otimes \mu)),$$

defined by $\psi_w^{\vartheta}(x) = \psi_w(x \otimes \vartheta)$, factors through $H^0(\tilde{X}_w, \mathcal{L})$ giving rise to a map $\overline{\psi}_w^{\vartheta} : H^0(\tilde{X}_w, \mathcal{L}) \to H^{\ell}_{\tilde{X}_w/\partial\tilde{X}_w}(\widetilde{G/B}, \mathcal{L}(\lambda \otimes \mu))$, and moreover the map $\overline{\psi}_w^{\vartheta}$ is injective (cf. Lemma 2.15). But, as proved in [Ku1], the canonical restriction map : $H^0(\widetilde{G/B}, \mathcal{L}) \approx V(\lambda+\rho)^* \otimes V(\mu+\rho)^* \to H^0(\tilde{X}_w, \mathcal{L})$ is surjective and moreover $H^0(\tilde{X}_w, \mathcal{L})$ contains a unique copy E_w of the $\Delta(\mathfrak{g})$-module $V(\overline{\mu_w - \lambda})$. We next prove that the image of E_w under the map $\overline{\psi}_w^{\vartheta}$ lands inside the irreducible submodule N_w of $H^{\ell}_{\tilde{X}_w/\partial\tilde{X}_w}(\widetilde{G/B}, \mathcal{L}(\lambda \otimes \mu))$ and in fact is the minimal $\Delta(\mathfrak{g})$-type of N_w. (It may be mentioned that we do not use the known information about the minimal $\Delta(\mathfrak{g})$-type of N_w, instead we deduce it as a consequence of the Beilinson-Bernstein realization of irreducible Harish-Chandra modules and our work.)

Acknowledgements. This work was done in the academic year 1988-89, while the second author was visiting The Institute for Advanced Study, Princeton; hospitality of which is gratefully acknowledged.

1. Notation and preliminaries

(1.1) **Notation.** The notation G is reserved to denote a semi-simple connected simply-connected complex algebraic group with a fixed Borel subgroup B and a (complex) maximal torus $T \subset B$. Let $\mathfrak{g} \supset \mathfrak{b} \supset \mathfrak{h}$ be the (real) Lie algebras of $G \supset B \supset T$ resp. Of course these Lie algebras have canonical complex structures coming from the corresponding groups.

Let $\{\alpha_1, ..., \alpha_\ell\} \subset \mathfrak{h}^*$ (where $\mathfrak{h}^* := \mathrm{Hom}_{\mathbb{C}}(\mathfrak{h}, \mathbb{C}))$ be the simple roots for the positive root system determined by $\mathfrak{b}$, and let $\{\alpha_1^\vee, ..., \alpha_\ell^\vee\}$ be the corresponding simple co-roots. Define the set of integral weights $\mathfrak{h}^*_{\mathbb{Z}} := \{\lambda \in \mathfrak{h}^* : \lambda(\alpha_i^\vee) \in \mathbb{Z}, \text{ for all } 1 \leq i \leq \ell\}$. The set of dominant integral weights D is by definition $\{\lambda \in \mathfrak{h}^*_{\mathbb{Z}} : \lambda(\alpha_i^\vee) \geq 0, \text{ for all } i\}$. As usual ρ is the element of D, defined by $\rho(\alpha_i^\vee) = 1$, for all $1 \leq i \leq \ell$. Denote by $D - \rho$ the set $\{\lambda \in \mathfrak{h}^*_{\mathbb{Z}} : \lambda + \rho \in D\}$.

Let $W \approx N(T)/T$ denote the Weyl group, where $N(T)$ is the normalizer of T in G. The group W, which has a canonical representation in $\mathfrak{h}^*$, is generated (as a Coxeter group) by the 'simple' reflections $\{r_i\}_{1 \leq i \leq \ell}$; where $r_i \in \mathrm{Aut}\, \mathfrak{h}^*$ is defined by $r_i(\lambda) = \lambda - \lambda(\alpha_i^\vee)\alpha_i$. In particular, we can talk of the length $\ell(w)$ of any $w \in W$. For any $\lambda \in D$, let $W_\lambda := \{w \in W : w\lambda = \lambda\}$ be the stabilizer of λ. Then W_λ is again a Coxeter group, generated by a certain subset of simple reflections $\{r_i\}$.

We also fix a maximal compact subgroup $K \subset G$, with Lie algebra $\mathfrak{k}$. The complexified Lie algebra $\mathfrak{g}^{\mathbb{C}} := \mathfrak{g} \otimes_{\mathbb{R}} \mathbb{C}$ can be identified with the direct sum (complex) Lie algebra $\tilde{\mathfrak{g}} := \mathfrak{g} \oplus \mathfrak{g}$, under the complex Lie algebra isomorphism $\varphi : \mathfrak{g}^{\mathbb{C}} \to \tilde{\mathfrak{g}}$ (uniquely) defined by $\varphi(X) = (\overline{X}, X)$ for $X \in \mathfrak{g}$; where the bar denotes the conjugate-linear isomorphism of $\mathfrak{g}$ determined by the compact form $\mathfrak{k}$. Clearly $\varphi(\mathfrak{k} \otimes_{\mathbb{R}} \mathbb{C})$ is the diagonal subalgebra $\Delta(\mathfrak{g})$ of $\tilde{\mathfrak{g}}$. *From now onwards, instead of the pair* $(\mathfrak{g}^{\mathbb{C}}, \mathfrak{k}^{\mathbb{C}})$, *we will only consider the isomorphic pair* $(\tilde{\mathfrak{g}}, \Delta(\mathfrak{g}))$ (*under* φ).

(1.2) **Definition.** Let $\mathfrak{g}_1$ be a complex Lie algebra with a complex reductive subalgebra $\mathfrak{k}_1$. A $\mathfrak{g}_1$-module (in a complex vector space) M is called a $(\mathfrak{g}_1, \mathfrak{k}_1)$-*module* (also called *Harish-Chandra module*) if it is locally $\mathfrak{k}_1$-finite and is semi-simple as a $\mathfrak{k}_1$-module. It is called an *admissible* $(\mathfrak{g}_1, \mathfrak{k}_1)$-*module* if all the isotypical components of M (under $\mathfrak{k}_1$) are finite dimensional. If the $(\mathfrak{g}_1, \mathfrak{k}_1)$- module M is irreducible as a $\mathfrak{g}_1$-module, it is called an *irreducible Harish-Chandra module* (for the pair $(\mathfrak{g}_1, \mathfrak{k}_1)$).

Since the centre of the universal enveloping algebra $U(\tilde{\mathfrak{g}})$ can canonically be identified with $Z(\mathfrak{g}) \times Z(\mathfrak{g})$ (where $Z(\mathfrak{g})$ is the centre of $U(\mathfrak{g})$), the *infinitesimal character* of (say) an irreducible $\tilde{\mathfrak{g}}$-module is given by an element $(\lambda, \mu) \in \mathfrak{h}^* \times \mathfrak{h}^*$, and moreover λ and μ can be assumed to be dominant. We follow the standard convention that the trivial (one dimensional) $\tilde{\mathfrak{g}}$-module has infinitesimal character (ρ, ρ).

(1.3) **Definitions.** We denote by $\widetilde{G/B}$ the product flag variety $G/B \times G/B$. The group G acts on $\widetilde{G/B}$ diagonally. For any $w \in W$, we define the *Schubert variety* $X_w \subset G/B$ (resp. the *G-Schubert variety* $\tilde{X}_w \subset \widetilde{G/B}$) as the closure of the B-orbit $\mathcal{B}_w := BwB/B \subset G/B$ (resp. the closure of the G-orbit $\tilde{\mathcal{B}}_w := G(e,w) \subset \widetilde{G/B}$). As is easy to see $\{X_w\}_{w \in W}$ (resp. $\{\tilde{X}_w\}_{w \in W}$) are precisely the B-orbit closures in G/B (resp. G-orbit closures in $\widetilde{G/B}$). We also set $\partial X_w := X_w \setminus \mathcal{B}_w$ (resp. $\partial \tilde{X}_w := \tilde{X}_w \setminus \tilde{\mathcal{B}}_w$) and $Y_w := G/B \setminus \partial X_w$ (resp. $\tilde{Y}_w := \widetilde{G/B} \setminus \partial \tilde{X}_w$). It is easy to see that ∂X_w (resp. $\partial \tilde{X}_w$) is closed in G/B (resp. $\widetilde{G/B}$).

For any $\lambda \in \mathfrak{h}^*_{\mathbb{Z}}$ there is defined a *line bundle* $\mathcal{L}(\lambda)$ on G/B; which is associated to the principal B-bundle: $G \to G/B$ by the 1-dimensional representation $\mathbb{C}_{-\lambda}$ (determined by the character $e^{-\lambda}$ of B). More generally, given an algebraic B-module M (cf. Definition 2.9), we can consider the corresponding *vector bundle* $\mathcal{L}(M) := G \times_B M$ over G/B. For any $\lambda, \mu \in \mathfrak{h}^*_{\mathbb{Z}}$, we define the line bundle $\mathcal{L}(\lambda \otimes \mu)$ on $\widetilde{G/B}$ as the external tensor product of the line bundles $\mathcal{L}(\lambda)$ and $\mathcal{L}(\mu)$ respectively. The restriction of $\mathcal{L}(\lambda)$ to X_w (resp. $\mathcal{L}(\lambda \otimes \mu)$ to $\tilde{X}_w$) is denoted by $\mathcal{L}_w(\lambda)$ (resp. $\mathcal{L}_w(\lambda \otimes \mu)$).

For any topological space X, closed subspaces $Z \subseteq Y \subseteq X$, and an abelian sheaf $\mathcal{S}$ on X, $H^*_{Y/Z}(X, \mathcal{S})$ (resp. $\mathcal{H}^*_{Y/Z}(X, \mathcal{S})$) denotes the *local cohomology* (resp. *local cohomology sheaf*) introduced by Grothendieck ([H_1; page 219, variation 2]). If Z is the empty set ϕ, $H^*_{Y/Z}(X, \mathcal{S})$ (resp. $\mathcal{H}^*_{Y/Z}(X, \mathcal{S})$) is abbreviated to $H^*_Y(X, \mathcal{S})$ (resp. $\mathcal{H}^*_Y(X, \mathcal{S})$). The cases of our interest will be when X is an algebraic variety over $\mathbb{C}$ with the Zariski topology and $\mathcal{S}$ is an $\mathcal{O}_X$-module (where $\mathcal{O}_X$ denotes the structure sheaf of X).

For a smooth algebraic variety X over $\mathbb{C}$, $\mathcal{D}_X$ denotes the sheaf of algebraic differential operators on X. A $\mathcal{D}_X$*-module* is, by definition, a sheaf $\mathcal{S}$ of left $\mathcal{D}_X$-modules, which is quasi-coherent as an $\mathcal{O}_X$-module.

We recall the following algebraic analogue of a result of Brylinski-Kashiwara :

(1.4) **Proposition** [BK; Proposition 8.5]. *Let Y be a closed subvariety,* *of pure codimension ℓ, of a smooth algebraic variety X, and let $Z \subset Y$ be a nowhere dense closed subvariety of Y which contains the singular locus of Y. Then there exists a unique holonomic $\mathcal{D}_X$-module with regular singularities* (cf. [BK;§ 1]) $\mathcal{F} = \mathcal{F}(Y, X)$ *($\mathcal{F}$ does not depend upon the choice of Z) satisfying :*

$$(P_1) \qquad \mathcal{F}\mid_{X\setminus Z} \approx \mathcal{H}^{\ell}_{Y\setminus Z}(X\setminus Z, \mathcal{O}_{X\setminus Z})$$

and

$$(P_2) \qquad \mathcal{H}^0_Z(X, \mathcal{F}) = \mathcal{H}^0_Z(X, \mathcal{F}^*) = 0,$$

where

$$\mathcal{F}^* := \mathcal{H}om_{\mathcal{O}_X}(\Omega_X, \mathcal{E}xt^{\dim_{\mathbb{C}} X}_{\mathcal{D}_X}(\mathcal{F}, \mathcal{D}_X)),$$

and Ω_X is the canonical bundle of X.

We also recall the following two results from local cohomology, for their use in Section (2).

(1.5) **Lemma** [K; § 11]. (a) *Let K be an affine algebraic group over $\mathbb{C}$ with Lie algebra $\mathfrak{k}$, let X be a K-variety over $\mathbb{C}$, and let $\mathcal{S}$ be a quasi-coherent K-module on X (also called K-linearized $\mathcal{O}_X$-module). Then, for any closed subspaces $Y \supseteq Z$ of X, the local cohomology $H^p_{Y/Z}(X, \mathcal{S})$ admits a natural $\mathfrak{k}$-module structure, which is functorial in the following sense:*

Let X' be another K-variety over $\mathbb{C}$ with a quasi-coherent K-module $\mathcal{S}'$ on X', a K-morphism $f : X' \to X$, and a K-equivariant sheaf morphism $\hat{f} : f^*(\mathcal{S}) \to \mathcal{S}'$. Then, for any closed subspaces $Y' \supseteq Z'$ of X' such that $Y' \supseteq f^{-1}(Y)$ and $Z' \supseteq f^{-1}(Z)$, the induced map $H^p_{Y/Z}(X, \mathcal{S}) \to H^p_{Y'/Z'}(X', \mathcal{S}')$ (cf. [K; Lemma 11.3]) *is a $\mathfrak{k}$-module map.*

(b) *If we assume, in addition, (in the first paragraph of a) that Y and Z are both K-stable, then the $\mathfrak{k}$-module structure on $H^p_{Y/Z}(X, \mathcal{S})$ "integrates" to give a canonical K-module structure.*

(1.6) **Lemma.** *Let $\mathbf{A}^d$ be the affine space of dimension d over a field k. Then :*

(a) $H^p_{\{0\}}(\mathbf{A}^d, \mathcal{O}_{\mathbf{A}^d}) = 0$, *for all $p \neq d$, and*

(b) $H^d_{\{0\}}(\mathbf{A}^d, \mathcal{O}_{\mathbf{A}^d})$ *is canonically isomorphic with* $\sum_{n_1,\dots,n_d<0} k x_1^{n_1}\cdots x_d^{n_d}$,

as k-vector spaces; where 0 is the origin of $\mathbf{A}^d$, and $(x_1, \dots, x_d)$ are the coordinate functions on $\mathbf{A}^d$.

2. Formulation of the main result and its proof

(2.1). *In this whole section we fix once and for all* $\lambda, \mu \in D - \rho$ (cf. § 1.1), *and* $w \in W$. *Put* $\ell = \ell(w_0) - \ell(w)$, *where* w_0 *is the longest element of* W. We set

$$\begin{aligned} \mathcal{F}_w &= \mathcal{F}(X_w, G/B) \\ \widetilde{\mathcal{F}}_w &= \mathcal{F}(\widetilde{X}_w, \widetilde{G/B}) \\ \mathcal{F}_w(\lambda) &= \mathcal{F}_w \otimes_{\mathcal{O}_{G/B}} \mathcal{L}(\lambda) \\ \widetilde{\mathcal{F}}_w(\lambda \otimes \mu) &= \widetilde{\mathcal{F}}_w \otimes_{\mathcal{O}_{\widetilde{G/B}}} \mathcal{L}(\lambda \otimes \mu) \end{aligned}$$

where $\mathcal{F}(,)$ is as defined in Proposition (1.4). Since X_w is B-stable (resp. $\widetilde{X}_w$ is G-stable, under the diagonal G-action) and the line bundle $\mathcal{L}(\lambda)$ is B-equivariant (resp. the line bundle $\mathcal{L}(\lambda \otimes \mu)$ is G-equivariant), by the uniqueness of $\mathcal{F}$, we obtain that $\mathcal{F}_w$ is a quasi-coherent B-module (resp. $\widetilde{\mathcal{F}}_w$ is a quasi-coherent G-module).

Now we recall the following fundamental result due to Beilinson and Bernstein. (Even though we do not need, a more general result is proved by them.)

(2.2) **Theorem** [BB]. *The map* $w \mapsto H^0(\widetilde{G/B}, \widetilde{\mathcal{F}}_w(\lambda \otimes \mu))$ *sets up a bijective correspondence from* $W'_{\lambda+\rho,\mu+\rho}$ *to the set of isomorphism classes of irreducible* $(\tilde{\mathfrak{g}}, \Delta(\mathfrak{g}))$*-modules with infinitesimal character* $(\lambda + \rho, \mu + \rho)$; *where* $W'_{\lambda+\rho,\mu+\rho} := \{w \in W : w$ *is the (unique) element of minimal length in its double coset* $W_{\lambda+\rho} w W_{\mu+\rho}\}$, *and* $W_{\lambda+\rho}$ *is as defined in* § 1.1.

If $w \notin W'_{\lambda+\rho,\mu+\rho}$, *then* $H^0(\widetilde{G/B}, \widetilde{\mathcal{F}}_w(\lambda \otimes \mu)) = 0$.

As a preparation to prove (or even to formulate) our main result, we prove the following lemmas.

(2.3) **Lemma.** *The canonical restriction map*

$$H^0(\widetilde{G/B}, \widetilde{\mathcal{F}}_w(\lambda \otimes \mu)) \to H^0(\widetilde{Y}_w, \widetilde{\mathcal{F}}_w(\lambda \otimes \mu))$$

is injective, where $\widetilde{Y}_w$ *is as defined in* § 1.3.

Proof. From the long exact sequence for the local cohomology (cf. [H_2; Chap. III, Exercise 2.3]), it suffices to prove that $H^0_{\partial \widetilde{X}_w}(\widetilde{G/B}, \widetilde{\mathcal{F}}_w(\lambda \otimes \mu)) = 0$: By the defining property (P_2) of $\widetilde{\mathcal{F}}_w$ (cf. Proposition 1.4), the sheaf $\mathcal{H}^0_{\partial \widetilde{X}_w}(\widetilde{G/B}, \widetilde{\mathcal{F}}_w(\lambda \otimes \mu)) = 0$. In particular, by [G; page 5, Proposition 1.4], $H^0_{\partial \widetilde{X}_w}(\widetilde{G/B}, \widetilde{\mathcal{F}}_w(\lambda \otimes \mu)) = 0$. □

(2.4) **Lemma.** *There is a canonical isomorphism*

$$\theta_w : H^{\ell}_{\widetilde{X}_w/\partial\widetilde{X}_w}(\widetilde{G/B}, \mathcal{L}(\lambda \otimes \mu)) \to H^0(\widetilde{Y}_w, \widetilde{\mathcal{F}}_w(\lambda \otimes \mu)).$$

Proof. By the defining property (P_1) (cf. Proposition 1.4), the sheaf $\widetilde{\mathcal{F}}_w(\lambda \otimes \mu) \mid_{\widetilde{Y}_w}$ is the local cohomology sheaf $\mathcal{H}^{\ell}_{\widetilde{\mathcal{B}}_w}(\widetilde{Y}_w, \mathcal{L}(\lambda \otimes \mu))$. Further $\mathcal{H}^{i}_{\widetilde{\mathcal{B}}_w}(\widetilde{Y}_w, \mathcal{L}(\lambda \otimes \mu)) = 0$ for all $i \neq \ell$, since $\widetilde{\mathcal{B}}_w$ is a smooth closed subvariety of $\widetilde{Y}_w$ of codimension ℓ. Now the lemma follows from [G; page 5, Proposition 1.4] together with [K; Lemma 7.7]. □

(2.5) **Remark.** Exactly the same proof as above gives an isomorphism: $H^{\ell}_{X_w/\partial X_w}(G/B, \mathcal{L}(\mu)) \xrightarrow{\sim} H^0(Y_w, \mathcal{F}_w(\mu))$, where Y_w is as defined in § 1.3. Similarly the restriction map : $H^0(G/B, \mathcal{F}_w(\mu)) \to H^0(Y_w, \mathcal{F}_w(\mu))$ is injective (cf. Lemma 2.3).

(2.6) **Lemma.** $H^{\ell}_{\widetilde{X}_w/\partial\widetilde{X}_w}(\widetilde{G/B}, \mathcal{L}(\lambda \otimes \mu))$ *is canonically isomorphic with*

$$H^0(G/B, \mathcal{L}(\lambda) \otimes \mathcal{L}(H^{\ell}_{X_w/\partial X_w}(G/B, \mathcal{L}(\mu)))),$$

where $\mathcal{L}()$ *is as in* § 1.3 *and* $X_w, \partial X_w$ *being* B*-stable,* $H^{\ell}_{X_w/\partial X_w}(G/B, \mathcal{L}(\mu))$ *has a canonical* $\mathfrak{g}$*-module structure which restricted to* $\mathfrak{b}$ *integrates to give a* B*-module structure* (cf. Lemma 1.5).

Proof. By the spectral sequence [K; Lemma 8.5(d)], connecting the local cohomology sheaves to local cohomology groups, we get :

$$H^{\ell}_{\widetilde{X}_w/\partial\widetilde{X}_w}(\widetilde{G/B}, \mathcal{L}(\lambda \otimes \mu)) \approx H^0(\widetilde{G/B}, \mathcal{S}),$$

where $\mathcal{S}$ is the local cohomology sheaf $\mathcal{H}^{\ell}_{\widetilde{X}_w/\partial\widetilde{X}_w}(\widetilde{G/B}, \mathcal{L}(\lambda \otimes \mu))$. (The spectral sequence degenerates because $\mathcal{H}^{i}_{\widetilde{X}_w/\partial\widetilde{X}_w}(\widetilde{G/B}, \mathcal{L}(\lambda \otimes \mu)) = 0$, for all $i \neq \ell$; see the proof of Lemma 2.4.)

Further by the definition of the direct image sheaf, applied to the projection on the first factor $\pi_1 : \widetilde{G/B} \to G/B$, we get

$$H^0(\widetilde{G/B}, \mathcal{S}) \approx H^0(G/B, \pi_{1*}(\mathcal{S})).$$

We next assert that the direct image sheaf $\pi_{1*}(\mathcal{S})$ on G/B is isomorphic with $\mathcal{L}(\lambda) \otimes \mathcal{L}(H^{\ell}_{X_w/\partial X_w}(G/B, \mathcal{L}(\mu)))$:

First of all, the sheaf $\pi_{1*}(\mathcal{S})$ is a G-linearized sheaf of $\mathcal{O}_{G/B}$-modules. This is clear because the map π_1 is G-equivariant (under the diagonal action of G on $\widetilde{G/B}$), $\widetilde{X}_w, \partial\widetilde{X}_w$ are G-stable, and $\mathcal{L}(\lambda \otimes \mu)$ is a G-equivariant line bundle. Let us now compute the stalk of $\pi_{1*}(\mathcal{S})$ at the base point $\underline{e} \in G/B$: Consider the affine open subset $U^-\underline{e} \subset G/B$, where U^- is the unipotent subgroup of G with Lie algebra $\oplus_{\alpha\in\Delta_-}\mathfrak{g}_\alpha$, where Δ_+ is the set of roots for $\mathfrak{b}, \Delta_- := -\Delta_+$, and $\mathfrak{g}_\alpha$ is the root space corresponding to the root α. Define a map $m : \pi_1^{-1}(U^-\underline{e}) = U^-\underline{e} \times G/B \to G/B$ by $m(g\underline{e}, x) = g^{-1}x$, for $g \in U^-$ and $x \in G/B$. Then m is an affine morphism. Also, as is easy to see, $m^{-1}(X_w) = \tilde{X}_w \cap \pi_1^{-1}(U^-\underline{e})$ and $m^{-1}(\partial X_w) = \partial\tilde{X}_w \cap \pi_1^{-1}(U^-\underline{e})$. In particular, by the spectral sequence [G; Proposition 5.5 and Corollary 5.6] together with [H_2; Chapter III, Exercise 8.2], we get $H^\ell_{\pi_1^{-1}(U^-\underline{e})\cap\tilde{X}_w/\pi_1^{-1}(U^-\underline{e})\cap\partial\tilde{X}_w}(\pi_1^{-1}(U^-\underline{e}), \mathcal{L}(\lambda\otimes\mu)) \approx H^\ell_{X_w/\partial X_w}(G/B, m_*\mathcal{L}(\lambda\otimes\mu))$. From this it is not difficult to deduce the assertion that $\pi_{1*}(\mathcal{S}) \approx \mathcal{L}(\lambda) \otimes \mathcal{L}(H^\ell_{X_w/\partial X_w}(G/B, \mathcal{L}(\mu)))$, and hence the lemma is proved. □

(2.7) **Definition.** The Lie algebra $\mathfrak{g}$ admits a unique complex linear involution τ such that $\tau\,|_{\mathfrak{h}} = -1$ and it sends the α-th root space $\mathfrak{g}_\alpha$ to $\mathfrak{g}_{-\alpha}$ for any root α. Given a $\mathfrak{g}$-module M, we get another $\mathfrak{g}$-module structure on M by twisting the original $\mathfrak{g}$-module structure by τ. We denote the twisted $\mathfrak{g}$-module by M^τ.

Let $\tilde{\mathcal{O}}$ be the category of finitely generated $U(\mathfrak{g})$-modules, which are locally finite as $U(\mathfrak{b})$-modules. Any $N \in \tilde{\mathcal{O}}$ satisfies $N = \oplus_{\lambda\in\mathfrak{h}^*}N_\lambda$, where N_λ is the λ-th generalized weight space. Set $N^\vee = \{f \in \mathrm{Hom}_{\mathbb{C}}(N, \mathbb{C}) : f(N_\lambda) = 0$, for all but finitely many $\lambda\}$. Then $N^\vee$ has a canonical $\mathfrak{g}$-module structure. Finally we set $N^\sigma := (N^\vee)^\tau$. It is easy to see that $N^\sigma \in \tilde{\mathcal{O}}$ and moreover $ch(N) = ch(N^\sigma)$, where $ch(N) := \sum(\dim N_\lambda)e^\lambda$ is the formal character of N.

The following lemma is well known (see,e.g., [BK; § 5]), but we recall the proof as it will be used in the proof of Lemma (2.14).

(2.8) **Lemma.** *$H^\ell_{X_w/\partial X_w}(G/B, \mathcal{L}(\mu)) \approx M(\mu_w)^\sigma$, as $\mathfrak{g}$-modules, where $\mu_w := -w(\mu+\rho) - \rho$.*

Proof. Consider the T-equivariant biregular isomorphism (cf. [KL; § 1.4]) $\xi = \xi_w : U_w \times U'_w \xrightarrow{\sim} wU^-B/B$, given by $(g,h) \mapsto ghwB$ (for $g \in U_w$ and $h \in U'_w$); where U_w (resp. U'_w) is the unipotent subgroup of G with Lie algebra $\oplus_{\alpha\in\Delta_+\cap w\Delta_-}\mathfrak{g}_\alpha$(resp. $\oplus_{\alpha\in\Delta_-\cap w\Delta_-}\mathfrak{g}_\alpha$), and T acts by conjugation on U_w and U'_w.

As can be easily seen, there is a nowhere vanishing section s of the line bundle $\mathcal{L}(\mu)\,|_{wU^-B/B}$, which transforms under the canonical T-action via the weight $-w\mu$. Further $\xi(U_w \times e) = \mathcal{B}_w$ and $\mathcal{B}_w$ is closed in the open subset $\xi(U_w \times U'_w)$ of G/B. Hence by [K; Lemmas 7.7 and 7.9],

$$\begin{aligned} H^\ell_{X_w/\partial X_w}(G/B, \mathcal{L}(\mu)) &\approx H^\ell_{\mathcal{B}_w}(Y_w, \mathcal{L}(\mu)) \\ &\approx H^\ell_{U_w \times e}(U_w \times U'_w, \mathcal{O}_{U_w \times U'_w}) \otimes s \end{aligned}$$

$$(I_1)\ldots\qquad H^\ell_{X_w/\partial X_w}(G/B, \mathcal{L}(\mu)) \approx H^\ell_{\{e\}}(U'_w, \mathcal{O}_{U'_w}) \otimes \mathbb{C}\,[U_w] \otimes s,$$

by [G; Proposition 5.5],

where $\mathbb{C}\,[U_w]$ is the ring of regular functions on U_w. So, by Lemma (1.6),

$$\begin{aligned} ch\, H^\ell_{X_w/\partial X_w}(G/B, \mathcal{L}(\mu)) &= ch\, H^\ell_{\{e\}}(U'_w, \mathcal{O}_{U'_w}) \cdot ch\, \mathbb{C}\,[U_w] \cdot e^{-w\mu} \\ &= e^{-\sum_{\alpha \in \Delta_+ \cap w\Delta_+} \alpha} \cdot \Big(\prod_{\alpha \in \Delta_+ \cap w\Delta_+} (1 - e^{-\alpha})^{-1} \Big) \\ &\quad \cdot \Big(\prod_{\beta \in \Delta_+ \cap w\Delta_-} (1 - e^{-\beta})^{-1} \Big) \cdot e^{-w\mu} \\ &= e^{\mu_w} \cdot \prod_{\alpha \in \Delta_+} (1 - e^{-\alpha})^{-1} \\ &= ch\, M(\mu_w) \\ &= ch\, (M(\mu_w)^\sigma), \quad \text{(cf. §2.7)}. \end{aligned}$$

So both the modules of the lemma have the same character. From this it is not difficult to establish that they are isomorphic as $\mathfrak{g}$-modules (cf. [BK; § 5] or [Ku$_2$; § 3]). □

(2.9) **Definition.** A B-module M is called *algebraic* if the action of B on M is locally finite and any finite dimensional B-submodule of M is an algebraic B-module.

The following result can easily be deduced from Peter-Weyl theorem. (In fact a more general result is proved by Bott [B; Theorem I].)

(2.10) **Proposition.** *Let M be an algebraic B-module. Then $H^0(G/B, \mathcal{L}(M))$ is G-module isomorphic with $\oplus_{\theta \in D}(V(\theta)^* \otimes_{\mathbb{C}} [M \otimes V(\theta)]^B)$, where we put the trivial G-module structure on the space of B-invariants $[M \otimes V(\theta)]^B$, $V(\theta)$ is the irreducible G-module with highest weight θ and $V(\theta)^*$ is its dual.*

(2.11) **Corollary.** *As G-modules*

$$H^\ell_{\tilde{X}_w/\partial\tilde{X}_w}(\widetilde{G/B}, \mathcal{L}(\lambda \otimes \mu)) \approx V(\overline{\mu_w - \lambda}) \oplus \Big(\bigoplus_{\substack{\theta \in D \\ \|\theta\| > \|\lambda - \mu_w\|}} V(\theta)^* \otimes [V(\theta)]_{\lambda - \mu_w} \Big),$$

where μ_w is as in Lemma (2.8), $[V(\theta)]_{\lambda-\mu_w}$ denotes the $(\lambda - \mu_w)$-th weight space in $V(\theta)$ and, for any $\chi \in \mathfrak{h}^*$, $\overline{\chi}$ denotes the unique dominant element *in the* W*-orbit of* χ.

Proof. Combining Lemmas (2.6) and (2.8) we get :

$$\begin{aligned}
H^{\ell}_{\tilde{X}_w/\partial\tilde{X}_w}(\widetilde{G/B},\mathcal{L}(\lambda\otimes\mu)) &\approx H^0(G/B,\mathcal{L}(\lambda)\otimes\mathcal{L}(M(\mu_w)^\sigma))\\
&\approx \oplus_{\theta\in D}(V(\theta)^*\otimes[V(\theta)\otimes\mathbb{C}_{-\lambda}\otimes M(\mu_w)^\sigma]^B),\\
&\qquad \text{by Proposition (2.10)}\\
&\approx \oplus_{\theta\in D}(V(\theta)^*\otimes[V(\theta)\otimes M(\mu_w-\lambda)^\sigma]^B)\\
&\approx \oplus_{\theta\in D}(V(\theta)^*\otimes \operatorname{Hom}_{\mathfrak{b}}(M(\mu_w-\lambda)^\tau,V(\theta)))\\
&\qquad \text{(cf. Definition 2.7)}\\
&\approx \oplus_{\theta\in D}(V(\theta)^*\otimes[V(\theta)]_{\lambda-\mu_w}),\\
&\qquad \text{since } M(\mu_w-\lambda)^\tau \text{ is } U(\mathfrak{n})\text{-free.}
\end{aligned}$$

We next observe that if any $\chi\in\mathfrak{h}^*$ occurs as a weight in $V(\theta)$, then $\|\chi\|\le\|\theta\|$ and equality occurs if and only if $\overline{\chi}=\theta$:

We can assume, without loss of generality, that χ is dominant. Write $\theta=\chi+\beta$ for some $\beta\in\sum_{i=1}^{\ell}\mathbb{Z}_+\alpha_i$, where $\mathbb{Z}_+$ is the set of non-negative integers. Then $\|\theta\|^2=\|\chi\|^2+\|\beta\|^2+2<\chi,\beta>$. In particular, $\|\chi\|\le\|\theta\|$ and equality occurs if and only if $\beta=0$. This proves the assertion and hence the corollary. □

The following is an immediate consequence of the above corollary.

(2.12) **Corollary.** *For any* $w\in W$, $H^{\ell}_{\tilde{X}_w/\partial\tilde{X}_w}(\widetilde{G/B},\mathcal{L}(-\rho\otimes-\rho))$ *has a unique (up to scalar multiples)* G*-invariant, where* ℓ *is as in* § 2.1.

(2.13) **The basic map.** For any $w\in W$ and $\lambda,\mu\in D-\rho$, there is defined a canonical Kunneth map (got by taking the tensor product)

$$\begin{aligned}
\psi_w=\psi_w^{\lambda,\mu}: H^0(\widetilde{G/B},\mathcal{L}(\lambda+\rho\otimes\mu+\rho))\otimes H^{\ell}_{\tilde{X}_w/\partial\tilde{X}_w}(\widetilde{G/B},\mathcal{L}(-\rho\otimes-\rho))\\
\to H^{\ell}_{\tilde{X}_w/\partial\tilde{X}_w}(\widetilde{G/B},\mathcal{L}(\lambda\otimes\mu)),
\end{aligned}$$

where ℓ is as in § 2.1. (Observe that $H^{p}_{\tilde{X}_w/\partial\tilde{X}_w}(\widetilde{G/B},\mathcal{L}(\lambda\otimes\mu))=0$, for all $p\neq\ell$.)

By naturality, the map ψ_w is a $\tilde{\mathfrak{g}}$-module map, where we put the tensor product $\tilde{\mathfrak{g}}$-module structure on the domain (cf. Lemma 1.5). By the above corollary, $H^{\ell}_{\widetilde{X}_w/\partial\widetilde{X}_w}(\widetilde{G/B}, \mathcal{L}(-\rho \otimes -\rho))$ contains a unique G-invariant ϑ. Hence by restricting ψ_w, (since ϑ is G-invariant) we get a G-module map

$$\psi_w^{\vartheta} : H^0(\widetilde{G/B}, \mathcal{L}(\lambda + \rho \otimes \mu + \rho)) \to H^{\ell}_{\widetilde{X}_w/\partial\widetilde{X}_w}(\widetilde{G/B}, \mathcal{L}(\lambda \otimes \mu)),$$

given by $\psi_w^{\vartheta}(x) = \psi_w(x \otimes \vartheta)$.

Now we have the following crucial :

(2.14) **Lemma.** *The map ψ_w^{ϑ} factors through $H^0(\widetilde{X}_w, \mathcal{L}_w)$, i.e., there exists a map $\overline{\psi}_w^{\vartheta} : H^0(\widetilde{X}_w, \mathcal{L}_w) \to H^{\ell}_{\widetilde{X}_w/\partial\widetilde{X}_w}(\widetilde{G/B}, \mathcal{L}(\lambda \otimes \mu))$ making the following diagram commutative:*

$$\begin{array}{ccc} H^0(\widetilde{G/B}, \mathcal{L}) & \xrightarrow{\psi_w^{\vartheta}} & H^{\ell}_{\widetilde{X}_w/\partial\widetilde{X}_w}(\widetilde{G/B}, \mathcal{L}(\lambda \otimes \mu)) \\ r_w \searrow & & \nearrow \overline{\psi}_w^{\vartheta} \\ & H^0(\widetilde{X}_w, \mathcal{L}_w) & \end{array}$$

where r_w is the canonical restriction, and $\mathcal{L} := \mathcal{L}(\lambda + \rho \otimes \mu + \rho)$ ($\mathcal{L}_w$ has a similar meaning).

Proof. From the naturality of the Kunneth map, we get that the following diagram $(\mathcal{D})$ is commutative :

$$\begin{array}{ccc} H^0(\widetilde{G/B}, \mathcal{L}) \otimes H^{\ell}_{\widetilde{X}_w/\partial\widetilde{X}_w}(\widetilde{G/B}, \mathcal{L}(-\rho \otimes -\rho)) \to & H^{\ell}_{\widetilde{X}_w/\partial\widetilde{X}_w}(\widetilde{G/B}, \mathcal{L}') \\ \downarrow\wr & \downarrow\wr \\ H^0(\widetilde{G/B}, \mathcal{L}) \otimes H^0(\widetilde{Y}_w, \widetilde{\mathcal{F}}_w(-\rho \otimes -\rho)) \to & H^0(\widetilde{Y}_w, \widetilde{\mathcal{F}}_w(\lambda \otimes \mu)) \end{array}$$

where $\mathcal{L}' := \mathcal{L}(\lambda \otimes \mu)$, and $\widetilde{Y}_w$ is as defined in § 1.3 and the vertical isomorphisms are induced by the isomorphism of Lemma (2.4).

Define a subsheaf $\mathcal{K}_w = \{x \in \mathcal{F}_w : \mathcal{I}_{X_w} x = 0\}$ (resp. $\widetilde{\mathcal{K}}_w = \{x \in \widetilde{\mathcal{F}}_w : \mathcal{I}_{\widetilde{X}_w} x = 0\}$), where $\mathcal{I}_{X_w}$ (resp. $\mathcal{I}_{\widetilde{X}_w}$) denotes the ideal sheaf of X_w in G/B (resp. of $\widetilde{X}_w$ in $\widetilde{G/B}$). Set $\mathcal{K}_w(-\rho) = \mathcal{K}_w \otimes_{\mathcal{O}_{G/B}} \mathcal{L}(-\rho)$ and $\widetilde{\mathcal{K}}_w(-\rho \otimes -\rho) = \widetilde{\mathcal{K}}_w \otimes_{\mathcal{O}_{\widetilde{G/B}}} \mathcal{L}(-\rho \otimes -\rho)$.

By the very definition, $\psi_w(Q_w \otimes H^0(\widetilde{Y}_w, \widetilde{\mathcal{K}}_w(-\rho \otimes -\rho))) = 0$, where Q_w is the kernel of the restriction map r_w. But, by Kumar [Ku1; Theorem 1.5], the map r_w is surjective and hence, to prove the lemma, it suffices to show that $\vartheta \in H^0(\widetilde{Y}_w, \widetilde{\mathcal{K}}_w(-\rho \otimes -\rho))$:

We first observe that

$$(I_2) \cdots H^0(\widetilde{Y}_w, \widetilde{\mathcal{K}}_w(-\rho \otimes -\rho)) \approx H^0(G/B, \mathcal{L}(-\rho) \otimes \mathcal{L}(H^0(Y_w, \mathcal{K}_w(-\rho)))),$$

where Y_w is as defined in § 1.3. By Remark (2.5),

$$H^\ell_{X_w/\partial X_w}(G/B, \mathcal{L}(-\rho)) \approx H^0(Y_w, \mathcal{F}_w(-\rho)).$$

Further by (I_1) (cf. proof of Lemma 2.8)

$$H^0(Y_w, \mathcal{K}_w(-\rho)) \approx \{x \in H^\ell_{\{e\}}(U'_w, \mathcal{O}_{U'_w}) : fx = 0, \text{ for all } f \in \mathbb{C}\,[U'_w] \text{ with } f(e) = 0\} \otimes \mathbb{C}\,[U_w] \otimes s.$$

Hence

$$(I_3)\cdots \qquad H^0(Y_w, \mathcal{K}_w(-\rho)) \approx (x_1^{-1}\cdots x_\ell^{-1}) \otimes \mathbb{C}\,[U_w] \otimes s,$$

by Lemma (1.6), where $\{x_1, ..., x_\ell\}$ are the coordinate functions on $U'_w \approx$ Lie U'_w (Lie U'_w denotes the Lie algebra of U'_w). In particular, $H^0(Y_w, \mathcal{K}_w(-\rho)) \neq 0$. Now $H^0(Y_w, \mathcal{K}_w(-\rho))$ is a B-stable subspace of $H^0(Y_w, \mathcal{F}_w(-\rho)) \approx M(-\rho)^\sigma$ (cf. Remark 2.5 and Lemma 2.8). As is easy to see, any B-stable non-zero subspace of $M(\lambda)^\sigma$ (for any $\lambda \in \mathfrak{h}^*$) contains the λ-th weight space. So $H^0(Y_w, \mathcal{K}_w(-\rho))$ contains the $(-\rho)$-th weight space. (This can also be obtained from I_3 .) This proves, by (I_2) and Proposition (2.10), that $\vartheta \in H^0(\widetilde{Y}_w, \widetilde{\mathcal{K}}_w(-\rho \otimes -\rho))$; thus proving the lemma. □

(2.15) **Lemma.** *The map*

$$\overline{\psi}^{\vartheta}_w : H^0(\widetilde{X}_w, \mathcal{L}_w(\lambda + \rho \otimes \mu + \rho)) \to H^\ell_{\widetilde{X}_w/\partial\widetilde{X}_w}(\widetilde{G/B}, \mathcal{L}(\lambda \otimes \mu))$$

(defined in the above lemma) is injective.

Proof. The sheaf $\widetilde{\mathcal{K}}_w\,|_{\widetilde{Y}_w}$ is supported in the G-orbit $\widetilde{\mathcal{B}}_w$ (cf. Definition 1.3) and moreover (by definition) it is a sheaf of $\mathcal{O}_{\widetilde{\mathcal{B}}_w}$-modules. Since the section $\vartheta \in H^0(\widetilde{Y}_w, \widetilde{\mathcal{K}}_w(-\rho \otimes -\rho))$ is G-invariant, $\vartheta(x) \neq 0$ (as an element of the stalk $\widetilde{\mathcal{K}}_w(-\rho \otimes -\rho)_x$) for any $x \in \widetilde{\mathcal{B}}_w$. Now take any $t \neq 0 \in H^0(\widetilde{X}_w, \mathcal{L}_w(\lambda + \rho \otimes \mu + \rho))$. Then there exists a $x_0 \in \widetilde{\mathcal{B}}_w$ such that $t(x_0) \neq 0$. But then, by the commutative diagram $(\mathcal{D})$ (of Lemma 2.14), $\overline{\psi}^{\vartheta}_w(t)(x_0) \neq 0$. In particular, $\overline{\psi}^{\vartheta}_w(t) \neq 0$. □

We recall the following result due to Kumar.

(2.16) **Theorem** [Ku$_1$; Theorem 2.10 and Proposition 2.9]. *The G-module $H^0(\widetilde{X}_w, \mathcal{L}_w(\lambda + \rho \otimes \mu + \rho))$ contains a unique copy of the irreducible G-module $V(\overline{\mu_w - \lambda})$; where μ_w is as in Lemma (2.8), and the bar is as in Corollary* (2.11).

Now we come to the main result of this paper :

(2.17) **Theorem.** *Let G be a semi-simple connected simply-connected complex algebraic group and fix $\lambda, \mu \in D - \rho$ (cf. § 1.1). Then, for any $w \in W'_{\lambda+\rho,\mu+\rho}$,*

$$\overrightarrow{\psi}^{\vartheta}_w(V(\overline{\mu_w - \lambda})) \subset H^0(\widetilde{G/B}, \widetilde{\mathcal{F}}_w(\lambda \otimes \mu)),$$

where $W'_{\lambda+\rho,\mu+\rho}$ is as in Theorem (2.2), μ_w is as in Lemma (2.8), and $\overrightarrow{\psi}^{\vartheta}_w$ is the G-module map defined in Lemma (2.14).

In particular, $\overrightarrow{\psi}^{\vartheta}_w(V(\overline{\mu_w - \lambda}))$ occurs with multiplicity exactly one in the irreducible Harish-Chandra module $N_w := H^0(\widetilde{G/B}, \widetilde{\mathcal{F}}_w(\lambda \otimes \mu))$ (cf. Theorem 2.2) and is its minimal $\Delta(\mathfrak{g})$-type.

(Recall that, by Lemmas (2.3) and (2.4), N_w canonically embeds inside $H^{\ell}_{\widetilde{X}_w/\partial\widetilde{X}_w}(\widetilde{G/B}, \mathcal{L}(\lambda\otimes\mu))$, and moreover the map $\overrightarrow{\psi}^{\vartheta}_w$ is injective by Lemma (2.15).)

Proof. By Corollary (2.11), any irreducible G-submodule $V(\theta)$ of N_w (in fact of $H^{\ell}_{\widetilde{X}_w/\partial\widetilde{X}_w}(\widetilde{G/B}, \mathcal{L}(\lambda \otimes \mu)))$ satisfies either $\| \theta \| > \| \lambda - \mu_w \|$ or $\theta = \overline{\mu_w - \lambda}$, and in the later case it occurs with multiplicity one in $H^{\ell}_{\widetilde{X}_w/\partial\widetilde{X}_w}(\widetilde{G/B}, \mathcal{L}(\lambda \otimes \mu))$. So the proof of the theorem will be completed, if we show that $V(\overline{\mu_w - \lambda})$ does occur as a component in N_w :

In view of Lemma (2.4) and the long exact local cohomology sequence [H_2; Chap. III, Exercise 2.3] (cf. proof of Lemma 2.3), it suffices to show that $H^1_{\partial\widetilde{X}_w}(\widetilde{G/B}, \widetilde{\mathcal{F}}_w(\lambda \otimes \mu))$ does not contain $V(\overline{\mu_w - \lambda})$ as a component; which is content of the next lemma. This completes the proof of the theorem (modulo the next lemma). □

(2.18) **Lemma.** *The irreducible G-module $V(\overline{\mu_w - \lambda})$ is not a component of $H^1_{\partial\widetilde{X}_w}(\widetilde{G/B}, \widetilde{\mathcal{F}}_w(\lambda \otimes \mu))$, for any $w \in W'_{\lambda+\rho,\mu+\rho}$.*

Proof. By the defining property (P_2) of the sheaf $\mathcal{F}_w$ (cf. Proposition 1.4), $\mathcal{H}^0_{\partial X_w}(G/B, \mathcal{F}_w(\mu)) = 0$. So, by an analogue of Lemma (2.6),

$$(I_4) \cdots H^1_{\partial\widetilde{X}_w}(\widetilde{G/B}, \widetilde{\mathcal{F}}_w(\lambda\otimes\mu)) \approx H^0(G/B, \mathcal{L}(\lambda)\otimes\mathcal{L}(H^1_{\partial X_w}(G/B, \mathcal{F}_w(\mu)))).$$

Consider the following exact sequence $(\mathcal{T})$:

$$H^0_{\partial X_w}(G/B, \mathcal{F}_w(\mu)) = 0 \to H^0(G/B, \mathcal{F}_w(\mu)) \to H^0(Y_w, \mathcal{F}_w(\mu))$$
$$\to H^1_{\partial X_w}(G/B, \mathcal{F}_w(\mu)) \to H^1(G/B, \mathcal{F}_w(\mu)) = 0,$$

where the vanishing of $H^1(G/B, \mathcal{F}_w(\mu))$ is due to [BB;§ 2]. Further, by [BB] (see also [Ka]), $H^0(G/B, \mathcal{F}_w(\mu))$ is the irreducible highest weight $\mathfrak{g}$-module $L(\mu_w)$ with highest weight μ_w (use the fact that w is of smallest length in its coset $wW_{\mu+\rho}$, since $w \in W'_{\lambda+\rho,\mu+\rho}$ by assumption). Hence, by combining Lemma (2.8) with Remark (2.5), we get (by the exact sequence $\mathcal{T}$)

$$H^1_{\partial X_w}(G/B, \mathcal{F}_w(\mu)) \approx M(\mu_w)^\sigma / L(\mu_w).$$

But then, by (I_4) and Proposition (2.10), we get

$$(I_5) \cdots H^1_{\partial \widetilde{X}_w}(\widetilde{G/B}, \widetilde{\mathcal{F}}_w(\lambda \otimes \mu)) \approx \oplus_{\theta \in D}(V(\theta)^* \otimes [\mathbb{C}_{-\lambda} \otimes K(\mu_w) \otimes V(\theta)]^B),$$

as G-modules, where $K(\mu_w) := M(\mu_w)^\sigma / L(\mu_w)$. So, to complete the proof of the lemma, we need to show that

$$\mathcal{C} := [\mathbb{C}_{-\lambda} \otimes K(\mu_w) \otimes V(\overline{\lambda - \mu_w})]^B = 0 :$$

As is easy to see

$$\mathcal{C} \approx \mathrm{Hom}_{\mathfrak{b}}(A^\tau \otimes \mathbb{C}_{\lambda}, V(\overline{\lambda - \mu_w})),$$

where A is the kernel of the map : $M(\mu_w) \to L(\mu_w)$. So

$$(I_6) \cdots \qquad \mathcal{C} \approx \mathrm{Hom}_{\mathfrak{b}^-}(A \otimes \mathbb{C}_{-\lambda}, V(\overline{\mu_w - \lambda})),$$

where $\mathfrak{b}^-$ is the opposite Borel subalgebra of $\mathfrak{g}$.

Next we claim that $\mu_{w'} - \lambda$ does not occur as a weight in $V(\overline{\mu_w - \lambda})$, for any $w' \in W$ such that

$$(I_7) \cdots \qquad \mu_{w'} = \mu_w - \beta, \text{ for some } \beta \neq 0 \in \textstyle\sum_{i=1}^{\ell} \mathbb{Z}_+ \alpha_i :$$

We first obtain

$$\| \mu_{w'} - \lambda \|^2 = \| \mu_w - \lambda \|^2 + 2 < \beta, \lambda + \rho > .$$

So if $\mu_{w'} - \lambda$ does occur as a weight in $V(\overline{\mu_w - \lambda})$, then

$$(I_8) \cdots \qquad < \beta, \lambda + \rho >= 0 \text{ (cf. proof of Corollary 2.11)}.$$

Rewriting (I_7) we get

$$(I_9) \cdots \qquad w^{-1}w'(\mu + \rho) - (\mu + \rho) = w^{-1}\beta.$$

But, by assumption, $w \in W'_{\lambda+\rho,\mu+\rho}$; in particular, $vw > w$ for any $v \in W_{\lambda+\rho}$. This, together with (I_8), gives that

$$(I_{10}) \cdots \qquad w^{-1}\beta \in \sum \mathbb{Z}_+ \alpha_i \text{ ,and of course } w^{-1}\beta \neq 0.$$

Further by (I_9)

$$(I_{11})\cdots \qquad -w^{-1}\beta \in \sum \mathbb{Z}_+\alpha_i.$$

Now (I_{10}) and (I_{11}) contradict each other, proving the assertion that $\mu_{w'}-\lambda$ does not occur as a weight in $V(\overline{\mu_w - \lambda})$. This proves the vanishing of C, by (I_6). □

References

[B] Bott, R. , Homogeneous vector bundles, Ann. Math. 66 (1957), 203-248.

[BB] Beilinson, A. and Bernstein, J. , Localisation de $\mathfrak{g}$-modules, C.R. Acad. Sc. Paris 292 (1981), 15-18.

[BK] Brylinski, J.L. and Kashiwara, M. , Kazhdan-Lusztig conjecture and holonomic systems, Invent. Math. 64 (1981), 387-410.

[D] Duflo, M. , Representations irreductibles des Groupes semi-simples complexes, Springer lecture notes in Mathematics no. 497 (1975), 26-88.

[G] Grothendieck, A. , *Local Cohomology*, Lecture notes in Mathematics no. 41, Springer-Verlag , Berlin-Heidelberg-New York, 1967.

[H_1] Hartshorne, R. , *Residues and Duality*, Lecture notes in Mathematics no. 20, Springer-Verlag, Berlin-Heidelberg-New York , 1966.

[H_2] Hartshorne, R. , *Algebraic Geometry*, Springer-Verlag , Berlin-Heidelberg-New York , 1977.

[Ka] Kashiwara, M. , Representation theory and D-modules on flag varieties, RIMS (Kyoto) Publications , 1988.

[K] Kempf, G. , The Grothendieck-Cousin complex of an induced representation, Adv. in Mathematics 29 (1978), 310-396.

[Ku_1] Kumar, S. , Proof of the Parthasarathy - Ranga Rao - Varadarajan conjecture, Invent. Math. 93 (1988), 117-130.

[Ku_2] Kumar, S., Bernstein-Gelfand-Gelfand resolution for arbitrary Kac-Moody algebras, Maths. Ann. 286 (1990), 709-729.

[KL] Kazhdan, D. and Lusztig, G., Schubert varieties and Poincaré duality, Proc. Symp. Pure Math. 36 (1980), 185-203.

B.K. : Department of Mathematics, M.I.T., Cambridge,MA. 02139,USA.
S.K. : School of Mathematics, T.I.F.R., Colaba, Bombay 400005,India.

Contemporary Mathematics
Volume 139, 1992

Towards a Generalized Bruhat Decomposition

LEX E. RENNER[1]

0. Introduction

The purpose of this paper is to survey what is known about the problem of generalized Bruhat decompositions. To frame the general idea we need a definition: Let G be a connected reductive group, and suppose $\mu : G \times X \to X$ is a regular action of G on the normal, irreducible variety X. We say the action is spherical (and X is called a spherical embedding) if, for some Borel subgroup $B \subseteq G$, and some $x \in X$, $B \cdot x \subseteq X$ is dense. Everybody's most familiar example is, of course $X = G/B$; and this one is closely related to $G \times G$ acting on G with the two sided action (which is also a spherical action).

If $G \times X \to X$ is spherical, then a theorem of Brion [1] asserts that B has only finitely many orbits on X.

After some reflection one is naturally led to the following basic problems:

(1) Classify all $H \subseteq G$ such that $G \times G/H \to G/H$ is spherical; and identify important special cases.

(2) Classify all spherical varieties X. (Note: Any such X has a dense G-orbit isomorphic to some G/H as in (1).) This turns out to be a discrete problem.

(3) If X is spherical,
  (a) Enumerate the G-orbits on X.
  (b) Enumerate the B-orbits on X.

---

[1]1991 Mathematics Subject Classification. Primary 14M17, 14L30.
Supported by a grant from NSERC.
This paper is in final form and no version of it will be submitted for publication elsewhere.

(4) Develop a structure theory of spherical embeddings whereby one can "read" the structure via discrete invariants (Weyl groups, toroidal embeddings,...).

I will discuss the recent progress on some of these problems, including the work of Brion, Luna and Vust. The reader is referred to DeConcini's ICM survey [5] for an overview of the theory of embeddings.

Very recently, Solomon [20] has initiated the study of a generalized Iwahori-Hecke algebra that arises naturally from the Bruhat decomposition on algebraic monoids (see 1.4). Algebraic monoids are, among other things, spherical varieties of the form $j : G \subset M$ where j is open and dense, and the two sided action of G on itself extends over M.

Putcha [13] has uncovered a natural description of the $R_{\theta,\sigma}$'s of Kazhdan-Lusztig theory. Briefly, he starts by generalizing Solomon's Iwahori algebra (with canonical basis $\{\mathfrak{X}_{\theta,\sigma}\}$), proves it is semisimple, and finds the equation

$$1 = \sum R_{\theta,\sigma} \cdot \mathfrak{X}_{\theta,\sigma}.$$

As is well known, the Kazhdan-Lusztig polynomials are defined recursively from the $R_{\theta,\sigma}$'s.

The reader may find this survey a little disjointed. On the one hand, there is some discussion of the basic theory of spherical embeddings, developed largely by Brion, Luna and Vust [3]. This is the most natural context for discussing analogues of the Bruhat decomposition. However, the B-orbit structure is not yet fully understood at this generality. On the other hand, there is the self contained subtheory of reductive monoids [12]. Here, the generalized Bruhat decomposition is already quite well understood [17], with many potential applications in sight [13], [20].

We hope this approach will give the reader an opportunity to develop a useful overview of the problems and results in this area.

1. Spherical Embeddings

Let G be a connected, reductive group. A closed subgroup $H \subseteq G$ is called <u>spherical</u> if a Borel subgroup of G has a dense orbit on G/H. G/H is then called a <u>spherical homogeneous space</u>. If H is reductive (and the characteristic is zero) this is equivalent to the condition $k[G/H] = \bigoplus_{\lambda \in S} V_\lambda$, where $\{V_\lambda\}_{\lambda\in S}$ is a collection of pairwise nonisomorphic simple G modules.

A spherical embedding for G is an irreducible, normal algebraic variety X together with a regular action $G \times X \to X$ such that a Borel subgroup of G has a dense orbit on X. This is equivalent to saying that X is a normal, equivariant, partial compactification of some spherical homogeneous space G/H.

It is appropriate to distinguish several classes of spherical embeddings. Brion [2] has classified all irreducible spherical homogeneous spaces.

1.1. Complete Homogeneous Spaces. These are so interesting and familiar, nothing else should be said here.

1.2. Toroidal Embeddings. Here $T = B = G$, and so any X with a dense G orbit is spherical. As is well known [4], the classification and structure theory of toroidal embeddings boils down to an analysis of certain polyhedral decompositions of the lattice of characters of a torus. It is certain that toroidal embeddings will be involved in any major classification theorem about spherical embeddings (see 2.1.2 for the reason).

1.3. Affine Symmetric Spaces. Let $\sigma : G \to G$ be an automorphism such that $\sigma^2 = \mathrm{id}$, and let $H = G^\sigma = \{x \in G \mid \sigma(x) = x\}$. Then by [23], G/H is a spherical homogeneous space. A spherical variety X with dense orbit isomorphic to G/G^σ is called a symmetric variety. These are discussed at length in [6,7], yielding a rigorous justification of some predictions of Schubert.

1.4. Algebraic Monoids. A reductive monoid is an irreducible, normal, algebraic variety M together with an associative morphism $\mu : M \times M \to M$ and an identity element $1 \in M$ for μ. We assume also that $G = G(M) = \{x \in M \mid x^{-1} \in M\}$ (which is always affine and open in M) is a reductive group. M is a spherical variety for the action $\mu : G \times G \times M \to M$ defined by $\mu((g,h), x) = gxh^{-1}$. The axiomatic theory of reductive monoids is in a fairly complete state due to the efforts of M.S. Putcha and the author. Putcha's monograph [12] is accessible to anyone with a background in algebraic groups and a willingness to entertain the notion of a semigroup.

2. Orbit Structure of Spherical Embeddings

Let $G \times X \to X$ be a spherical embedding for G and let $O \subseteq X$ be the unique, dense G-orbit. In this section we describe some of the most general properties about the orbit structure of spherical embeddings. The guiding philosophy is to describe the orbit structure of X as completely as possible in terms of toroidal embeddings, Weyl groups and other numerical or combinatorial invariants.

2.1. Basic properties of Spherical Embeddings.

2.1.1[2]. Let $B \subseteq G$ be a Borel subgroup. Then B has only finitely many orbits on X.

2.1.2[3]. There exist $x \in O$ and $T \subseteq G$, a maximal torus such that $\overline{T \cdot x} \subseteq X$ (Zariski closure) intersects every G-orbit.

2.1.3[11]. X has rational singularities.

I have not done the general theory of spherical embeddings justice with this paltry sketch. My purpose here is to give a "satellite picture" of the landscape. Luna [9] has written a detailed survey of the general theory of spherical embeddings, but I don't know if it has been published. The interested reader should consult [5] for a proper survey up to 1986.

2.2. The Enumerative Problem for Orbits. By 2.1.1, any spherical embedding X has a finite number of B orbits. In case $X = G/P$ for some parabolic subgroup P of G, the B orbits on X are canonically indexed by W/W_P where W and W_P are the Weyl groups of G and P respectively.

As another example, let $\sigma : G \to G$ be an involution, and let $H = \{x \in G \mid \sigma(x) = x\}$. Then by [23], $X = G/H$ is spherical for G. But in this case, one can say much more about B orbits than 2.1.1. In [21], Springer has shown that the B - orbits on G/H are in one to one correspondence with the set

$$\{[T,B] \mid T \subseteq B,\ T \text{ a maximal torus, } B \text{ Borel, } \sigma(T) = T\}$$

where $[T,B] = [T',B']$ if there exists $s \in H$ such that $sTs^{-1} = T'$ and $sBs^{-1} = B'$.

On the other hand, the problem of computing G-orbits in a nonhomogeneous spherical variety is far from trivial. Let M be a reductive monoid (1.4) with unit group G, and assume that M has a zero element, and further that there is a unique minimal non zero $G \times G$ orbit $J \subseteq M$ (e.g. If $M = M_n(k)$ then $J = \{x \in M \mid \text{rank}(x) = 1\}$). Such a reductive monoid is called <u>J-irreducible</u>. The simplest way to construct such a monoid is as follows: Let G_0 be semisimple and let $\rho : G_0 \to G\ell_n(k)$ be an irreducible representation. Define $M_\rho = \overline{\rho(G_0)\cdot Z(G\ell_n(k))}$ (Zariski closure). By [16], M_ρ is J-irreducible.

2.2.1. THEOREM [14]. *Let* M *be J-irreducible, and let* $x \in J$. *Then* $G\backslash M/G$ *is a finite lattice* (*with* $GxG \leq GyG$ *if* $GxG \subseteq \overline{GyG}$), *which is uniquely deyermined by the parabolic subgroup* $P = \{g \in G \mid gx = \alpha x$ *for some* $\alpha \in Z(G)^0\}$. *We say* <u>M *is of type* P</u>.

Figure 1 depicts all cases where G_0 is simple and $M = M_\rho$ is obtained from $\rho =$ the adjoint representation. The circled nodes indicate the fundamental weights involved in each adjoint representation. By 2.2.1 these nodes determine the lattice $G\backslash M/G$. Incidently, these lattices also represent the lattice of centers of unipotent radicals of standard parabolics.

But we have digressed somewhat. The above description of the $G \times G$ orbits on a reductive monoid is offered as a warming up exerciese. What we really want is to describe the B-orbits on a spherical variety. To be precise let $\mu : G \times X \to X$ be a spherical variety. One would like to find a finite poset $(\mathcal{S}, \leq)$ canonically associated with X, such that

(i) $\varphi : B\backslash X \cong \mathcal{S}$

(ii) $Bx \subseteq \overline{By}$ iff $\varphi(x) \leq \varphi(y)$

(iii) For any $x \in X$ one can identify (through φ) the B-orbits on X that intersect ρBx, where ρ is a simple reflection.

In the next section we describe what is known for reductive monoids. Notice that (iii) asks for an analogue of Tits' well known axiom for BN pairs. Springer [21] has solved (iii) for an important class of spherical <u>homogeneous</u> spaces, namely the affine symmetric spaces.

3. Bruhat Decomposition for Reductive Monoids

Let M be a reductive monoid, and let $T \subseteq G$ be a maximal torus, where $G = \{x \in M \mid x^{-1} \in M\}$ is the unit group of M. Let

$R = \overline{N_G(T)}$ (Zariski closure in M) and let

$\mathcal{R} = R/T$, the set of T orbits on R.

Notice that, for $x \in R$, $Tx = xT$ so one can multiply the elements of $\mathcal{R}$. If $T \subseteq B$ is a Borel subgroup, and $\sigma = xT \in \mathcal{R}$

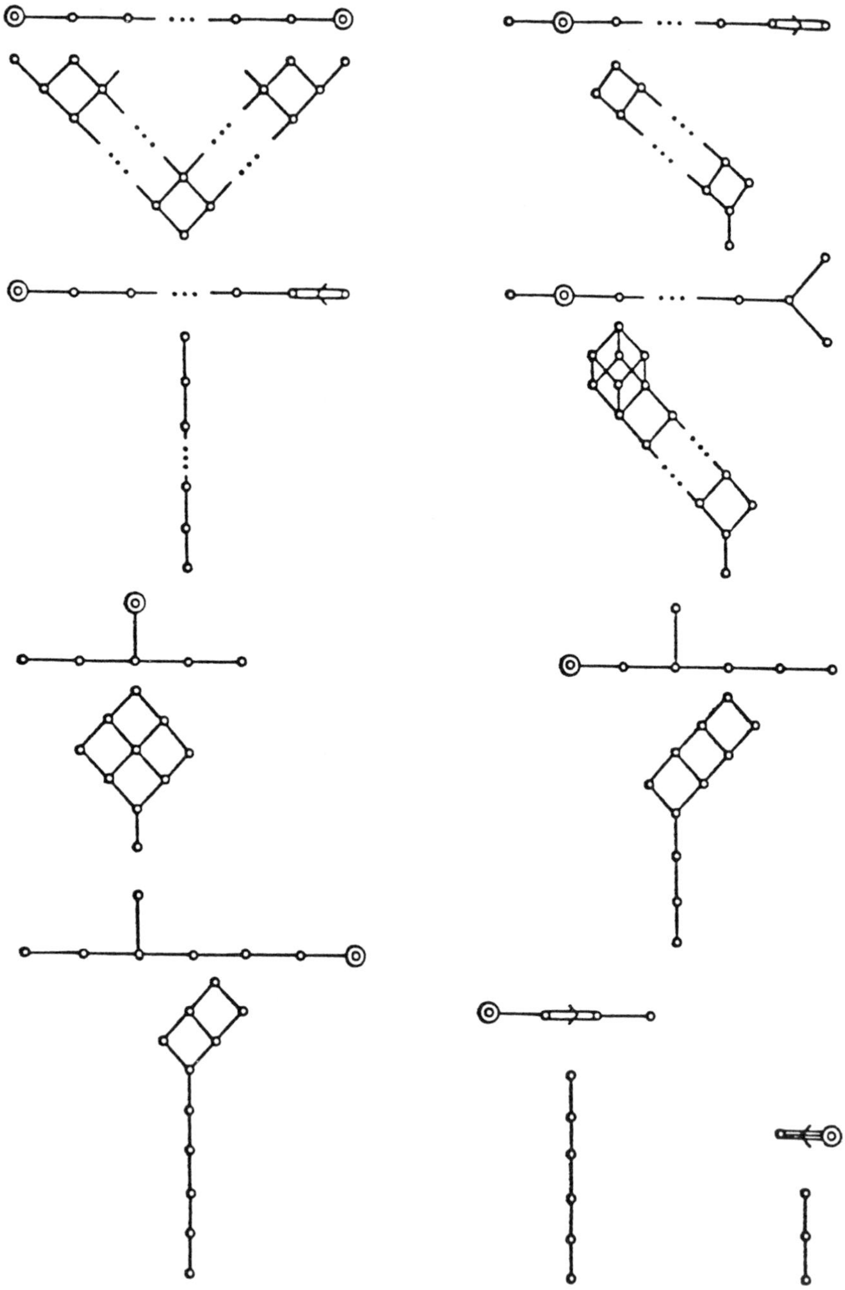

Fig. 1. The adjoint representations of simple algebraic groups. Here the lattice also represents the lattice of centers of unipotent radicals of standard parabolics.

we write $B\sigma B$ for BxB.

3.1. THEOREM [17].

(1) *$\mathcal{R}$ is a finite monoid with unit group* W.

(2) $\mathcal{R} = W \cdot \Lambda \cdot W$ *where* $\Lambda = \{e \in T \mid e^2 = e,\ Be \subseteq eB\}$

(3) $M = \bigcup_{e \in \Lambda} G\, e\, G$ (*disjoint union*)

(4) $M = \bigcup_{\sigma \in \mathcal{R}} B\, \sigma\, B$ (*disjoint union*)

(5) *If* $\rho \in W$ *is a simple reflextion relative to* B *and* T *and* $\sigma \in \mathcal{R}$, *then* $\rho B \sigma \subseteq B\sigma B \cup B\rho\sigma B$.

$\mathcal{R}$ can be partially ordered by the relation $\sigma \leq \tau$ if $B\sigma B \subseteq \overline{B\tau B}$, extending the Bruhat ordering on W. If $M = M_2(k)$ then $\mathcal{R}$ can be identified with the set $\{O, e, f, n, \bar{n}, i, \rho\}$ where $O = \begin{bmatrix} 0 & 0 \\ 0 & 0 \end{bmatrix}$, $e = \begin{bmatrix} 1 & 0 \\ 0 & 0 \end{bmatrix}$, $f = \begin{bmatrix} 0 & 0 \\ 0 & 1 \end{bmatrix}$, $n = \begin{bmatrix} 0 & 1 \\ 0 & 0 \end{bmatrix}$, $\bar{n} = \begin{bmatrix} 0 & 0 \\ 1 & 0 \end{bmatrix}$, $i = \begin{bmatrix} 1 & 0 \\ 0 & 1 \end{bmatrix}$, and $\rho = \begin{bmatrix} 0 & 1 \\ 1 & 0 \end{bmatrix}$. The poset structure on $\mathcal{R}$ is depicted in figure 2.

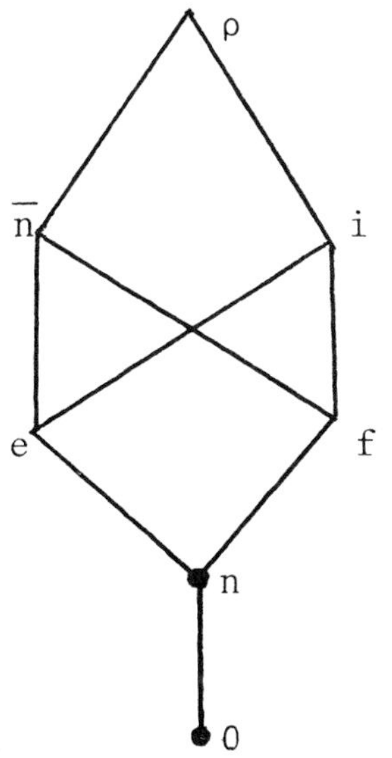

Fig. 2.

For $M = M_n(k)$, $\mathcal{R} = \mathcal{R}_n$ can be identified with the set of 0 - 1 matrices with at most one non zero entry in each row and column. If we let $r_n = |\mathcal{R}_n|$, we then obtain a recursive formula for r_n as follows

$$r_0 = 1, \; r_1 = 2$$
$$r_n = 2nr_{n-1} - (n-1)^2 r_{n-2}, \text{ for } n \geq 2.$$

If we let $r(x) = \sum_{n\geq 0} (r_n/n!)x^n$, then $r(x)$ converges to the function $(e^{x/(1-x)})/(1-x)$. See [18] for more details.

If $k = \mathbb{F}_q$, then Solomon [20] has defined a length function $\ell : \mathcal{R} \to \mathbb{N}$ (for $\mathcal{R} = \mathcal{R}_n$) and established the identity

$$|M_n(k)^r| = (q-1)^r q^{r(r-1)/2} \sum_{\sigma\in\mathcal{R}^r} q^{\ell(\sigma)} \qquad (1)$$

where

$$M_n(k)^r = \{x \in M_n(k) \,|\, \text{rank}(x) = r\}, \text{ and}$$
$$\mathcal{R}^r = \{\sigma \in \mathcal{R} \,|\, \text{rank}(\sigma) = r\}.$$

It is extremely likely that a generalization could be found to yield similar formulae for any finite reductive monoid. Notice that $|B_r| = (q-1)^r q^{r(r-1)/2}$, where B_r is the $r \times r$ upper triangular group of $G\ell_r(k)$.

4. Hecke Algebras and Kazhdan Lusztig Polynomials

Solomon [20] has also introduced an Iwahori-Hecke algebra contruction for the finite matrix monoids $M = M_n(k)$ $(k = \mathbb{F}_q)$. Let K be a field of characteristic zero and let $e = \frac{1}{|B|} \sum_{b\in B} b \in K[M]$, where $B \subseteq G\ell_n(k)$ is the upper triangular group and $K[M]$ is the monoid algebra. For each $D = BxB \in B\backslash M/B$ let $[D] = \sum_{y\in D} y \in eK[M]e$. Then it is easily checked that $\{[D] \,|\, D \in B\backslash M/B\}$ is a canonical basis for $eK[M]e$. On the other hand, if $\sigma \in \mathcal{R}^r$, then formula (1) suggests that

$$T_\sigma = (q-1)^{-r} q^{-r(r-1)/2} [B\sigma B]$$

is the appropriate normalized basis of $eK[M]e$. With this in mind Solomon defines the Iwahori algebra of (B,M) as

$$H(M,B) = \oplus_{\sigma\in R} \mathbb{Z}\, T_\sigma \subseteq eK[M]e$$

and proves the following theorem.

4.1. THEOREM [20]. *Let* $\nu = \begin{bmatrix} 0 & 1 & & & 0 \\ 0 & 0 & \ddots & & \\ & & \ddots & \ddots & \\ 0 & & & 0 & 1 \\ & & & 0 & 0 \end{bmatrix} \in \mathcal{R}$, *and let* $S \subseteq \mathcal{R}$ *be the set of simple reflections relative to* B *and* T. *Then* $H(M, B)$ *is the subring of* $eK[M]e$ *generated by* T_ν *and* T_s, $s \in S$. *Furthermore,*

$$T_s T_\sigma = \begin{cases} qT_\sigma & , \text{ if } \ell(s\sigma) = \ell(\sigma) \\ T_{s\sigma} & , \text{ if } \ell(s\sigma) = \ell(\sigma) + 1 \\ qT_{s\sigma} + (q-1)T_\sigma, & \text{ if } \ell(s\sigma) = \ell(\sigma) - 1 \end{cases}$$

$$T_\sigma T_s = \begin{cases} qT_\sigma & , \text{ if } \ell(\sigma s) = \ell(\sigma) \\ T_{\sigma s} & , \text{ if } \ell(\sigma s) = \ell(\sigma) + 1 \\ qT_{\sigma s} + (q-1)T_\sigma, & \text{ if } \ell(\sigma s) = \ell(\sigma) - 1 \end{cases}$$

$$T_\nu T_\sigma \equiv q^{\ell(\sigma) - \ell(\nu\sigma)}\, T_{\nu\sigma}$$
$$T_\sigma T_\nu = q^{\ell(\sigma) - \ell(\nu\sigma)}\, T_{\sigma\nu}$$

for all $\sigma \in \mathcal{R}$, $s \in S$.

REMARK. Recall from 3.1(5) that we have $sB\sigma \subseteq B\sigma B \cup Bs\sigma B$ for $s \in S$ and $\sigma \in \mathcal{R}$. For groups this leads to the dichotomy

$$\mathrm{BsB}\cdot\mathrm{B}\sigma\mathrm{B} = \begin{cases} \mathrm{Bs}\sigma\mathrm{B} & \text{if } \ell(\mathrm{s}\sigma) = \ell(\sigma) + 1 \\ \mathrm{Bs}\sigma\mathrm{B} \cup \mathrm{B}\sigma\mathrm{B} & \text{if } \ell(\mathrm{s}\sigma) = \ell(\sigma) - 1 \end{cases}.$$

However, there is a trichotomy for monoids which, according to Proposition 3.14 of [20], leads to

$$\mathrm{BsB}\cdot\mathrm{B}\sigma\mathrm{B} = \begin{cases} \mathrm{B}\sigma\mathrm{B} & \text{if } \ell(\mathrm{s}\sigma) = \ell(\sigma) \\ \mathrm{Bs}\,\sigma\mathrm{B} & \text{if } \ell(\mathrm{s}\sigma) = \ell(\sigma) + 1 \\ \mathrm{Bs}\,\sigma\mathrm{B} \cup \mathrm{B}\sigma\mathrm{B} & \text{if } \ell(\mathrm{s}\sigma) = \ell(\sigma) - 1 \end{cases}.$$

The reader may view the equation "$\ell(\mathrm{s}\sigma) = \ell(\sigma)$" with some suspicion, but there really isn't anything to worry about. After all, in a monoid, an equation of the form "$\mathrm{s}\sigma = \sigma$" is part of the fun.

Again, one sees an idea that will very likely apply to any finite reductive monoid. Finite reductive monoids are constructed from reductive monoids (over $\mathbb{F}_q$) in exactly the same way that Chevalley, Stienberg, Suzuki and Ree construct finite groups of Lie type from algebraic groups [22]. See [15] and [19] for some results in this direction.

Putcha [13] has already taken up the general notion of an Iwahori algebra for monoids (without the integral structure of 4.1), since he has found a semigroup theoretical way of describing the Kazhdan-Lusztig polynomials. The connection here is really quite surprising, so I will give a brief description.

Let G be a split, simple, algebraic group defined over $k = \mathbb{F}_q$, and let $\rho : G \to G\ell_n(k)$ be an irreducible representation of G with highest weight in general position. Define

$$M(G) = \overline{Z(G\ell_n(k))\rho(G)} \subseteq M_n(k)$$

where, as usual, the overscore denotes Zariski closure. It turns out that M(G) is a J-irreducible monoid of type B (see 2.2.1). Let $M_1 = G \cup J \cup \{0\}$ where J is the minimal non zero $G \times G$ orbit, and let

$$M = G(k) \cup J(k) \cup \{0\}.$$

Then M is a finite monoid with exactly three $G \times G$ orbits; G, J and 0. Furthermore, $J = GeG$ for some $e \in J$ with $Be = eBe$. Let

$J^0 = J \cup \{0\}$, and let

$$f = \frac{1}{|B|} \sum_{b \in B} b.$$

Define

$$H = f\, \mathbb{C}_0[J^0] f$$

where $\mathbb{C}_0[J^0] = \mathbb{C}[J^0]/(0)$ (semigroup algebra modulo the (one dimensional) ideal generated by the 0 of J^0).

REMARK. $\mathbb{C}_0[J^0]$ is referred to as the <u>contracted semigroup algebra</u> of J^0. But notice that, on the face of it, $\mathbb{C}_0[J^0]$ may not have an identity element, since J^0 is a semigroup without identity. On the other hand, by [10], $\mathbb{C}_0[J^0]$ is a semisimple algebra, so an identity element appears automatically for structural reasons.

Let $x, y \in G$. Then by the Bruhat and Birkhoff decompositions, $x \in B\sigma^{-1}U$ and $y \in U^-\theta B$ for some unique σ, $\theta \in W$. So $xey \in B\sigma^{-1}e\theta B$, and

$$A_{\sigma,\theta} = f\sigma^{-1}e\theta f = fxeyf$$

does not depend on the choice of x and y. Furthermore, $\{A_{\sigma,\theta}\} \subseteq H = f\, \mathbb{C}_0[J^0] f$ is a basis of H.

4.2. Theorem [13]. (a) *The unit element of* H *is given by*

$$u = \sum \tilde{Q}_{\theta,\sigma} \cdot A_{\theta,\sigma}$$

where $\tilde{Q}_{\theta,\sigma} = (-1)^{\ell(\theta)-\ell(\sigma)} \dfrac{|\theta B\sigma^{-1} \cap B^-B|}{|B|}$;

(b) $\tilde{Q}_{\theta,\sigma} = (-q)^{\ell(\theta)-\ell(\sigma)} R_{\theta,\sigma}(q)$, *where* $R_{\theta,\sigma}$ *is the one from Kazhdan-Lusztig theory* [8].

This is really quite a surprise. The $\tilde{Q}_{\theta,\sigma}$'s are derived from the "sandwich matrix", a construction arising naturally in abstract semigroup theory, while the $R_{\theta,\sigma}$'s are defined recursively in terms of the Bruhat order.

REFERENCES

1. M. Brion, Quelques propriétés des espàces homogènes sphériques, Manuscripta Math. **55** (1986), 191-198.
2. M. Brion, Classification des espàces homogènes sphériques, Compositio Mathematica **63** (1987), 189-208.
3. M. Brion, D. Luna, Th. Vust, Espàces homogènes sphériques, Invent. Math. **84** (1986), 617-632.
4. V. Danilov, The geometry of toric varieties, Russian Math. Surveys **33** (1978), 97-154.
5. C. DeConcini, Equivariant embeddings of homogeneous spaces, Proceedings of ICM, Berkely (1986), 367-376.
6. C. DeConcini, C. Procesi, Complete symmetric varieties, Lecture Notes in Math. Vol. **996**, Springer-Verlag, Berlin-New York, 1983, 1-44.
7. C. DeConcini, C. Procesi, Complete symmetric varieties II, Adv. Stud. Pure Math. **6** (1985), 482-513.
8. V. Deodhar, On some geometric aspects of Bruhat orderings I. A finer decomposition of Bruhat cells, Invent. Math. **79** (1985), 499-511.
9. D. Luna, Report on spherical varieties, unpublished.
10. J. Okninski, M. Putcha, Complex representations of matrix semigroups, Trans. Amer. Math. Soc., to appear.
11. V. Popov, Contraction of the actions of reductive algebraic goups, Math. USSR Sbornik **58** (1987), 311-335.
12. M. Putcha, "Linear Algebraic Semigroups", London Math. Soc. Lecture Notes Vol. **133**, Cambridge Univer. Press, 1988.
13. M. Putcha, Sandwich matrices, Solomon algebras and Kazhdan-Lusztig polynomials, preprint, 1991.
14. M. Putcha, L. Renner, The system of idempotents and the lattice of J-classes of reductive algebraic monoids, J. Algebra **116** (1988), 385-399.

15. M. Putcha, L. Renner, Canonical compactification of a finite group of Lie type, Trans. Amer. Math. Soc., to appear.

16. L. Renner, Classification of semisimple algebraic monoids, Trans. Amer. Math. Soc. **292** (1985), 193-223.

17. L. Renner, Analogue of the Bruhat decomposition for algebraic monoids, J. Algebra **101** (1986), 303-338.

18. L. Renner, Enumeration of injective partial transformations, Discrete Math. **73** (1989), 291-296.

19. L. Renner, Finite monoids of Lie type, in "Monoids and Semigroups with Applications" (J. Rhodes ed.), World Scientific (1991), 278-287.

20. L. Solomon, The Bruhat decomposition, Tits system and Iwahori ring for the monoid of matrices over a finite field, Geom. Ded. **36** (1990), 15-49.

21. T. Springer, Some results on algebraic groups with involutions, Adv. Stud. Pure Math. **6** (1985), 525-543.

22. R. Steinberg, Endomorphisms of linear algebraic groups, Mem. Amer. Math. Soc. **80** (1968).

23. Th. Vust, Opération des groupes réductifs dans un type de cônes presque homogènes, Bull. Soc. Math. France **102** (1974), 317-333.

DEPARTMENT OF MATHEMATICS
UNIVERSITY OF WESTERN ONTARIO
LONDON, ONTARIO N6A 5B7 CANADA

Recent Titles in This Series

(Continued from the front of this publication)

(See the AMS catalog for earlier titles)